Genetically Modified Organisms in Developing Countries

Risk Analysis and Governance

Bringing together the ideas of experts from around the world, this incisive text offers cutting-edge perspectives on the risk analysis and governance of genetically modified organisms (GMOs), supporting effective and informed decision-making in developing countries.

Comprised of four comprehensive sections, this book covers:

- **Integrated risk analysis and decision-making**, giving an overview of the science involved and examining risk analysis methods that impact decision-making on the release of GMOs, particularly in developing countries.
- **Diversification of expertise** involved in risk analysis and practical ways in which the lack of expertise in developing countries can be overcome.
- **Risk-analysis-based regulatory systems** and how they can be undermined by power relationships and socio-political interests, as well as strategies for improving GMO policy development and regulatory decision-making.
- **Case studies from developing countries** providing lessons based on real-world experience that can inform our current thinking.

Ademola A. Adenle has over 16 years of research and international experience, and was a research fellow and principal investigator at the United Nations University (UNU), Japan, where he coordinated the largest and most comprehensive study in the history of genetically modified (GM) agriculture in Africa, revealing a diverse range of challenges impeding the development of effective biosafety regulatory frameworks across the continent. Further to this research work, he specializes in science and technology policy addressing sustainable development challenges and uses his background in applied science to study social aspects of science and technology in agricultural development, biodiversity conservation, renewable energy, climate change adaptation and mitigation, and health innovation in developing countries.

Formerly Director of the African Centre for Gene Technologies and now an independent bioscience consultant, **E. Jane Morris** has over 25 years' experience working in the biotechnology field in Africa. As Chair of the South African advisory committee on genetic modification before the introduction of the GMO Act, she was instrumental in the introduction of GMO legislation in that country. She has a fellowship at the University of Leeds.

Denis J. Murphy is Professor of Biotechnology at the University of South Wales and advisor to the Food and Agriculture Organization. He has published numerous books and over 300 research articles on crop biotechnology and related areas.

Genetically Modified Organisms in Developing Countries

Risk Analysis and Governance

Edited by

ADEMOLA A. ADENLE

School of Global Environmental Sustainability,
Colorado State University, Fort Collins, CO, USA

E. JANE MORRIS

School of Biology, University of Leeds, Leeds, UK

DENIS J. MURPHY

Genomics and Computational Biology Research Group,
University of South Wales, Pontypridd, UK

CAMBRIDGE
UNIVERSITY PRESS

CAMBRIDGE
UNIVERSITY PRESS

University Printing House, Cambridge CB2 8BS, United Kingdom

One Liberty Plaza, 20th Floor, New York, NY 10006, USA

477 Williamstown Road, Port Melbourne, VIC 3207, Australia

4843/24, 2nd Floor, Ansari Road, Daryaganj, Delhi - 110002, India

79 Anson Road, #06-04/06, Singapore 079906

Cambridge University Press is part of the University of Cambridge.

It furthers the University's mission by disseminating knowledge in the pursuit of education, learning and research at the highest international levels of excellence.

www.cambridge.org
Information on this title: www.cambridge.org/9781107151918
DOI: 10.1017/9781316585269

First published 2017

Printed in the United Kingdom by Clays, St Ives plc

A catalogue record for this publication is available from the British Library

Library of Congress Cataloging-in-Publication data
Names: Adenle, Ademola A., 1976– editor. | Morris, E. Jane, editor. | Murphy, Denis J., editor.
Title: Genetically modified organisms in developing countries : risk analysis and governance / edited by Ademola A. Adenle, E. Jane Morris, Denis J. Murphy.
Description: Cambridge, United Kingdom : New York, NY : Cambridge University Press, 2017. | Includes bibliographical references and index.
Identifiers: LCCN 2017002752| ISBN 9781107151918 (hardback : alk. paper) | ISBN 9781316606292 (pbk. : alk. paper)
Subjects: LCSH: Transgenic organisms – Risk assessment – Developing countries. | Transgenic organisms – Government policy – Developing countries. | MESH: Organisms, Genetically Modified | Food, Genetically Modified | Developing Countries | Risk Assessment | Government Regulation | Socioeconomic Factors
Classification: LCC QH442.6 .G475 2017 | NLM QH 442.6 | DDC 636.08/21091724 – dc23
LC record available at https://lccn.loc.gov/2017002752

ISBN 978-1-107-15191-8 Hardback
ISBN 978-1-316-60629-2 Paperback

Contents

Part IV Case Studies from Developing Countries

Contributors

Ademola A. Adenle
School of Global Environmental Sustainability, Colorado State University, Fort Collins, CO, USA

Olalekan Akinbo
NEPAD Agency African Biosafety Network of Expertise, Kampala, Uganda

Klaus Ammann
University of Bern, Bern, Switzerland

Hossein Azadi
Department of Geography, Ghent University, Ghent, Belgium

Dieter Blancquaert
Laboratory of Functional Plant Biology, Department of Physiology, Ghent University, Ghent, Belgium

Helaine Carrer
University of São Paulo, São Paulo, Brazil

Wendy Craig
International Centre for Genetic Engineering and Biotechnology, Trieste, Italy (craig@icgeb.org)

Bertrand Dagallier
Environment, Health and Safety Programme, Organisation for Economic Co-operation and Development, Paris, France

Maria Lúcia Zaidan Dagli
University of São Paulo, São Paulo, Brazil

Paulo Paes de Andrade
Federal University of Pernambuco, Recife, Brazil

Gutemberg Delfino De Souza
CTNBio, Ministry of Science, Technology and Innovation, Brasília, Brazil

Hans De Steur
Department of Agricultural Economics, Faculty of Bioscience Engineering, Ghent University, Ghent, Belgium

Julius Ecuru
Uganda National Council for Science and Technology, Kampala, Uganda

José Falck-Zepeda
The International Food Policy Research Institute, Washington, D.C., USA

Xavier Gellynck
Department of Agricultural Economics, Faculty of Bioscience Engineering, Ghent University, Ghent, Belgium

Marnus Gouse
Bureau for Food and Agricultural Policy, University of Pretoria, Pretoria, South Africa

Eric M. Hallerman
Department of Fish and Wildlife Conservation, Virginia Polytechnic Institute and State University, Blacksburg, VA, USA

Wilna Jansen van Rijssen
FoodNCropBio, Monument Park, South Africa

Peter Kearns
Environment, Health and Safety Programme, Organisation for Economic Co-operation and Development, Paris, France

Muffy Koch
Global biosafety consultant, Ottawa, ON, Canada

John Komen
Komen Bioscience Consultancy, Haarlem, the Netherlands

Willy Lambert
Laboratory of Toxicology, Department of Bioanalysis, Ghent University, Ghent, Belgium

Godwin Lemgo
African Biosafety Network of Expertise, Ouagadougou, Burkina Faso

Yunhe Li
State Key Laboratory for Biology of Plant Diseases and Insect Pests, Institute of Plant Protection, Chinese Academy of Agricultural Sciences, Beijing, China

Yanhui Lu
State Key Laboratory for Biology of Plant Diseases and Insect Pests, Institute of Plant Protection, Chinese Academy of Agricultural Sciences, Beijing, China

Phil Macdonald
Plant Health Science Services, Canadian Food Inspection Agency, Mississauga, ON, Canada

Diran Makinde
NEPAD Agency African Biosafety Network of Expertise, Midrand, South Africa

E. Jane Morris
School of Biology, University of Leeds, Leeds, UK

Denis J. Murphy
Genomics and Computational Biology Research Group, University of South Wales, Pontypridd, UK

Rubens José Nascimento
CTNBio, Ministry of Science, Technology and Innovation, Brasília, Brazil

Dennis Obonyo Ndolo
International Centre for Genetic Engineering and Biotechnology, Cape Town, South Africa (ndolo@icgeb.org)

Narender Nehra
Donald Danforth Plant Science Center, St Louis, MO, USA

Takahiko Nikaido
Environment, Health and Safety Programme, Organisation for Economic Co-operation and Development, Paris, France

Yufa Peng
State Key Laboratory for Biology of Plant Diseases and Insect Pests, Institute of Plant Protection, Chinese Academy of Agricultural Sciences, Beijing, China

Peter W. B. Phillips
Johnson Shoyama Graduate School of Public Policy, University of Saskatchewan, Saskatoon, SK, Canada

Tassiana Fronza Pinho
CTNBio, Ministry of Science, Technology and Innovation, Brasília, Brazil

Hector Quemada
Donald Danforth Plant Science Center, St Louis, MO, USA

Monica Racovita
Institute of Science, Technology and Society Studies, Alpen-Adria University, Graz, Austria

Sandra Sharry
Faculty of Agricultural and Forestry Sciences, La Plata University, Buenos Aires, Argentina

Stuart J. Smyth
Department of Agricultural and Resource Economics, University of Saskatchewan, Saskatoon, SK, Canada

Andrea Sonnino
National Agency for New Technologies, Energy and Sustainable Economic Development (ENEA), Rome, Italy

Christophe Stove
Laboratory of Toxicology, Department of Bioanalysis, Ghent University, Ghent, Belgium

Krishna Ravi Srinivas
Research and Information System for Developing Countries (RIS), New Delhi, India

Mark Tepfer
Institut Jean-Pierre Bourgin, INRA, AgroParisTech, CNRS, University of Paris-Saclay, Versailles, France (mark.tepfer@inra.fr)

Eduardo Trigo
Grupo CEO S.A., Buenos Aires, Argentina

Dominique Van Der Straeten
Laboratory of Functional Plant Biology, Department of Physiology, Ghent University, Ghent, Belgium

Wynand J. van der Walt
FoodNCropBio, Monument Park, South Africa

Edivaldo Domingues Velini
São Paulo State University, São Paulo, Brazil

Carmen Vicién
Faculty of Agronomy, University of Buenos Aires, Buenos Aires, Argentina

Mauro Vigani
University of Gloucestershire, Cheltenham, UK

Kongming Wu
State Key Laboratory for Biology of Plant Diseases and Insect Pests, Institute of Plant Protection, Chinese Academy of Agricultural Sciences, Beijing, China

Wen Xiang
Centre for Public Regulation and Administration, Faculty of Law, University of Copenhagen, Copenhagen, Denmark

Foreword

The last three decades have witnessed spectacular progress in the plant sciences, especially regarding our knowledge at the molecular level. It was only possible within such a short period of time because of genetic engineering. The plant gene engineering endeavour started with the finding that soil bacteria like *Agrobacterium tumefaciens* that were known to induce crown galls were also able to force a plant to synthesise exudates in these galls, containing a high concentration of a class of odd amino acids. These novel compounds are nutrients for these infecting Agrobacteria, hence this has been called "genetic colonisation". The next step was the unravelling of the molecular mechanism that made this possible. Surprisingly it was a transfer and integration into the plant genome of a bacterial DNA segment, part of a large plasmid harboured by this very *Agrobacterium* strain. These findings, from the 1970s, culminated in 1983 in the demonstration that it was possible to turn these *Agrobacterium* strains into efficient and reliable systems for delivery and expression of foreign genes into healthy plants.

Most research centres directly accepted this new technology in their attempts to discover the molecular basis of plant physiology and plant growth and development. They also started to make the first prototypes of novel crops through genetic engineering. As the development and production of commercially valuable novel crops is not the task of public research institutes, new actors appeared. Spin-off or start-up companies, often situated in or near the universities where the innovative science was generated, could raise the money and attract the post-docs and skilled technicians to construct the first genetically modified (GM) crops.

Agrochemical companies were very well aware of this opening up of the plant sciences. The prospect of obtaining more efficiently improved cultivars through this genetic engineering technology convinced them to join the race. They bought up the best-performing seed companies, and laid the foundation of Crop Science departments worldwide. They became the bases for developing a successful intensive global agriculture.

From the beginning of gene cloning, concerned scientists asked to discuss the 'safety' of the recombinant DNA technology. The Asilomar conference (February 1975) triggered decades-long debates on 'biosafety procedures' that generated helpful inputs as well as distortions, among which the Cartagena Protocol is one related to plant biotechnology. The first publication of the Cartagena Protocol appeared in 2000 with concerns based on the lack of experience of the recombinant DNA technology applied to plants of the middle 1970s and without any reflection of the experience and knowledge gained in the intervening three decades. Some political parties, particularly in Europe (mostly belonging to the 'Green movement') and globally well-structured NGOs such as Greenpeace and Friends of the Earth saw in the Cartagena Protocol an opportunity to mobilise people respectful of nature and active in organisations expressing environmental concerns. They took advantage of the absence, in most of the schools of agriculture, of basic training in molecular genetics to spread misinformation. In those days the structure and function of genomes were not yet well documented and it was not difficult to fuel fear of the 'unknown'.

On the other hand, scientists so well trained in rational thinking and fact finding in their own discipline have not been able to effectively balance the misinformation with science-based facts. Deprived of science communication strategies, isolated from social, political and economic sciences, and with poor exchanges with the civil society, plant scientists were astonished that they were not listened to when they praised the technology.

All humans, scientists or not, have during many tens of thousands of years of recent evolution developed a neurobiology that stimulates their social behaviour and the awareness of the consequences of their actions. This 'emotional thinking' is deeply rooted in our neurophysiology and we should not expect that some hundred years of 'scientific enlightenment' would alter the trend.

We have made enormous progress in the life sciences. But we all know that our knowledge in medicine, agriculture and environmental care is rudimentary compared with what we will know within a few decades. The question that arises is how deep our understanding of the technology has to become before we can declare it safe. But, to convince society, we should not count on explaining science. Scientists are needed in broader public debates. To build up trust in a new technology, it is necessary to have a continuous, open, two-way exchange of information and opinion on the impact of the technology on societies and the earth's environment.

The regulatory hurdles are holding back innovative developments, especially those of concern to agricultural systems in low-income countries and emerging economies. There is an urgent need for a regulatory framework that fosters rather than discourages innovation. Only through open debate can we create confidence that we all, sincerely, want and can offer better living conditions for fellow citizens of this planet. More than 90% of the specialists in the field are convinced that we are moving in the right direction. To the best of our knowledge the new tools are not more hazardous, for humans and animals, than the traditional crop improvement techniques. On the contrary, the new approaches can bring major contributions for many environmental, social and health problems. Industry meanwhile has demonstrated the economic advantages. In many developing countries farmers have accepted the technology and 10% of the world's agriculture is growing transgenic plants.

It is also our responsibility to have a sound regulatory framework. Technological advances will be pivotal if we want to democratically improve global societal and environmental welfare. This book is a valuable contribution towards the implementation of an internationally harmonised scientifically based biosafety framework. I congratulate the authors for their efforts in communicating their unique experience in the field and raising relevant issues that influence biosafety decision-making processes in developing countries. The lessons learned from the biosafety case studies and regulatory models in different countries and organisations should serve in the current debate about a thorough renewal of GMO crop regulation worldwide.

Marc Van Montagu
Emeritus Professor
VIB – International Plant Biotechnology Outreach (IPBO)
World Food Prize Laureate 2013

1 Introduction

Ademola A. Adenle, E. Jane Morris and Denis J. Murphy

Background to the Book

The predicted growth of the world population to over 9 billion people by the year 2050 will pose a great challenge for developing countries, where most of the population growth is occurring. The Food and Agriculture Organization of the United Nations (FAO) estimates that global demand for food, feed and fibre will nearly double in the first half of this century, while increasingly crops may also be used for bioenergy and other industrial purposes. At the same time food production alone will need to increase by 70% over this period, and this will have to be achieved largely through the application of technology to increase crop yields rather than through an increase in the area under cultivation. Meeting this challenge will require developing countries in particular to embrace modern agricultural technologies.

Genetically modified organisms (GMOs) form a potentially vital tool if this increase in production is to be achieved. However, most developing countries have not yet developed integrated, fit-for-purpose risk analysis and decision-making frameworks that could guide the sustainable development, use and regulation of GMOs. At the same time, we have already seen the explosive growth of genomic technologies over the past decade, and the twenty-first century is likely to witness the emergence of new and potentially disruptive biological developments related to agriculture. Research and new frontiers in genomic biosciences will revolutionise not only medicine, but also the way we develop and use biological resources and undertake agricultural production. The appropriate governance of GMOs and a suite of new technologies such as genome editing needs to be addressed with urgency both in the developing world and in the developed world.

GMOs are currently defined in different ways by different regulatory bodies, and there is an emerging debate over the utility of such a broad-brush term in the context of risk assessment and regulation. There is also a dichotomy between GMO risk analysis based largely on the process of creating new crop varieties (typified by the approach in Europe, which has also been adopted by most developing countries that are signatories to the Cartagena Protocol on Biosafety) and risk management based largely on the nature of the new crop varieties themselves (the product-based approach commonly applied in North America). Added to this confusion, the development of new technologies is bringing into question the validity of some of the historic risk assessment and regulatory models currently in use. Some groups of developed countries, particularly in the European Union, have up to now imposed severe restrictions on genetically modified (GM) crops that have effectively precluded their cultivation in these regions, and this negativity has been reflected in the attitudes of many developing countries. However, it is difficult to see how developing countries will be able to meet the rising demand for food without embracing all the technology options available, some of which may well involve the use of new varieties developed using GM-related technologies. Innovations are needed that can contribute to more productive (in terms of both quality and quantity) and resource-efficient crop and agricultural systems for developing countries that use water, nutrients, energy and agricultural land in a more sustainable manner. For example, technological innovation including GM technology can play a significant role in achieving some of the sustainable development goals such as Goal 2, end hunger, achieve food security and improve nutrition, and promote sustainable agriculture; and Goal 13, take urgent action to combat climate change and its impact.

If the potential for GM technology is to have maximum impact, it will also be necessary to develop GM crops that are relevant for the needs of developing countries; this will require significant innovation, since to date the technology has been commercially introduced only into the globally important commodity crops such as maize, soya and cotton (these three species currently account for >90% of all GM crops grown around the world). Moreover, the traits introduced are not necessarily the most important ones for developing country agriculture. Regulatory barriers and the cost of overcoming regulatory hurdles form a significant disincentive to the innovative development of GM crops for which there is a smaller market and/or for which the introduced trait meets local rather than global needs.

At the same time, there are major requirements to increase the skills and capacity in developing countries on a number of fronts. Firstly, it is important that regulators and policymakers fully understand the scientific basis of modern breeding technologies. Secondly, the capacity to undertake risk assessment and to reach informed decisions is essential. Many developing countries lack sufficient procedures for testing and evaluating any health issues, let alone assessing the impact of novel crops on the environment. The skills to analyse and monitor for the presence of GMOs are essential if the regulatory framework requires labelling. Finally, the capacity for scientific innovation needs to be fostered within developing countries if they are to produce new GMOs that are tailored to their own needs.

This book addresses how risk analysis (composed of the elements of risk assessment, risk management and risk communication) of GMOs can be governed effectively by bringing together chapters that address a variety of issues that will contribute to a much improved approach to decision-making aimed towards sustainable development, regulation and safe use of GMOs in developing countries. The current challenges facing the application of GMOs in developing countries have encouraged a broad spectrum of experts to contribute a unique collection of chapters that specifically look into biosafety experience-based strategies, regulatory models and related projects in different countries and organisations to facilitate the implementation of functional biosafety frameworks and GMO policy both at the national level and at the international level.

One of the key objectives of this book is to assess the current status of biosafety strategies and the impact of impediments with particular reference to the precautionary principle. We examine how these impediments influence biosafety decision-making processes in developing countries and identify possible solutions to the problem. Since one of the most significant obstacles to the development of functional biosafety systems has been linked to a very strict interpretation of the precautionary principle, some of the chapters in this book focus on the underlying challenges and state of national biosafety systems, with examples from representative countries and regions. Another main objective is to bring together the novel ideas of a wide range of experts, case studies and current approaches (e.g., proposed methodologies and frameworks) on the risk analysis and governance of GMOs as they inform effective decision-making in developing countries. A third important objective is to identify opportunities to address current challenges, and what often seems like an impasse in the GMO dialogue, in constructive and meaningful ways.

The book provides an interesting overview of approaches and frameworks in risk analysis of GMOs at the national, regional and international levels. We shed some light on how a lack of consensus for different types of approaches or frameworks is undermining the development of biosafety systems as well as impeding new innovation for tackling low agricultural productivity, food insecurity and health problems in developing countries. We conclude the book with some suggestions for lasting solutions to the problem.

The Structure of the Book

Subsequent to this introduction, the book is divided into four main parts, which are followed by a concluding chapter. The parts address the following issues.

Part I: Risk Analysis Methodology and Decision-Making

Starting with an overview of GMO technology and recent developments in genomics, this part examines the factors that make current methodological approaches deficient for the effective risk analysis of GMOs in developing countries. Chapters in this part discuss relevant, fit-for-purpose and integrated risk analysis methods that may play an important role in decision-making for the release of various GM products. How can the new methods be applied in conducting case-by-case risk assessment under different geographical and social conditions? What are the critical elements that may facilitate decision-making among relevant stakeholder groups such as scientists, policymakers, farmers and environmental activists? How important is it to inform consumers through product labelling? How can these elements lead to a consensus for policy implementation?

Part II: Diversification of Expertise

The second part of the book focuses on diversification of expertise such as that of natural and social scientists in the risk–benefit and socio-economic analysis of GMOs. In Chapters 8–12 the complexities of building appropriate capacity amongst scientists and regulators in a developing country context are discussed. The issues range from lack of expertise, and limited financial and infrastructure capacity, to shortcomings in political and organisational structures. After reading this part, readers will have a clear understanding about the debate regarding the inclusion of socio-economic considerations in the decision-making process. Readers will also learn about the importance of facilities, technical scientific capacity and expertise in risk assessment of GMOs as well as opportunities for international collaboration.

Part III: Risk-Analysis-Based Regulatory Systems

This part (Chapters 13–17) describes the development of risk analysis methods, and outlines issues for risk analysis and risk communication that may improve regulatory policy and decision-making. What have we learnt from past experience that should help to determine the way forward? After reading this part, readers will appreciate how GMO regulatory mechanisms can be undermined by socio-political interests and through inappropriate implementation of international protocols such as the Cartagena Protocol on Biosafety (CPB). Readers will also appreciate a variety of strategies outlined in each of these chapters to improve GMO policy development and regulatory decision-making, especially in developing countries.

Part IV: Case Studies from Developing Countries

In Chapters 18–24 a number of case studies from developing countries are presented, illustrating challenges but also successes that might be emulated by other developing countries. China, India, South America (Brazil and Argentina) and both east and west Africa are represented. Despite differences between the situations of individual countries, some commonalities are evident.

The Chapters

Part I

In Chapter 2, Recent Scientific Developments in Genetic Technologies: Implications for Future Regulation of GMOs in Developing Countries, Denis J. Murphy highlights the history of modern biotechnology, including GMOs and the emergence of GM-related technologies like genome editing, from a developing country perspective. This chapter discusses the impediments for existing GMO biosafety regulatory frameworks at European–USA level, the implication for developing countries and the potential benefits of genome editing in the latter. The author argues that genome editing technologies may need less regulation in view of the fact that the new techniques do not necessarily involve insertion of novel genes as in GMOs, thereby paving the way for new innovations for improving agricultural productivity and food security in developing countries.

In Chapter 3, A Strategy for Integrating Science into Regulatory Decision-Making for GMOs, Craig *et al.* present the use of problem formulation in risk assessment of

GMOs and highlight six steps in the risk assessment as used in workshops organised by the International Centre for Genetic Engineering and Biotechnology (ICGEB). They demonstrate the application of the methodology using a case study from one of their workshops, and discuss the integration of the risk assessment in a broader framework of risk analysis. They conclude that careful evaluation of the pathway for the risk assessment of GMOs can help facilitate decision-making for the application of GMOs.

In Chapter 4, Governance of Benefits and Risks of GMOs in Developing Countries, E. Jane Morris examines the dimensions of risk and benefits of GMOs on the basis of available evidence in the literature. She highlights some examples of potential benefits of GMOs in developing countries and points to the fact that the consideration of benefits is largely absent from European legislation as well as from international regulatory instruments such as the CPB. She presents a survey of 38 developing countries to assess to what extent their legislation accommodates a consideration of benefits. She argues for the need to develop a structured approach to risk–benefit analysis at international level.

In Chapter 5, The Canadian Experience with the Creation and Implementation of Regulatory Frameworks for the Environmental Release of GM Crops as a Model for Developing Countries, Phil Macdonald of the Canadian Food Inspection Agency analyses the extensive Canadian experience of risk assessment and management of large-scale commercial GM crop releases over the past two decades, the lessons learned and the implications for developing countries. This chapter describes how product-based risk assessments were formulated and how they have evolved in the Canadian setting, which is dominated by grain crops such as canola/rapeseed and where much of the crop product is subsequently exported. Important elements in the risk assessment process include the monitoring of initial small-scale releases, either in a confined or in an unconfined setting, and of the wider environmental effects, either actual or potential.

In Chapter 6, Labelling of Food from GMOs: Options to Consider by Developing Countries, Jansen van Rijssen and van der Walt provide an overview of labelling of GM food and products with a specific focus on the requirements of international regulatory bodies. They stress that labelling has to be regarded as a consumer right rather than a safety issue. They describe a number of challenges for labelling, particularly for developing countries, including increased costs for consumers. The authors discuss the lack of harmonisation of rules on GM labelling and the concern that the proliferation of standards can lead to trade barriers. The authors conclude with four recommendations for governments to take action.

In Chapter 7, Building Human Capacity and Skills in Biosafety: Lessons Learned and Emerging Best Practices, Komen and Koch review capacity development programmes on biosafety in developing countries with a focus on sub-Saharan Africa, and highlight that many African countries are making slow progress in developing functional biosafety frameworks despite international support. The authors argue that short-term training and workshops have failed to deliver organisational capacity, and that a more strategic, long-term and needs-driven approach is necessary, focusing on providing regulators with practical skills. There should be better coordination between organisations to prevent duplication of efforts, and more regional collaboration. Monitoring and evaluation systems should gauge the performance of capacity-development programmes.

Part II

In Chapter 8, Regulation of GMOs in Developing Countries: Why Socio-Economic Considerations Matter for Decision-Making, Falck-Zepeda and Gouse provide real-life examples of socio-economic impact assessment of GMOs and how their consideration in decision-making may facilitate or hinder adoption and sustainable use of this technology. The authors describe the divergent views on the need for inclusion of socio-economic considerations (SEC) in biosafety decision-making. Many developing countries have incorporated SEC into GMO legislation without a clear definition of what doing so entails, leading to an unworkable system with increased costs and regulatory delays. *Ex-post* studies identified some of the unpredictable factors that hinder adoption of GM technology in a sustainable manner, which can be difficult to capture in *ex-ante* studies, thereby making SEC difficult for decision-making. However, this chapter does not rule out inclusion of SEC in biosafety decision-making as this can have a positive impact, depending on how it is implemented.

In Chapter 9, Food Safety Assessment of Genetically Modified Crops in Developing Countries: The Experience in Africa, Lemgo *et al.* demonstrate the current regulatory and technical challenges that are associated with food safety assessment in developing nations, with a focus on countries in Africa. In addition to limited technical capacity, there is often a lack of coordination among key ministries, and representatives with inappropriate expertise are included in technical committees. Obtaining *de novo* food safety data for GM food security crops will be challenging. Regional or continental harmonisation of food safety risk assessment could reduce the regulatory burden for individual countries.

In Chapter 10, Being Scientific about Socio-Economics in GMO Decision-Making in Developing Countries, Monica Racovita states that any regulatory framework for GMOs including SEC for developing countries should be based on the scientific method and legitimate decisions. As mentioned in Chapter 8, this chapter also discusses the influence of the CPB and how the introduction of SEC into the CPB increases the potential for conflict with the World Trade Organization (WTO) science-based approach. She mentions that many developing countries are clamouring for inclusion of SEC despite the lack of a clear definition of the term, and the concomitant increase in cumbersome regulatory processes. For developing countries that do wish to include SEC, she describes a pragmatic evidence-based scientific approach to SEC that employs problem formulation.

In Chapter 11, The Sharing of Information in Risk Assessment: How National Authorities Cooperate, Kearns *et al.* describe cooperative activities in GMO risk assessment from the perspective of the Organisation for Economic Co-operation and Development (OECD). The OECD's consensus documents on the biology of a wide range of crops and traits provide a common basis for environmental risk assessment. Likewise, consensus documents for food safety assessment provide compositional data on a variety of crops, against which a GM derivative can be compared. A further document describes molecular characterisation of plants, and other documents are in development. Developing countries have been fully involved in the preparation of these documents, and they are freely available to regulators in developing countries.

In Chapter 12, Should GM Rice with Nutrition Benefits Be Deployed? Findings from Biotech and Socio-Economic Research, De Steur *et al.* seek to answer this question by looking at the evidence based on projected health impacts and consumer studies with a focus on vitamin A-enriched 'Golden Rice' and Folate Biofortified Rice in developing countries. Through *ex-ante* studies, these authors discuss the positive attitudes of consumers towards biofortified rice and their willingness to pay for the product. The authors of this chapter point out that regulatory constraints are a major hurdle, arguing for the need to shift socio-economic research on biofortified rice towards a stakeholder approach that will help unravel some of the challenges that may further delay the introduction of the new innovation.

Part III

In Chapter 13, The Risk Analysis Framework and Biosafety Policy, Phillips and Smyth describe the history of risk analysis and its linkage to biosafety regulatory systems. The authors of this chapter discuss the articulation and balancing of errors in the risk analysis system and state that, while regulators fear making Type I errors (acceptance of unsafe products), the unchecked precautionary principle in the CPB has caused Type II errors (rejection of product). This has been a major hindrance to development and acceptance of GMO products particularly in developing countries, as well as in the European Union. They point out that several attempts to coordinate and harmonise regulatory protocols governing GMOs have been unsuccessful to date. To facilitate the introduction of science-based regulatory systems, the authors call for an increase in scientific technical capacity in developing countries, particularly at regional level, while encouraging individual countries to improve their governance systems.

In Chapter 14, The Precautionary Principle as a Barrier to GMO Risk Analysis: Elicitation of Experts' Viewpoints, Ademola Adenle describes a series of interviews with experts who gave their opinions on the impact on developing countries of the precautionary principle in the CPB and the controversy between the US and European approaches to GMO risk analysis. In Europe, decisions are made on a political basis rather than a scientific basis, and this trend is followed in many developing countries. Some experts felt that the UNEP-GEF capacity building programmes under the CPB had encouraged a restrictive approach to GMOs, while others blamed the lack of scientific expertise in many developing countries. Experts were concerned that new gene-editing techniques would fall under the same regulatory constraints as GMOs. On the basis of these experts' opinions, the author recommends a review of the CPB and argues that there is no need for international regulation of GMOs. There should be more transparency in risk analysis, and, in the case of developing countries, risk assessment should be undertaken at a regional rather than a national level.

In Chapter 15, How Do Socio-Political Disputes Concerning GMOs Influence Decision-Making in Developing Countries?, Azadi *et al.* provide more explanation of the controversy between the EU and US approaches and the impact of this conflict on developing nations. Developing countries that have adopted GM crops have demonstrated economic and environmental benefits, but other developing countries have adopted a 'go-slow' approach, often citing the EU position as a reason. Policymakers in developing countries need to consider a number of objectives, including market access, agricultural productivity, food availability and biosafety risks, when deciding whether or not to accept GMOs. The authors stress that fragmentation at international level has increased the complexity and made it more difficult for countries to reach a decision.

In Chapter 16, Biosafety Communication: Beyond Risk Communication, Sonnino and Sharry stress the need for improved transparency and public participation in biosafety decision-making. International frameworks specify the need for risk communication and public participation. However, the ongoing controversies demonstrate that communication approaches adopted to date have not worked. They advocate a participatory approach to biosafety communication as key to engaging all the relevant stakeholders, including the public, scientists and media, in the decision-making process. Decisions should be socially robust as well as scientifically sound. Biosafety should be communicated by authoritative and neutral sources without vested interests.

In Chapter 17, The Role of Mass Media and Lobbies in the Formulation of GMO Regulations, Mauro Vigani discusses the impact of mass media in the formulation of GMO regulation through the lens of social science and political economy approaches. Under the social science approach, the author describes how communication strategies through mass media influence the perceived risks and benefits of GMOs among the public. Media bias has contributed to negative public perceptions, and benefits are not sufficiently communicated. From the political economic standpoint, biased media coverage can also influence government policies that respond to public concerns. The structure of the media market and the strength of lobbying groups can influence the outcome. The influence of lobbying groups could also affect regulatory approaches to new genome-editing techniques.

Part IV

In Chapter 18, Risk Regulation of Agricultural GMOs in China: Challenges and Prospects, Wen Xiang examines risk regulation of agricultural GMOs and identifies some of the challenges related to adoption of GM rice in China. The author explains the processes involved in risk assessment and the roles of various stakeholders in GMO regulation at the national level. The system has some weaknesses, including a lack of transparency and public participation. Following public concern, the commercialisation of Bt rice has been suspended. Problems include confusing labelling requirements, lack of clarity on socio-economic evaluation, delays in decision-making, and a top-down process that does not incorporate input from farmers and consumer groups. The author concludes that risk regulation of GMOs is not only a scientific issue but also a social issue, and therefore government should be more accountable by engaging the broader public in decision-making.

In Chapter 19, Commercial Use and Governance of Bt Cotton in China, Li *et al.* provide a detailed analysis of the impact of the successful introduction of Bt cotton in terms of economic and environmental benefits. Opposition has focused on evolution of pest resistance and secondary pest outbreaks. The authors reiterate concerns raised in Chapter 18 regarding low public awareness, public mistrust of the technology and inadequate communication by scientists. Empirical evidence of long-term benefits could help resolve the public debate.

In Chapter 20, The Regulatory Regime of Genetically Modified Crops in India, Krishna Ravi Srinavas explains the complex regulatory system for biotechnology and biosafety in India, which involves a balance between the promotional Ministry of Science and Technology and the more precautionary Ministry of Environment, Forestry and Climate Change. So far Bt cotton is the only crop approved for commercial cultivation. Challenges to an application for commercialisation of Bt brinjal have exposed polarised attitudes and resulted in the imposition of a moratorium on its release. GM mustard and chickpea, both developed by Indian scientists, face similar problems. The regulatory system has been challenged in the Supreme Court, which appointed a committee of technical experts. This committee has recommended not only a revamp of the regulatory regime but also the introduction of extremely stringent and expensive long-term toxicity-testing measures. Until a new and independent regulatory structure is established, the current logjam will continue.

In Chapter 21, The Argentinian GMO Biosafety System: An Evolving Perspective, Vicién and Trigo describe the GMO regulatory framework and its evolution. Subsequent to an assessment of food and feed safety, the effects of commercial release on trade and markets are separately assessed under the so-called 'mirror policy'. Commercial approval normally requires prior approval of an event in Argentina's main markets. The authors state that the biosafety regulatory framework is responsive and minimally bureaucratic, and is widely respected. They argue for international harmonisation of regulatory oversight to minimise problems arising from asynchronous approvals and to facilitate trade in commodities. Regulatory systems need to function properly and decision-making processes must be clear and transparent.

In Chapter 22, The Brazilian GMO Regulatory System: A Historical View and Perspective, Velini *et al.* discuss how, following an initially disrupted start with regulatory challenges to the legal framework, a new regulatory system was introduced in 2005. Subsequent to this, Brazil has become the second-largest grower of GM crops after the USA. Scientific risk assessment is carried out by the National Biosafety Technical Commission, while optional socio-economic assessment is the responsibility of the National Biosafety Council. However, despite the widespread adoption of GM crops, hardly anything developed within Brazil's public sector has come to market, due to the high cost of meeting regulatory requirements. Moreover, public opinion is generally unfavourable to GMOs. The authors stress the need for better risk communication, focusing on safety and benefits of the technology.

In Chapter 23, An Effective Regulatory Regime Supported by Research and Development Is Key to Adoption of GM Technology in West Africa: Burkina Faso and Nigeria as Case Studies, Akinbo *et al.* explain that in West Africa only three countries (Burkina Faso, Ghana and Nigeria) have functional biosafety laws that allow field trials with GMOs, and only Burkina Faso has experience with commercialisation. The adoption of Bt cotton in Burkina Faso represents a learning experience for the region, despite setbacks and ongoing problems with cotton lint quality. The authors urge West African countries to increase investment in biotechnology R&D in order to place them on a growth trajectory similar to that of Brazil or China. There is a significant hurdle to move from approval of a fibre crop (cotton) to approval of feed and food crops. The region needs regulatory provisions that encourage less expensive and quick and safe evaluation of local GM crops, and a harmonised regulatory approach in the region could facilitate this.

In Chapter 24, A Pathway for Biosafety Regulation of GMOs in Sub-Saharan Africa, Julius Ecuru starts with the premise that biosafety regulation should be viewed as an integral element of the gene technology innovation system in Sub-Saharan Africa (SSA). He discusses the concept of innovation in terms of factors that facilitate GMO development and diffusion in SSA. He argues that application and diffusion of GMOs as guided by biosafety regulation should be a shared responsibility between the developers, government and other key stakeholders. GMOs are being developed in SSA for subsistence crops through publicly funded programmes, but also separately for commercial advantage. Taking the example of Uganda, field trials of food security crops are under way, but the legislation that would allow commercialisation has not yet been passed. Where there are strong commercial interests, regulation tends to proceed more quickly. For the most part, the regulation of the technology has been reactive rather than proactive, and preventative rather than facilitating. The author recommends collective engagement of key stakeholders to develop functional biosafety systems that can enhance GMO development and introduction.

Conclusion

The book's concluding chapter highlights some key messages from the book, and presents a series of recommendations that could effectively contribute to the successful development of GMOs in developing countries.

Part I

Risk Analysis Methodology and Decision-Making

2 Recent Scientific Developments in Genetic Technologies: Implications for Future Regulation of GMOs in Developing Countries

Denis J. Murphy

Introduction

In the early 1970s the first man-made transgenic organisms, popularly referred to as genetically modified organisms (GMOs) were created. At the time, this technology was regarded as a potentially revolutionary scientific development that could create entirely new, and possibly harmful, life forms. The resulting transgenic organisms contained combinations of DNA sequences either copied from existing organisms or created synthetically in a form that had never previously existed in the biological world. At the time, many leading researchers were concerned that novel biohazards might arise from this new technology. As a result, the Committee on Recombinant DNA Molecules was set up and an international conference convened at Asilomar, California in 1975 (Berg, 2008). Following this conference, a series of non-binding guidelines relating to GMOs was issued to inform future scientific research and the regulation of GMO use. As Paul Berg from Stanford University later stated,

> Participants agreed on the final day of the conference that research should continue, but under stringent restrictions. The recommendations formed the basis of the official US guidelines on research involving recombinant DNA, issued in July 1976. They have proved remarkably effective. (Berg, 2008)

During the late 1970s, different types of systems were drawn up by governments in the USA and Europe to regulate GM microbes. As the scope of GM technology expanded, further regulations were enacted during the 1980s and 1990s to cover more complex multicellular organisms, including plants and animals. These original systems are still the basis for the more recently established regulatory systems set up in some developing countries. Indeed, all the regulatory systems in use in 2016 are still based on decades-old GM technologies dating from the twentieth century. Moreover, there is still a fundamental dichotomy between the product-based regulatory paradigm, in countries such as the USA and Canada, and the process-based paradigm prevalent in Europe, where no account is taken of potential benefits of the technology (see Chapter 4). Since the turn of the 2010s both of these regulatory models have come under increased scrutiny and challenge in the face of recent developments in GM technology. This is especially true for some of the new gene-editing methods discussed in this chapter. For example, in contrast to classical twentieth-century GM technologies, gene editing can sometimes leave no detectable trace in the recipient genome and therefore no evidence that the organism has been subjected to any form of genetic modification.

In this chapter, recent scientific findings concerning transgenic organisms and the development of new GM-related technologies will be reviewed, particularly in the context of developing countries. Their potential impacts on the development of new products and processes, and the wider context of such technologies, will be assessed. The current state of the art in terms of the deployment of new and old GM technologies in developing countries will be outlined. Finally, the implications of these developments for the future regulation and management of GMOs in developing countries will be discussed.

New Scientific Discoveries Are Changing the Nature of GM-Related Technologies and Our Perspectives about Gene Transfer as a Natural Biological Process

Since the turn of the twenty-first century, fundamental scientific research has resulted in the discovery of several new ways in which the genetic composition of organisms could be deliberately and precisely modified to create useful new phenotypes that may not be subject to existing GMO-related regulations in some jurisdictions, such as the USA. Some recent examples of these emerging GM technologies include cisgenics, synthetic amino acids and gene editing via CRISPR-related techniques (CRISPR stands for 'clustered, regularly interspaced, short palindromic repeats'). The so-called 'cisgenic' methods of gene transfer, developed in the 2010s, are claimed by some scientists to be sufficiently different from transgenic methods that they should not be subject to the same regulatory constraints as the latter (Telem *et al.*, 2013).

In 2015, a radical new GM technology relating to modified microbial organisms and involving the use of synthetic amino acids was published in *Nature* (Dolgin, 2015). The authors claim that this technology could decisively address consumer and regulatory concerns about containment and biosafety of GMOs that carry potentially problematic traits such as expression of bioactive pharmaceutical agents. Also, as discussed below and in Figure 2.1, the Swedish Board of Agriculture recently published findings from field trials with the conclusion that plants mutated via CRISPR-Cas9 technology did not contain 'foreign' DNA and were 'exempted from the GM legislation' (Eklöf, 2015). The potentially game-changing nature of new GM-related technologies was demonstrated in late 2015 when, in an echo of the Asilomar conference on GMOs in 1975, a special meeting of researchers, regulators, ethicists and other interested parties was convened in Washington, D.C., to discuss the implications of the new gene-editing technologies, especially as applied to humans (National Academies, 2015).

In addition to the many technical advances in GM-related methods since the 1970s, there have also been fundamental changes in our knowledge of inter-organism and inter-species gene transfer in the so-called 'natural' world. Prior to the 2000s, it was believed that the transfer of functional DNA sequences between different species was limited to

The *Arabidopsis thaliana* PsbS mutant

The same mutant produced five times, but which ones are within the scope of the European GMO legislation?

What is the function of PsbS?
PsbS is a protein which is involved in photosynthetic light harvesting and has been characterised as a 'safety valve'. Plants that lack the protein show reduced fitness and seed production under natural conditions. Mutant plants that fully lack the protein or produce a dysfunctional protein have been obtained in different ways.

A. The radiation mutant

The first PsbS mutant was made by exposing Arabidopsis plants to fast neutrons. The fast neutrons generate damage in the DNA that is repaired by the cells' own DNA-repair machinery. During this repair, the whole PsbS gene is deleted and PsbS is therefore not present in the plant. However, changes in other genes may also occur following the radiation.

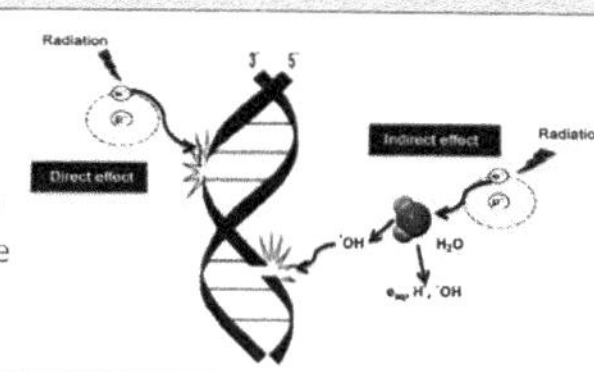

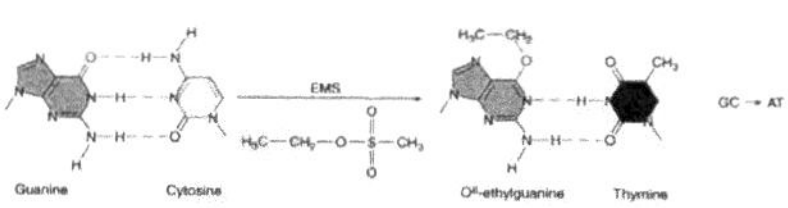

B. The chemically induced mutant

The second PsbS mutant was made by exposing Arabidopsis plants to the chemical mutagen EMS. One letter in the gene for PsbS was changed, leading to a dysfunctional PsbS protein.

C. The T-DNA mutant

The third PsbS mutant was made by transferring so-called T-DNA from the soil bacterium *Agrobacterium tumefaciens* to Arabidopsis plants. The T-DNA was inserted into the PsbS gene, leading to a disruption of the gene. The result is that the PsbS gene is no longer functional.

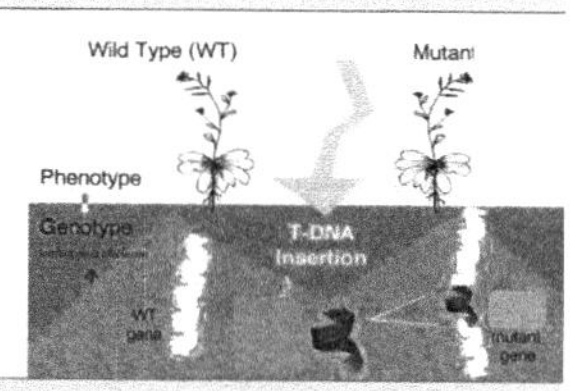

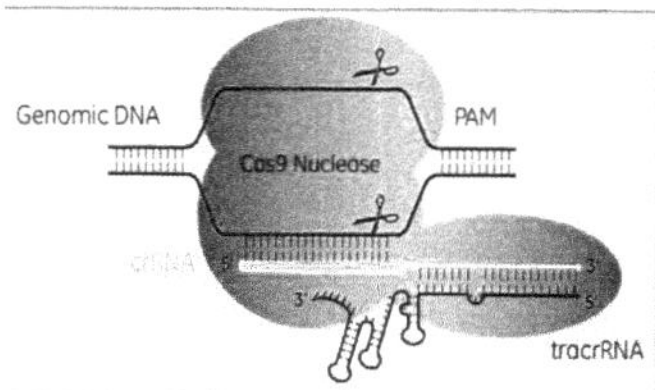

D. & E. The modern genome-edited mutant

The most recent technology to generate PsbS mutants is the so-called CRISPR/Cas-mediated genome editing. This CRISPR/Cas system generates two double-strand breaks close to each other at predetermined locations in the PsbS gene. The DNA-repair machinery repairs the break, deleting the DNA between the breaks, leading to a dysfunctional PsbS gene. The intermediate mutant that still contains the DNA for producing the CRISPR/Cas complex that generates the double-strand break is called mutant D. The final mutant E is produced from mutant D after a round of spontaneous fertilisation. One quarter of the offspring no longer contain the genes for the complex and these are selected as mutant E. They contain no foreign DNA and only differ from wild-type Arabidopsis by a small deletion in the PsbS gene.

A GMO or not a GMO?

Mutants A and B are not within the scope of the European GMO legislation. Mutant C is, even though T-DNA sequences have been shown to naturally occur in crops like tobacco and sweet potato, considered GMOs. What about mutants D and E? Mutant D still contains foreign DNA and is therefore considered a GMO. Mutant E does not contain foreign DNA and only lacks a number of DNA base pairs in the PsbS gene. Does this removal of a few base pairs constitute a novel combination of genetic material? Probably not. When compared with mutants A and B it would be illogical to subject the genome-edited mutant E to the requirements of the GMO legislation.

Figure 2.1. Comparison of various methods that can be usd to produce a new mutation in a plant and the potential regulatory implications. Adapted from a diagram courtesy of the Swedish Board of Agriculture.

very simple organisms such as bacteria. In contrast, it was (and still is in many quarters) believed that the genomes of more complex organisms such as higher plants and animals were relatively fixed within individual species and that the insertion of exogenous DNA was therefore 'unnatural'. This perspective has now entirely changed and, as discussed below, there are many recently discovered examples of gene transfer between completely unrelated plants, animals and microbes, leading to the creation of new genotypes and phenotypes that sometimes have evolutionarily significant selective advantages (Gao *et al.*, 2014; Soucy *et al.*, 2015). To quote again from Berg (2008),

> In the 33 years since Asilomar, researchers around the world have carried out countless experiments with recombinant DNA without reported incident. Many of these experiments were inconceivable in 1975, yet as far as we know, none has been a hazard to public health. Moreover, the fear among scientists that artificially moving DNA among species would have profound effects on natural processes has substantially disappeared with the discovery that such exchanges occur in nature.

In addition to naturally occurring DNA transfer between species, there are reports of soil bacteria acting as 'natural' genetic engineers by inserting novel genes into plants, including crops (Kyndt *et al.*, 2015). The progeny of such plants will inherit the bacterial genes and pass them on in turn to their descendants. This results in the non-human-mediated creation of a completely new transgenic plant variety – in effect a naturally occurring GMO. In the case reported by Kyndt *et al.* (2015), a transgenic version of the sweet potato plant with four extra bacterial genes was found. The sweet potato has been grown by farmers as a food crop for millennia but, interestingly, wild relatives of sweet potato did not contain the bacterial genes. This means that humans have inadvertently selected a transgenic version of the sweet potato to grow as a crop in preference to non-transgenic versions of the plant. In the light of these discoveries, it is difficult to maintain the stance that GMOs themselves, and even the creation of GMOs (whether by humans or bacteria), are completely aberrant and fundamentally 'unnatural' instances of human interference in the wider biological world.

The Status of GM Crops and Technologies in Developing Countries in 2016

Commercial GM crops have been grown on an increasingly widespread scale since 1996, when 4.3 million hectares (Mha) were planted, nearly all in the USA. By the 2014 growing season, the global area of GM crops had increased to 181.5 Mha, of which 96 Mha were being grown in developing countries (James, 2015). The major developing countries growing GM crops in 2014 were Brazil (42.2 Mha), Argentina (24.3 Mha), India (11.6 Mha), China (3.9 Mha), Paraguay (3.9 Mha), Pakistan (2.9 Mha), South Africa (2.7 Mha), Uruguay (1.6 Mha), Bolivia (1.0 Mha), the Philippines (0.8 Mha), Burkina Faso (0.5 Mha), Myanmar (0.3 Mha) and Mexico (0.2 Mha). Over the first two decades since their introduction, the portfolio of GM crop species has remained very narrow, with just four crops, namely soybean, maize, cotton and rapeseed, accounting for the vast majority of the global area. In some developing countries only one GM crop was under cultivation in 2014. For example, in India, Pakistan, Burkina Faso and Myanmar, only GM cotton was grown. The number of genetic traits that were modified in this limited catalogue of GM crops is also very small, with only two traits, namely herbicide tolerance and/or Bt-based insect resistance, accounting for >99.9% of the total area (James, 2015).

In all of these cases, the types of GM technology used to create the transgenic crops were similar to the original methods used to make the earliest GM plants in the 1980s. For example, such transgenes are inserted randomly into a recipient crop genome either by using *Agrobacterium*-mediated transformation of plant tissues or via biolistic delivery using small DNA-coated particles that are propelled into plant tissues from a so-called gene gun (Murphy, 2011). In both cases the site at which the transgene-containing DNA is inserted into the recipient genome is entirely random. It is also common to have multiple copies of either fully intact or partially fragmented pieces of exogenous DNA inserted at various locations in the recipient genome. The location of the insertion site might also determine the expression levels or stability of the transgene expression, which complicates the selection of suitable transformants. Therefore, it can be argued that, while such 'first-generation' GM tools can be an improvement on many earlier plant-breeding methods, they are still relatively crude and limited in their scope for creating radically new phenotypes.

Another feature of such twentieth-century GM crops is that they were normally created using selectable marker genes that confer antibiotic resistance or herbicide tolerance. In some commercial GM crops in use in 2015, these selectable marker genes are still present in the crop genome. While there have been no documented cases of adverse consequences arising from the presence of selectable marker genes, it is generally agreed that it would be better if the transgenic plants did not contain such genes (Murphy, 2011). To summarise, by the time of writing in early 2016, developing countries accounted for well over 50% of global GM crops by area. However, this is well under 10% of the total arable area in developing countries, so GM crops are by no means a predominant form of plant husbandry. We can now regard these crops as an early form of GM technology that has remained very restricted in terms of the range of species and genetic traits available to farmers. Arguably, these 'first-generation' GM crops are the products of a now-outdated set of technologies dating back as much as thirty years. Moreover, as we saw above, many of the regulatory processes devised for the management of such crops are based on outdated scientific knowledge and are not fit for purpose in addressing the 'second generation' of GM crops that will soon arise from newly emerging technologies of gene editing.

Reception of First-Generation GM Crops and Technologies in Developing Countries

As discussed in Chapter 4, GM crops and technologies have had very mixed receptions in developing countries. As in many industrialised countries, some of the objections to GM crops in developing countries have been based on inaccurate knowledge of the underpinning science. Probably the major commercial distinction between GM and other modern so-called 'non-natural' breeding methods is that new GM varieties can be protected by patent legislation, which is stronger then the plant breeders' rights available for non-GM varieties. For example, the developer of a new GM variety has ownership of the plants and their progeny and can stop them being propagated for reuse by growers.

There is still a widespread perception that GM crops are overwhelmingly a 'Western' technology developed by private sector USA- or Europe-based multinational companies as part of a for-profit intensification of agriculture. Therefore, although more than 50% of GM crops grown in 2015 are being cultivated in developing countries, there is little sense of ownership of the technology and even the regulatory frameworks, if they exist at all in developing countries, tend to be closely modelled on those that were drafted in most Western countries several decades ago. However, this situation is gradually changing as developing countries begin to establish their own research programmes that are able to use modern genomics-based breeding technologies, including some that are GM-related. During the 2010s, the impact of genomic technologies on agricultural systems in developing countries was rather mixed. In some of the more rapidly expanding economies such as India, China, Brazil, Malaysia and South Africa, uptake of genomic technologies has been relatively rapid, but in much of sub-Saharan Africa it was much slower.

However, as we have seen from other areas such as DNA sequencing, the costs and resource-intensity of these genomics-related technologies will steadily decline with time. Such technologies will also become increasingly easy to use, which will greatly facilitate their dissemination to regions with less advanced know-how and infrastructure. Dissemination of all forms of plant-related genomic technologies will be especially important for those parts of the developing world, such as sub-Saharan Africa, that face the most urgent food security and environmental threats and where there are potentially great gains to be won in terms of increasing crop yield and quality. As discussed below, some of the new gene-editing technologies have the promise to enable the creation of more radical genetic alterations that would be directly relevant to poorer farmers. These technologies also require less investment in biotechnological equipment and expertise, and therefore have the potential to disseminate and democratise the ability to use such advanced technologies, especially for public good rather than commercial ventures.

The New Genome-Editing Technologies – A Game Changer for Genetic Modification?

Since the mid 2010s several new and potentially revolutionary forms of plant and animal (including human) GM technologies, collectively known as 'gene editing', have been developed. Probably the most powerful of these is the CRISPR system (Bhaya *et al.*, 2011; Hsu *et al.*, 2014; Mao *et al.*, 2013). In 2015, the CRISPR/Cas9 system was described in a *Nature* article as 'the biggest game changer to hit biology since PCR' (Ledford, 2015). Applying this method to crop improvement has opened up many new possibilities for radical genome modifications (Belhaj *et al.*, 2015; Zhang *et al.*, 2014). Gene editing will greatly accelerate crop and livestock breeding by precise and predictable genetic modifications directly in selected individuals and will also enable simultaneous modification of multiple traits.

In terms of crop breeding, this means that it will soon be possible to advance from the random insertion of single genes or small numbers of genes into a genome (as in traditional GM) to the highly precise insertion at a defined location of large numbers of genes, chromosome segments or pseudo-segments encoding entire metabolic pathways into virtually any plant species. Methods such as TALEN (which stands for 'transcription activator-like effector nucleases') and CRISPR/Cas9 can be used for gene knockouts, e.g., to eliminate unwanted genes that adversely affect food quality or confer susceptibility to pathogens, or that divert metabolism away from valuable end products (Davis, 2015). An example reported in 2014 was the use of both TALEN and CRISPR/Cas9 to target the genes of the mildew-resistance locus in wheat. This resulted in the production of plants resistant to powdery mildew disease, which is a serious crop disease (Wang *et al.*, 2014). Genome editing and other methods will be increasingly used to create transgenic livestock with improved quality and yield as well as for reducing environmental footprints (such as methane emissions) and for production of valuable products, including pharmaceuticals, from transgenic crops and livestock (Murphy, 2007; Laible *et al.*, 2015).

Precise nucleotide exchanges using oligonucleotide donor sequences can also be used to modify the regulatory sequences upstream of genes that determine agricultural performance, leading to improved crop yields. In some cases it will be possible to re-programme genes so that they are expressed in different organs at much higher rates. For example, seed-specific genes regulating storage lipid accumulation could be re-programmed to function in leaves or roots in order to generate high levels of oil in such vegetative tissues (Murphy, 2014a; Vanhercke *et al.*, 2013). The oil could be used for food or fodder, as a source of renewable oleochemicals for industry, or even for biofuels. Within the next decade, it is likely that entire metabolic pathways will be transferred to the plant and tissue of choice so that some crops could become low-cost production systems for industrial green chemicals, including high-value specialty chemicals (Murphy, 2014a), nutraceuticals such as omega-3 oils (Napier *et al.*, 2015) and pharmaceuticals such as vaccines, antibodies and drugs (Murphy, 2007).

Since the first plant genome sequence was published in 2000 (The Arabidopsis Genome Initiative, 2000), many dozens of species, including nearly all of the major crops, have been sequenced and their genomic data uploaded into public databases. In many cases, there is still a great deal to do in terms of improving the quality of the existing genomic data and the identification of key genes that regulate important agronomic processes such as seed development, yield, photosynthetic efficiency, stress/disease responses and climatic resilience (Murphy, 2014b). However, good progress has been made and, once key genes involved in developmental and metabolic regulation have been identified, the new gene-editing technologies will enable these genes to be precisely altered in order to achieve the desired agronomic improvement (Edwards and Batley, 2010; Ricroch and Hénard-Damave, 2015). The identification and editing of such sets of genes will enable breeders to consider the manipulation of complex traits, such as plant architecture (Foster *et al.*, 2015), salt tolerance (Fita *et al.*, 2015) and photosynthetic efficiency (Covshoff and Hibberd, 2012), in food crops.

Implications of New Technologies for Regulatory Systems

As discussed earlier in this chapter and elsewhere in this book (see Chapter 4), the current systems designed to regulate GMOs, including transgenic crops, were largely drawn up in the wake of the Asilomar Conference and subsequent developments during the period from the 1970s to the 1990s. Quite apart from the fact that these regulatory systems are inconsistent among countries, and in many cases lack a scientific basis, they have also signally failed to keep abreast with developments in genetic technologies, particularly in the 2010s. As early as 2009, concern was expressed about the regulatory implications of oligonucleotide-mediated mutagenesis (Breyer *et al.*, 2009), which had been shown to be a useful tool for the creation of site-specific DNA changes in several crops, including rice (Okuzaki and Toriyama, 2004; Dong *et al.*, 2006).

There have been several recent authoritative reviews of the implications of gene-editing technologies (Oye *et al.*, 2014; Petherick *et al.*, 2015; Roberts *et al.*, 2015), but these have yet to result in any regulatory changes, even in the USA, which has often led the way in GM regulation (National Academies, 2015). In Europe, legislators have been very slow to respond to such developments. As early as 2007 the European Commission appointed an expert panel to advise it on the ever-expanding plant-breeding toolbox, but the panel's report, submitted in 2012, was never officially published, although a draft version is available online (EU New Techniques Working Group, 2012). Even in 2016, there is little evidence of awareness of the potentially momentous nature of these new technologies amongst policymakers (Abbott, 2015; Anon., 2015a, 2015b). This is despite the many scientific publications that have highlighted the need for a reconsideration of the regulatory environment, particularly for GM crops, in the light of new technologies such as, but by no means limited to, gene editing (ACRE, 2012; Ainsworth, 2015; Nyman, 2014; Wolt *et al.*, 2016).

Another significant aspect of the new genome-editing systems is that it can be virtually impossible to detect resultant modifications in a genome. This is in contrast to conventional late-twentieth-century GM methods, for which the presence of novel DNA can be readily detected (Ledford, 2015; Lunshof, 2015). Therefore some new GM plants may not carry any proof of whether they were produced via a GM-type technology or via one of several non-GM technologies that could have been used instead. This development has the potential to undermine the entire framework that is currently used for the regulation of GMOs because, for example, it would become impossible to distinguish between a new plant variety that has arisen via spontaneous mutation from a plant produced via deliberate mutagenesis in the lab and a GM plant that has been modified by genome editing.

In late 2015, an assessment concerning whether plants created via the CRISPR/Cas9 system should be classified as 'GM' was published by the Swedish Board of Agriculture

(SBA) (Eklöf, 2015). The conclusion was that, while individual cases might differ, the Arabidopsis plants examined in their study should not be regarded as GMOs (see also Figure 2.1). It was stated that

> The SBA makes the interpretation that those plants in the description mutated by CRISPR / Cas9 and which do not contain any foreign DNA are exempted from the GM legislation. (Eklöf, 2015)

Despite such clear and explicit evidence-based verdicts from this and other expert studies, it is difficult to be optimistic that European regulations will be modified in the near future to recognise the new scientific landscape arising from gene editing and related technologies. In contrast, in the USA there are intensive consultations involving scientists, policymakers and legislators that seem more likely to give rise to some sort of amended regulatory structures within the next year or so. Given the enormous potential of gene editing for crops in developing countries (Erbentraut and Shapiro, 2015), it seems more likely that those countries will tend to model their own regulatory systems on US systems that will enable the technologies to be disseminated more rapidly to address real-life agronomic challenges.

Potential Impacts of New GM-Related Technologies in Developing Countries

In 2015, over half of all GM crops were being grown in developing countries, including several countries in Africa. There has been an increasing tendency for such crops to have been developed by public-sector institutions in developing countries and to be distributed at low cost to poorer farmers. An example is Golden Rice, developed by the International Rice Research Institute (IRRI) in the Philippines and scheduled for release to farmers in 2016–17 (Kowalski, 2015; Moghissi *et al.*, 2016). As developing countries produce new varieties of GM crops, with traits such as improved nutritional quality, drought tolerance and disease resistance, it will become increasingly difficult to defend the 1990s-era argument that GM crops represent a tool for the imposition of global food hegemony by Western governments and multinationals (Engdahl, 2007). Providing that realistic and evidence-based regulatory mechanisms are adopted, the new gene-editing technologies, in conjunction with genomics and other advanced breeding systems, have the potential to democratise the process of crop improvement in developing countries. This could enable those countries to better address their urgent needs related to priorities such as food security and climate change, as we will now briefly discuss.

According to the 2015 report by the FAO, IFAD and WFP, almost 800 million people, or one in nine in the world, continue to experience hunger (FAO, IFAD & WFP, 2015). While real progress has been made in recent decades, this has been patchy in some regions, such as sub-Saharan Africa. One issue that is especially acute in Africa is that many of the most important staples are grown by smallholders who cultivate so-called 'orphan crops', such as cassava. These crops have yet to benefit from many aspects of modern breeding technologies (Dawson and Jaenicke, 2006). Although orphan crops suffer from low yields and may be severely impacted by environmental changes, they have considerable scope for significant genetic improvement for yield, quality and resilience using new genomic approaches (Armstead *et al.*, 2009).

An example of a way forward is the African Orphan Crops Consortium. This is an international effort to improve the nutrition, productivity and climatic adaptability of some of Africa's most important food crops, helping to decrease the malnutrition and stunting rife among the continent's rural children. With the endorsement of the African Union, the goal is to sequence, assemble and annotate the genomes of 100 traditional African food crops and to make the results available in public databases. These 'orphan crops' are species that were neglected by researchers and industry because they are not economically important on the global market. By focusing on such orphan crops it should be possible to make significant gains in overall food production in sub-Saharan Africa by the 2050s.

Conclusions and Future Prospects

Over recent decades there has been a great deal of bad news and pessimism about developing country agriculture, especially in sub-Saharan Africa. For a variety of reasons the region was unable to match the truly spectacular yield gains made in the rest of the developing world as part of the momentous Green Revolution in the twentieth century. More recently, however, there have been signs that the rhetoric may be changing and there may be grounds for cautious optimism for African breeders, farmers and consumers (Cernansky, 2015; Sanchez, 2015). This may be the first glimmer of a new dawn both for industrialised countries and for developing countries in terms of making full use of the burgeoning toolkit now available for breeders and agronomists. In particular, the use of genome editing and genomics technologies has the scope to vastly increase crop yield, quality and biodiversity.

Although some of these new technologies are based on genome engineering, they are completely different from the now-outdated twentieth-century 'GM' technologies in terms of both precision and potential benefits. Surely, therefore, it is now time to move on from the old sterile debates about GMOs and to enable the new genetic technologies to be disseminated and applied for the good of humankind, albeit in the context of robust and evidence-based regulatory systems.

References

Abbott, A. (2015). Europe's genetically edited plants stuck in legal limbo. Scientists frustrated at delay in deciding if GM regulations apply to precision gene editing. *Nature* **528**, 319–320.

ACRE (2012). ACRE advice: New techniques used in plant breeding. [Online]. Available from www.gov.uk/government/uploads/system/uploads/attachment_data/file/239542/new-techniques-used-in-plant-breeding.pdf

Ainsworth, C. (2015). Agriculture: A new breed of edits. *Nature* **528**, S14–S15.

Anon. (2015a). Seeds of change. The European Union faces a fresh battle over next-generation plant-breeding techniques. *Nature* **520**, 131–132.

Anon. (2015b). Crop conundrum. The EU should decide definitively whether gene-edited plants are covered by GM laws. *Nature* **528**, 307–308.

Armstead, I. *et al.* (2009). Bioinformatics in the orphan crops. *Briefs in Bioinformatics* **10**, 645–653.

Belhaj, K. *et al.* (2015). Editing plant genomes with CRISPR/Cas9. *Current Opinion in Biotechnology* **32**, 76–84.

Berg, P. (2008). Meetings that changed the world: Asilomar 1975: DNA modification secured. *Nature* **455**, 290–291.

Bhaya, D. *et al.* (2011). CRISPR–Cas systems in bacteria and archaea: versatile small RNAs for adaptive defense and regulation. *Annual Review of Genetics* **45**, 273–297.

Breyer, D. *et al.* (2009). Genetic modification through oligonucleotide-mediated mutagenesis. A GMO regulatory challenge? *Environmental Biosafety Research* **8**(2), 57–64.

Cernansky, R. (2015). Super vegetables. *Nature* **522**, 146–148.

Covshoff, S. and Hibberd, J. M. (2012). Integrating C_4 photosynthesis into C_3 crops to increase yield potential. *Current Opinion in Biotechnology* **23**, 209–214.

Davis, E. (2015). Genome editing: which should I choose, TALEN or CRISPR? Technical note, GeneCopoeia Inc. [Online]. Available from www.genecopoeia.com/resource/genome-editing-talen-or-crispr/

Dawson, I. K. and Jaenicke, H. (2006). Underutilised plant species: the role of biotechnology. Position Paper No. 1. Colombo, Sri Lanka: International Centre for Underutilised Crops.

Dolgin, E. (2015). GM microbes created that can't escape the lab. Engineered bacteria kept in check with a designer diet. *Nature* **517**, 423.

Dong, C. *et al.* (2006). Oligonucleotide-directed gene repair in wheat using a transient plasmid gene repair assay system. *Plant Cell Reports* **25**, 457–465.

Edwards, D. and Batley, J. (2010). Plant genome sequencing: applications for crop improvement. *Plant Biotechnology Journal* **8**, 2–9.

Eklöf, S. (2015). CRISPR/Cas9 mutated Arabidopsis. [Online]. Swedish Board of Agriculture, Jönköping. Available from www.umu.se/digitalAssets/171/171717_backgroud-psbs.pdf and www.upsc.se/documents/Information_on_interpretation_on_CRISPR_Cas9_mutated_plants_Final.pdf

Engdahl, W. F. (2007). *Seeds of Destruction – The Hidden Agenda of Genetic Manipulation.* Montreal: Global Research.

Erbentraut, J. and Shapiro, L. (2015). The genetic revolution could curb world hunger and pesticide use. *HuffPost Science* 12 December 2015.

EU New Techniques Working Group (2012). Final report. [Online]. Available from www.seemneliit.ee/wp-content/uploads/2011/11/esa_12.0029.pdf

FAO, IFAD & WFP (2015). *The State of Food Insecurity in the World 2015. Meeting the 2015 International Hunger Targets: Taking Stock of Uneven Progress.* Rome: FAO.

Fita, A. *et al.* (2015). Breeding and domesticating crops adapted to drought and salinity: a new paradigm for increasing food production. *Frontiers in Plant Science* **6**, 978.

Foster, T. M. *et al.* (2015). Two quantitative trait loci, Dw1 and Dw2, are primarily responsible for rootstock-induced dwarfing in apple. *Horticulture Research* **2**, 15001.

Gao, C. *et al.* (2014). Horizontal gene transfer in plants. *Functional and Integrative Genomics* **14**, 23–29.

Hsu, P. D. *et al.* (2014). Development and applications of CRISPR–Cas9 for genome engineering. *Cell* **157**, 1262–1278.
James, C. (2015). Global status of commercialized Biotech/GM Crops: 2015. ISAAA Brief No. 49. Ithaca, NY: ISAAA. [Online]. Available from http://isaaa.org/resources/publications/briefs/51/executivesummary/default.asp
Kowalski, S. P. (2015). Golden Rice, open innovation, and sustainable gobal food security. *Industrial Biotechnology* **11**, 84–90.
Kyndt, T. *et al.* (2015). The genome of cultivated sweet potato contains *Agrobacterium* T-DNAs with expressed genes: an example of a naturally transgenic food crop. *Proceedings of the National Academy of Sciences of the USA* **112**, 5844–5849.
Laible, G. *et al.* (2015). Improving livestock for agriculture – technological progress from random transgenesis to precision genome editing heralds a new era. *Biotechnology Journal* **10**, 109–120.
Ledford, H. (2015). CRISPR, the disruptor. *Nature* **522**, 20–24.
Lunshof, J. (2015). Regulate gene editing in wild animals. *Nature* **521**, 127.
Mao, Y. *et al.* (2013). Application of the CRISPR–Cas system for efficient genome engineering in plants. *Molecular Plant* **6**, 2008–2011.
Moghissi, A. A. *et al.* (2016) Golden Rice: scientific, regulatory and public information processes of a genetically modified organism. *Critical Reviews in Biotechnology* **36**(3), 535–541.
Murphy, D. J. (2007). Improved containment strategies in biopharming. *Plant Biotechnology Journal* **5**, 555–569.
Murphy, D. J. (2011). *Plants, Biotechnology, and Agriculture*. Egham: CABI Press.
Murphy, D. J. (2014a). Using modern plant breeding to improve the nutritional and technological qualities of oil crops. *Oilseeds & Fats Crops and Lipids* **21**, D607, DOI: 10.1051/ocl/2014038.
Murphy, D. J. (2014b). The future of oil palm as a major global crop: opportunities and challenges. *Journal of Oil Palm Research* **26**, 1–24.
Napier, J. A. *et al.* (2015). Transgenic plants as a sustainable, terrestrial source of fish oils. *European Journal of Lipid Science and Technology* **117**(9), 1317–1324.
National Academies (2015). www.nationalacademies.org/gene-editing/index.htm
Nyman, M. (2014). New techniques and the GMO-legislation, Swedish Gene Technology Advisory Board, Mistra Biotech Workshop. [Online]. Available from www.genteknik.se
Okuzaki, A. and Toriyama, K. (2004). Chimeric RNA/DNA oligonucleotide-directed gene targeting in rice. *Plant Cell Reports* **22**, 509–512.
Oye, K. A. *et al.* (2014). Biotechnology. Regulating gene drives. *Science* **345**(6197), 626–628.
Petherick. A. *et al.* (2015). Genome editing. *Nature* **528**, S1–S48.
Ricroch, A. E. and Hénard-Damave, M. C. (2015). Next biotech plants: new traits, crops, developers and technologies for addressing global challenges. *Critical Reviews in Biotechnology* **36**(4), 675–690.
Roberts, A. F. *et al.* (2015). Biosafety research for non-target organism risk assessment of RNAi-based GE plants. *Frontiers in Plant Science* **6**, 958.
Sanchez, P. A. (2015). En route to plentiful food production in Africa. *Nature Plants* **1**, 1–2.
Soucy, S. M. *et al.* (2015). Horizontal gene transfer: building the web of life. *Nature Reviews Genetics* **16**, 472–482.
Telem, R. S. *et al.* (2013). Cisgenics – a sustainable approach for crop improvement. *Current Genomics* **14**, 468–476.
The Arabidopsis Genome Initiative (2000). Analysis of the genome sequence of the flowering plant *Arabidopsis thaliana*. *Nature* **408**, 795–815.

Vanhercke, T. *et al.* (2013). Metabolic engineering of plant oils and waxes for use as industrial feedstocks. *Plant Biotechnology Journal* **11**, 196–210.

Wang, Y. *et al.* (2014). Simultaneous editing of three homoeoalleles in hexaploid breadwheat confers heritable resistance to powdery mildew. *Nature Biotechnology* **32**, 947–951.

Wolt, J. D. *et al.* (2016). The regulatory status of genome-edited crops. *Plant Biotechnology Journal* **14**(2), 510–518.

Zhang, H. *et al.* (2014). The CRISPR/Cas9 system produces specific and homozygous targeted gene editing in rice in one generation. *Plant Biotechnology Journal* **12**, 797–807.

3 A Strategy for Integrating Science into Regulatory Decision-Making for GMOs

Wendy Craig, Dennis Obonyo Ndolo and Mark Tepfer

Introduction

The first transgenic plants were described in 1983, and, since then, the phased commercialisation of transgenic (or 'genetically modified'; GM) organisms has increased to the extent that, 30 years later, 181.5 million hectares (Mha) of GM crops were cultivated (James, 2014; Figure 3.1). In addition, products with more and more complex gene–trait combinations (e.g., stacked events) are also now reaching the later stages of product development. Estimates show that 43.7 Mha were planted to stacked events (i.e., events with multiple trait-related transgenes) in 2012, accounting for more than 26% of the 170.3 Mha of GM crops planted worldwide that year (James, 2012).

The increasing number of GM traits in recent stacked events has set the trend for the next generation of GM crops. It is estimated that, by the end of 2015, for maize alone, 24 single-trait events will have been commercialised. This potentially translates to over 2,000 triple-stacks and over 12,000 quadruple-stacks that may enter the pipeline (Stein and Rodriguez-Cerezo, 2009). Future stacks are likely to involve not just multiple traits, but the combination of these traits to engineer metabolic pathways and even to introduce a number of pathways into an organism (Halpin, 2005; Naqvi *et al.*, 2009; Que *et al.*, 2010). This demonstrates how GM organisms (GMOs) are becoming more complex, and underlines the need to have efficient regulatory mechanisms to ensure timely decision-making.

In most countries, GMOs are subject to regulation. Each of the key GMO developments to date has therefore been overseen by an international regulatory environment which itself has undergone significant developments (Figure 3.2).

The international guidelines highlighted in Figure 3.2 are similar in their approach to what a risk assessment should comprise, in that they all agree that it should be a structured, reasoned approach, based on scientific/technical evidence, for considering the chance of harm from activities with a GMO. Risk assessments should aim to identify, characterise and evaluate risks to the health and safety of people or to the environment from activities with GMOs. To do so, they should identify risk by considering what could go wrong and how harm might occur, whilst the identified risk should be characterised by considering how serious the harm could be (consequences) and how likely it is that such harm could occur in the context of the case (Gray, 2012). This is consistent with a long-standing understanding in other domains that risk is the combination of the magnitude of the consequences of a hazard with the likelihood that the consequences

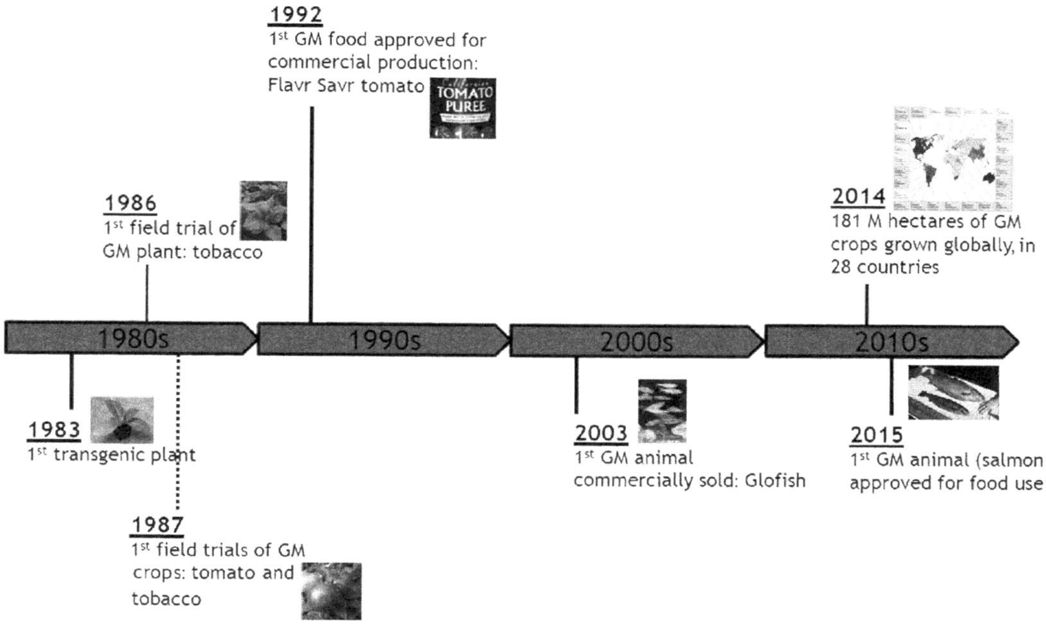

Figure 3.1. Timeline of GMO developments to date, featuring examples of key developments in the last four decades.

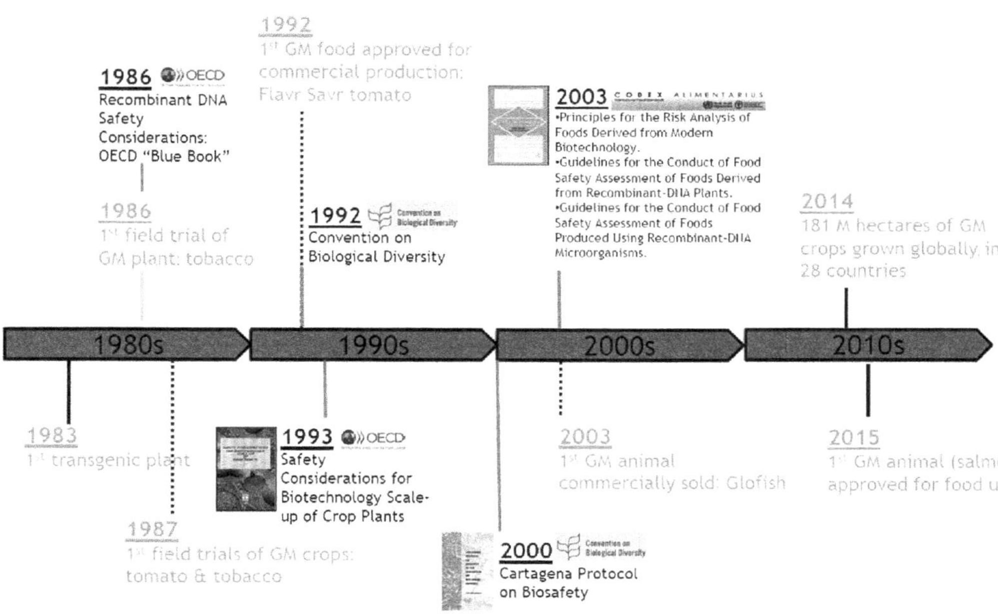

Figure 3.2. Timeline of development of international risk assessment guidelines, featuring examples of key developments in the last four decades.

will occur. By integrating consequences and likelihood, the level of risk can be evaluated, and the need for any measures to reduce it considered where pertinent. The list of 'Points to Consider' in Annex III of the Cartagena Protocol on Biosafety (Convention on Biological Diversity, 2000) is a good summary of the types of information that are regularly considered during a risk assessment:

- the biological and genetic characteristics of the recipient parental organism;
- the biological and genetic characteristics of the donor organisms;
- characteristics of the vector used for transformation;
- genetic characteristics of the nucleic acids inserted, and the functions encoded;
- the identity of the GMO and the differences in characteristics between it and the recipient parental organism;
- the recommended detection method;
- information relating to the intended use;
- the receiving environment.

In addition to the foregoing, it is essential that the reasons underpinning decisions based on risk analysis are shared with stakeholders and the general community; this is the role of risk communication. Risk communication also provides decision-makers with access to the relevant factual information and analyses, as well as awareness of the needs, values and concerns of stakeholders, for incorporation into the decision-making considerations. Finally, risk communication ensures that the risk analysis and decision-making processes are transparent both to applicants and to the broader community. Informed public participation (input) processes are essential to effective decision-making. Moreover, it is absolutely essential that all legitimate input bearing on the issues relevant to the decision-making process be accepted and seriously considered (Fitzpatrick *et al.*, 2009).

Given the need to enhance biosafety regulatory decision-making capacity, since the mid 1990s the International Centre for Genetic Engineering and Biotechnology (ICGEB) has organised 5-day workshops on the assessment of safety of GMOs. This long-term commitment to biosafety capacity enhancement has led to the development of a practical strategy for risk assessment, which has proved to be useful not only from a regulatory perspective, but also for the evaluation of potential risks early in the development of a GMO, and to summarise how a complex corpus of data can be presented to facilitate drawing conclusions regarding a potential risk in a manner that is clear and can be readily communicated (Julié-Galau *et al.*, 2014; Tepfer *et al.*, 2015; Tepfer and García-Arenal, 2015). This chapter presents the risk assessment methodology that has been continuously refined in the ICGEB workshops, including an outline of how this can be integrated into the broader context of risk analysis.

The Science of Risk Assessment

In many facets of daily life, the assessment of potential risks is a regular feature, often carried out informally and without detailed consideration. However, in the risk assessment of GMOs, such assessments are generally governed by a formalised framework and treated as strictly scientific processes. This is because, unlike risk assessments carried out in many other aspects of daily life, scientific expertise is required in the risk assessment and decision-making regarding the use of GMOs (Herman *et al.*, 2013). This is also because genetic engineering is a relatively new and rapidly evolving area, and therefore the methodology for analysing risks from genetic engineering may appear different from long-established methodologies, for example those used to assess hazardous chemicals or mechanical failure (Office of the Gene Technology Regulator, 2005). Consequently, strategies in which problem formulation has been a central feature have been developed to implement scientific risk assessment of GMOs, and to clarify how the process should be integrated into the broader risk analysis that underlies appropriate decision-making (Johnson *et al.*, 2007; Wolt *et al.*, 2010; Gray, 2012; García-Alonso, 2013; Tepfer *et al.*, 2013). In this context, risk analysis comprises three activities: risk assessment, which leads to a determination of the probability of specified harmful effects; development of any required risk management; and communication amongst relevant stakeholders.

Even though international guidelines have been developed, the assessment of risk has still tended to be rather complex in practice. This has led to an alarmingly rapid increase in the amount of information required in order to characterise the potential risks arising from the open cultivation of GM crops. For example the USA's environmental assessment of CZW-3 squash in June 1996 comprised only 31 pages (Animal and Plant Health Inspection Service, 1996), whilst current submissions in the EU now consistently comprise 8,000–9,000 pages. There are many possible reasons for why this has occurred, but amongst the most commonly proposed reasons are the following.

- *The assessment depends upon the interpretation of the baseline comparator.* The comparator may be the non-GMO counterpart, existing agricultural practices for the crop, agriculture in general, or any combination of the aforementioned.
- *The ability to detect differences.* For example, molecular tools (sequencing, -omics, etc.) able to generate data on a massive scale have now become more affordable.
- *Questionable utility of data submitted for risk assessment.* Regulatory guidance can be open to multiple interpretations, therefore leading to the submission by default of everything that is thought to have been requested and, just in case, everything else that was discovered, measured or evaluated in the course of events as well. Not every piece of information will serve the risk assessment, leading to the much-cited conundrum of separating out the 'Need to know' from the 'Nice to know'.

With special focus on the last of the above points, it can be a truly difficult exercise to discern those risks which are of highest pertinence among the many that can be conjectured. Also, regulatory authorities are bound by their mandates, so it is important that they can demonstrate that the risk assessment of the GMO in question has been considered in the light of their policy objectives. Fit-for-purpose, practical and integrated risk analysis methods are therefore required in order to ensure that effective and efficient regulatory decision-making is carried out, and to make rigorously scientific risk assessment and risk analysis more broadly accessible to diverse stakeholder groups.

Using Problem Formulation in Risk Assessment

One of the earliest proponents of a strategic approach to risk assessment was the USA's Environment Protection Agency; their 1998 guidelines explicitly integrated problem formulation into the formative risk assessment steps to help link the policy objectives that drive risk analysis to the assessment of the potential impact(s) of the dissemination of specific GMOs (EPA, 1998). Problem formulation is now being actively interpreted and adopted by other regulatory authorities, e.g., the European Food Safety Authority (EFSA, 2010).

A viable way to begin evaluating the safety of GMOs is to outline the potential risks that could occur as a result of their deployment, and then determine which of these are plausible and should warrant further investigation. This first step can be done through an approach that has been referred to as problem formulation. At its very simplest, problem formulation is about framing questions relevant to the risk assessment and coming up with a plan to answer them. It involves outlining potential risks (leading to the creation of risk hypotheses) and then determining which risks are plausible (testing and accepting or rejecting the risk hypotheses) (Raybould, 2007; Gray, 2012; Tepfer *et al.*, 2013). It may be worth emphasising that problem formulation is *not* risk assessment, is *not* a regulatory protocol and is *not* a legal instrument. It is, variously, an approach, a tool for risk assessment, a first step, a concept, a method, a way of thinking and an academic construct. It is a very useful approach to risk assessment in that it has great utility with regard to clarifying the exact nature of potential risks, determining what organisms or other features we don't want to be harmed, how they could be harmed, how likely it is that they would be harmed, and whether it matters if they were (Gray, 2012).

The importance of problem formulation in risk assessment is, however, often overlooked. Problem formulation identifies elements that are important to society and how these may potentially be harmed by a GMO. Poor problem formulation may compromise the entire risk assessment and impair subsequent decision-making (Fitzpatrick *et al.*, 2009). One outcome of this failure has been the production of data of limited relevance for risk assessment (Craig *et al.*, 2008), which then increases, rather than allays, concerns about the impacts of GMOs (Raybould, 2007; Johnson *et al.*, 2007). The simple question at the heart of problem formulation for regulatory risk assessments should ideally be 'What could go wrong if this GM crop is adopted or cultivated in this country?' Unfortunately, this question is often subtly but significantly modified to 'What will change if this GM crop is adopted or cultivated in this country?' This more open-ended question has led to seeking any form of biological or environmental effect, whereas risk assessment is concerned with a much smaller subset of effects, namely, those considered harmful (Fitzpatrick *et al.*, 2009). Clearly distinguishing between information that predicts change and that which predicts harm is central in ensuring regulatory efficiencies and avoiding the generation of irrelevant data (Raybould, 2007). Depending on the jurisdiction, the potential benefits of a GMO are taken into consideration at a later stage of the decision-making process.

The strength of the strategy described in this chapter lies in the fact that it follows natural thought processes as closely as possible and also minimises the use of technical jargon (Tepfer *et al.*, 2013). The strategy has the following key steps: (1) assembling a list of entities (i.e., individuals, species or commodities) that may possibly be impacted through the introduction of the GMO being considered and that may therefore need to be protected; (2) creating a list of possible harms based on the objectives above; (3) restating the harms as risk hypotheses that describe each harm in a simple declarative sentence; (4) elaborating in each case a pathway to harm that explains in a chain of cause and effect how deployment of the GMO would lead to the harm; (5) assessing in each case how scientific knowledge supports or invalidates each step of the pathway; and (6) considering any limitations to the conclusions, such as whether breaks in the pathway to harm constitute a formal refutation of the risk hypothesis.

Amongst the many case studies that we have used in our workshops, one was a hypothetical maize line transformed by biolistic methods to express an HSP60 protein to confer salt and drought tolerance, which was developed to be grown in the marginal rainfall areas of Kenya. In the following sections, we present the outcomes of the workshop sessions using this case study for the purpose of demonstrating the subsequent steps of the problem formulation strategy.

1. *Assembling a list of entities that may possibly be harmed through the introduction of the GMO being considered and that may therefore need to be protected*

The identification of potential harms is central to problem formulation, but the lack of consensus on how to value the entity which may potentially be harmed presents serious difficulties to GMO regulators and risk assessors (Tepfer *et al.*, 2013). Governments and societies will generally have entities that they consider useful and worthy of protection: for example, 'We, as a society, value and wish to protect X', where X may be agricultural sustainability, children's health, populations of rare and endangered species, etc. These are often outlined in policy objectives and may differ from government to government. A first step in problem formulation therefore involves assembling a list of entities, either specified by the prevailing legal mandates or arising from generic concerns, that may possibly be harmed through the introduction of the GMO being considered, and which therefore need to be protected. This step leads to the formulation of what are generally referred to as protection goals, which may include

- protecting the environment;
- protecting human health;
- protecting animal health;
- ensuring food security and alleviating poverty;
- preventing negative impacts on trade;
- ensuring compatibility with social and cultural values.

Considering that governmental regulation of GMOs and other agricultural technologies is designed to serve society, societal concerns and values should also be addressed and therefore are also captured at this stage. In the ICGEB workshops we have had participants of many nationalities, but the entities which they felt would be worth protecting are often in line with most of the existing policy objectives as elucidated in many government policy documents.

2. *Creating a list of possible harms based on the objectives above*

During brainstorming sessions, a list of potential undesirable harms (derived from the various protection goals) can easily be created, and this may include many that are clearly unrealistic or that require additional factors in order for them to actually occur. In the light of known information, these unrealistic scenarios can quickly be discounted and need not be explored further. Having discounted the unrealistic ones, the following potential harms were amongst those considered plausible by workshop participants:

- allergic reactions in humans;
- loss of genetic diversity of maize landraces;
- emergence of new virulent viruses;
- loss of income to farmers.

While these potential harms may not be specific to GMOs, the idea is to assess the potential harm from GMOs relative to the conventional counterpart/comparator.

3. *Restating the more plausible harms as risk hypotheses that describe the harm in a simple declarative sentence*

The harms are then restated as risk hypotheses, in the form of declarative sentences that describe exactly what could be harmed and how this would occur. In other words, the risk hypothesis serves to clarify the nature of the risk, and must clearly link ***an initiating action*** to *a potential negative or unwanted outcome*. The aforementioned plausible harms can therefore be restated as

- ***high consumption of the drought-tolerant (DT) maize*** leads to *allergic reactions in humans*;
- ***cultivation of DT maize*** leads to *loss of genetic diversity of maize landraces*;
- ***cultivation of DT maize*** leads to *emergence of new virulent viruses*;
- ***cultivation of DT maize*** leads to *loss of income to farmers*.

4. *Elaborating a pathway to harm that explains in a chain of cause and effect how deployment of the GMO would lead to the harm*

The pathway to harm is a chain of cause and effect that links the initial cause (i.e., dissemination of the GMO) with a potential harm (Figure 3.3). In some instances, there may be a number of alternative ways to link the cause with the potential harm, in which case the pathway may be branched. Each step must be formulated in a way that can be evaluated in the light of data that are either already available or can be generated. Even before considering what data may be pertinent to testing the risk hypothesis, careful scrutiny of the pathway to harm will identify which step(s) may be most decisive in attempting to break the pathway. A particularly useful feature of this stepwise analysis of the pathway to harm is that it suffices to break one step in the pathway decisively to invalidate the risk hypothesis and conclude that the likelihood that any harm will occur via that pathway is minimal.

Risk hypothesis: Cultivation of DT maize leads to loss of genetic diversity of maize landraces

Pathway to harm

1) DT maize and traditional landraces are grown in the same region

↓

2) Farmers prefer growing DT maize to conventional maize, including specific landraces

↓

3) With time farmers will stop growing certain/all landraces

↓

4) Farmers do not store or multiply their landraces to maintain viability

↓

5) Loss of landraces and hence of crop genetic diversity.

Figure 3.3. An example of a simple pathway to harm, describing key steps in linking one way in which cultivation of a hypothetical drought-tolerant (DT) maize could lead to the loss of genetic diversity in maize landraces.

5. *Assessing how scientific knowledge supports or invalidates each step of the pathway*

The steps in the pathway are then subjected to scientific scrutiny, through the application of existing knowledge or of new data generated from specifically designed experiments addressing the step in question. In the latter cases, knowledge of the pathway to harm assists the identification of data critical to challenging the specific step, thereby helping communication between the regulator and the applicant of the significance of the data to the risk assessment, and thus enriching subsequent discussions as to how best to obtain the data in the least time-consuming and most rigorous manner.

If data that are pertinent to the risk hypothesis, but which do not fit comfortably with a precise step in the pathway, are found, it may be useful to modify the manner in which the pathway to harm has been broken down into steps, in order to incorporate all data that may help test the specific risk hypothesis as clearly as possible.

Each step is then assessed in the light of the available data to determine the likelihood that it will occur and the resulting impact if it does occur (Figure 3.4). However, due to the complex, dynamic and variable nature of the biological systems in question, there will always be a degree of uncertainty concerning the mechanisms involved, thus making it virtually impossible to quantify the probability and magnitude of the potential harm occurring. Qualitative assessments that make use of relative descriptions for likelihoods and consequences can combine the maximum amount of data from disparate sources, and thus are usually the preferred evaluation method, which can be implemented by using a specific scale for each factor (Office of the Gene Technology Regulator, 2013).

Risk hypothesis: Cultivation of DT maize leads to loss of genetic diversity of maize landraces

Pathway to harm

1) DT maize and traditional landraces are grown in the same region (***highly likely***)

↓

2) Farmers prefer growing DT maize to conventional maize, including specific landraces (***likely***)

↓

3) With time farmers will stop growing certain/all landraces (***likely***)

↓

4) Farmers do not store or multiply their landraces to maintain viability (***unlikely***)

↓

5) Loss of landraces and hence of crop genetic diversity (***unlikely***)

Figure 3.4. A simple pathway to harm that has been subjected to scientific scrutiny through the application of existing knowledge and data.

Thus, for the likelihood of the harm expected to occur, the following qualitative scale can be envisaged:

- only in rare circumstances (***highly unlikely***);
- in some limited circumstances (***unlikely***);
- in many circumstances (***likely***);
- in most circumstances (***highly likely***).

Consequence assessment is done relative to a comparator (in this case a hybrid maize line). The following qualitative scale can be envisaged:

- minimal or no increase in harm (**marginal**);
- minor increase in harm that is readily treatable or reversible and limited in time, space or numbers affected (**minor**);
- significant increase in harm that requires specialised treatment or whose impact can be sufficiently mitigated or remedied (**intermediate**);
- significant and persistent increase in the severity of harm that is highly difficult to mitigate or remedy (**major**).

An overall conclusion to the pathway can then be made on the basis of the extent to which scientific knowledge invalidates one or more links in the causal chain or the degree to which the likelihood and magnitude of harm have been ascertained (Figure 3.4). It is useful to note that, if several steps have only a small chance of occurring, then the overall pathway has an extremely limited chance of completion due to the combination of several low-probability steps. Alternatively, one step may have hardly any chance of occurring, resulting in a very low overall probability even if all the other steps have a reasonably high chance of occurring.

		LEVEL OF RISK			
LIKELIHOOD ASSESSMENT	Highly likely	Low	Moderate	High	High
	Likely	Low	Low	Moderate	High
	Unlikely	Negligible	Low	Moderate	Moderate
	Highly unlikely	Negligible	Negligible	Low	Moderate
		Marginal	Minor	Intermediate	Major
		CONSEQUENCE ASSESSMENT			

Figure 3.5. An example of a risk matrix (Office of the Gene Technology Regulator, 2013).

6. *Considering any limitations on the conclusions, such as whether breaks in the pathway to harm constitute a formal refutation of the risk hypothesis*

The ability to make the distinction between major and minor impacts makes it possible for competent authorities to go forward with a decision regarding the planned activities with a GMO. Careful evaluation of the steps in the pathway to harm often makes it possible to show whether the potential impact is large or small. By examining the pathway, a risk matrix of likelihood and consequence assessments can be used at this stage to make an evaluation of the level of risk (Figure 3.5). For the example given in Figure 3.4, the overall level of risk occurring is **moderate** (**intermediate** consequence × ***unlikely*** likelihood).

The level of risk resulting from the matrix in Figure 3.5 then indicates the degree of concern and the extent to which mitigation or management actions are needed:

- **negligible** – risk is of no discernible concern, with no need to invoke mitigation;
- **low** – risk is of minimal concern, but may require mitigation beyond standard practices;
- **moderate** – risk is of marked concern, requiring demonstrably effective mitigation measures;
- **high** – risk is of considerable concern, which is unacceptable unless highly plausible and effective mitigation actions are implemented.

The overall strengths of this approach are that it maintains the risk assessment focus on causal links between the initial event and the potential harm; it identifies necessary data for the analysis; and it facilitates the use of existing data in the analysis, whilst ultimately shortening the analysis if the causal link is broken, thereby simplifying the conclusion of the risk assessment.

Risk assessment will therefore generate information upon which a risk management plan may then be developed if necessary. A risk management plan would make provisions on how to minimise any identified risks of moderate to high importance. As an example, the moderate-level risk identified for the hypothetical DT maize used in Figure 3.4 could be managed by local landraces being conserved in publicly accessible gene banks. This could be made a condition of the licence.

Integration of Risk Assessment in a Broader Risk Analysis

Even though risk assessment forms a scientific basis for the evaluation of risks, it might not be possible for it to capture all societal concerns for at least two reasons. First, risk acceptance is subject to cost–benefit trade-offs. It has been argued that, if a GM product has little demonstrable benefit, even a minimal risk associated with the product would not be acceptable to the general public and society. Furthermore, benefit to farmers might not be perceived as benefit by the general public. On the other hand, if the product has tremendous demonstrable benefit, some level of risk (even with uncertainty) would be acceptable to the general public. Second, it is difficult if not impossible to grasp, let alone compute rigorously, the will and concerns of the 'the general public and society' (Rajan and Letourneau, 2012). It is important, therefore, that such issues are taken into consideration in the decision-making process and that the recommendations of risk assessments are shared with wider stakeholder groups in order to obtain as much stakeholder input as possible. Such consultation would therefore be an integral component of the biosafety regulatory decision-making processes.

From the above, it is apparent that biosafety decision-making requires more than just a risk assessment; although risk assessment is crucial, it is but the first of the processes that would ultimately lead to decisions on activities involving GMOs. Risk management and risk communication are other important requirements. The three processes together, i.e., risk assessment, risk management and risk communication, constitute what has been referred to as risk analysis. While risk assessment identifies risks from plausible sets of circumstances that may result in harm to people or to the environment from a GMO, characterising the risks on the basis of their seriousness and the chance of harm, risk management selects and implements plans or actions to ensure that risks are appropriately managed. Risk communication, on the other hand, is the exchange of information, ideas and views between regulatory decision-makers and various stakeholders, but also conveys the rationale for decisions made by the decision-makers (Office of the Gene Technology Regulator, 2013). In a nutshell, therefore, risk analysis integrates the assessment, management and communication of risks posed by, or arising as a result of, adoption of a GMO. The effectiveness of biosafety regulatory decision-making is enhanced by requiring that all these processes be integrated.

Conclusion

As described above, there are solid pathways to carrying out scientifically valid assessments of potential risks associated with dissemination of GMOs. But as soon as the risk assessment is situated within the somewhat broader framework of risk analysis, many subjective factors come immediately into play, such as the degree of risk aversion in the jurisdiction in question and the degree of public trust in the objectivity of expert opinion, and more generally in the ability of the government to protect the common good. In the end, additional factors, such as cultural and socio-economic interests, may need to be considered in the decision-making process. Although scientists and government officials may be averse to considering the subjective elements involved in decision-making, for the excellent reason that it is extremely difficult to weight how they should be put into the balance, in the end, the decision-making process must be designed to respond appropriately to the needs of the diversity of stakeholder groups.

References

Animal and Plant Health Inspection Service (1996). USDA/APHIS Petition 95–352–01P for the Determination of Nonregulated Status for CZW-3 Squash. [Online]. Available from www.aphis.usda.gov/brs/aphisdocs2/95_35201p_com.pdf

Convention on Biological Diversity (2000). Cartagena Protocol on Biosafety to the Convention on Biological Diversity. Text and Annexes. [Online]. Available from www.cbd.int/doc/legal/cartagena-protocol-en.pdf

Craig, W. *et al.* (2008). An overview of general features of risk assessment of genetically modified crops. *Euphytica* **164**, 853–880.

EFSA (2010). Guidance on the environmental risk assessment of genetically modified plants. EFSA Panel on Genetically Modified Organisms (GMO), European Food Safety Authority (EFSA), Parma, Italy. *EFSA Journal* **8**(11), 1879–1990.

EPA (1998). Guidelines for Ecological Risk Assessment. EPA/630/R-95/002F. *Federal Register* **63**(93), 26846–26924.

Fitzpatrick, J. *et al.* (2009). Problem formulation in environmental risk assessment of genetically modified crops: a Brazilian workshop. *BioAssay* **4**, 5.

García-Alonso, M. (2013). Safety assessment of food and feed derived from GM crops: using problem formulation to ensure 'fit for purpose' risk assessment. *Collection of Biosafety Reviews* **8**, 72–101.

Gray, A. (2012). Problem formulation in environmental risk assessment for genetically modified crops: a practitioner's approach. *Collection of Biosafety Reviews* **6**, 14–65.

Halpin, C. (2005). Gene stacking in transgenic plants – the challenge for 21st century plant biotechnology. *Plant Biotechnology Journal* **3**, 141–155.

Herman, R. A. *et al.* (2013). Bringing policy relevance and scientific discipline to environmental risk assessment for genetically modified crops. *Trends in Biotechnology* **31**, 493–496.

James, C. (2012). Global Status of Commercialized Transgenic Crops. [Online]. Available from www.isaaa.org/resources/publications/briefs/44/executivesummary/default.asp

James, C. (2014). Global Status of Commercialized Transgenic Crops. [Online]. Available from www.isaaa.org/resources/publications/briefs/49/executivesummary/pdf/B49-ExecSum-English.pdf

Johnson, K. L. *et al.* (2007). How does scientific risk assessment of GM crops fit within the wider risk analysis? *Trends in Plant Science* **12**, 1–5.

Julié-Galau, S. *et al.* (2014). Evaluation of the potential for interspecific hybridization between *Camelina sativa* and related wild Brassicaceae in anticipation of field trials of GM camelina. *Transgenic Research* **23**, 67–74.

Naqvi, S. *et al.* (2009). When more is better: multigene engineering in plants. *Trends in Plant Science* **15**, 48–56.

Office of the Gene Technology Regulator (2005). Risk Analysis Framework 2005. [Online]. Available from www.ogtr.gov.au/internet/ogtr/publishing.nsf/content/raf-3/$FILE/raffinal2.2.pdf

Office of the Gene Technology Regulator (2013). Risk Analysis Framework 2013. [Online]. Available from www.ogtr.gov.au/internet/ogtr/publishing.nsf/content/raffinal5-toc

Que, Q. *et al.* (2010). Trait stacking in transgenic plants – challenges and opportunities. *GM Crops* **1**(4), 220–229.

Rajan, R. S. and Letourneau, D. K. (2012). What risk assessments of genetically modified organisms can learn from institutional analyses of public health risks. *Journal of Biomedicine and Biotechnology* 203093, 8 pages, doi:10.1155/2012/203093.

Raybould, A. (2007). Ecological versus ecotoxicological methods for assessing the environmental risks of transgenic crops. *Plant Science* **173**, 589–602.

Stein, A. J. and Rodriguez-Cerezo, E. (2009). The global pipeline of new GM crops: implications of asynchronous approval for international trade. [Online]. Available from http://ipts.jrc.ec.europa.eu/publications/pub.cfm?id=2420

Tepfer, M. *et al.* (2013). Putting problem formulation at the forefront of GMO risk analysis. *GM Crops & Food* **4**, 10–15.

Tepfer, M. *et al.* (2015). A critical evaluation of whether recombination in virus-resistant transgenic plants will lead to the emergence of novel viral diseases. *New Phytologist* **207**, 536–541.

Tepfer, M. and García-Arenal, F. (2015). Does recombination in virus-resistant transgenic plants lead to emergence of novel viral diseases? *ISB News Report*, June.

Wolt, J. D. *et al.* (2010). Problem formulation in the environmental risk assessment for genetically modified plants. *Transgenic Research* **19**(3), 425–436.

4 Governance of Benefits and Risks of GMOs in Developing Countries

E. Jane Morris

Background

The approach to governance of genetically modified organisms (GMOs) in most developing countries is based on the Cartagena Protocol on Biosafety (CPB) (Convention on Biological Diversity, 2000), which specifically deals with 'adverse effects' and 'risks' in accordance with the precautionary principle. Most developing countries therefore incorporate risk analysis as a central theme of their National Biosafety Frameworks (NBFs); many of these frameworks were developed with support provided under the capacity-building programmes of the CPB. The NBFs were generally not legally binding documents, but in some cases guided the process for the subsequent introduction of biosafety legislation. Other developing countries have independently introduced biosafety legislation without direct support through the instruments of the CPB, either as early adopters of the technology desiring a legislative basis for decision making, or through a desire to ban the technology completely.

Despite the focus on risk analysis, there have been calls for a balanced approach to assess both benefits and risks (Morris, 2011; Herman *et al.*, 2013). Research suggests that the perception of benefits and risks influences the acceptance of the technology. In fact the introduction of all technologies involves a risk–benefit trade-off (Arvidsson, 2015), but this relies on the benefits and risks being evaluated objectively and assumes that the benefits are in fact taken into account (Siegrist *et al.*, 2016). The author of this chapter previously published a methodology for a structured semi-quantitative approach to GMO risk–benefit analysis that could guide regulatory decision makers (Morris, 2011), but formal adoption of such an approach would require that countries incorporate consideration of benefits into their regulatory frameworks.

Interestingly, the draft Guidance on Risk Assessment published under the CPB (Convention on Biological Diversity, 2012) makes the statement that 'the recommendation on the acceptability of risk(s) should take into account any available scientific analysis of potential benefits for the environment, biodiversity, and human health' and include 'scientific benefit analyses, carried out using similar principles of sound science as those used throughout the risk assessment'. In response to the publication of the draft Guidance document the European Union feedback stated that 'The EU does not support an approach to risk assessment that balances risk acceptability with benefits'. On the other hand, the feedback from a few developing countries (e.g., Honduras and South Africa)

Table 4.1 Benefits of GM crops

Category	Potential benefits of relevance for developing countries
Environment	Reduced soil erosion from low-till agriculture, reduced need to bring new land into production, reduced need to use agrochemicals, increased biodiversity
Agriculture	Increased yield, tolerance to climate change, pest and disease resistance
Health	Improved nutritional value of food, decreased infection by bacterial and fungal pathogens, improved shelf life, increased food availability, reduced exposure of workers to potentially harmful agrochemicals, reduced toxicity
Socio-economic	Increased income and profitability for farmers, improved food security, increased competitiveness

emphasised the importance of also assessing benefits. In an updated version of the Guidance (Convention on Biological Diversity, 2015) the word 'should' is replaced by 'may', and a statement has been added to say that 'balancing risk acceptability with potential benefits is not laid out in the provisions of the Protocol'.

This chapter examines the extent to which the neglect of consideration of benefits has led to the current impasse in many developing countries, and what might be done to address this problem. It also brings into focus the European stance that excludes consideration of benefits, and examines how this may affect attitudes in developing countries.

Dimensions of Risk and Benefit

There are various dimensions of both risks and benefits for GMOs. To date (and also in this chapter), the main focus has been on GM crops, although GM microorganisms, insects, fish and mammals have also received attention. The first generation of GM crops incorporated two main traits, namely herbicide tolerance and insect resistance, with benefits accruing primarily to the farmer rather than the consumer. Increasingly, however, the range of traits is expanding to incorporate aspects such as adaptation to climate change and particularly nutritional improvement, where there will be clear benefits to the consumer. The approach to risk assessment of GM crops has been detailed in many publications from a purely scientific perspective (e.g., European Food Safety Authority, 2006), but scientific assessment of benefits is hardly mentioned. It is more common for both risks and benefits to be considered in publications written from a socio-economic perspective; however, the assessment methodologies are less well developed (Ludlow *et al.*, 2013; see also Chapters 8 and 10 in this book). The potential risks and benefits from the introduction of GM crops can be environmental, agricultural, health-related or socio-economic in nature. Morris (2011) outlined some examples of benefit in each of these categories; potential benefits that are of particular relevance for developing countries are summarised in Table 4.1.

How Do Developing Countries Take into Account Risks and Benefits?

Table 4.2 summarises the extent to which benefits are taken into account in a range of developing countries. The countries were selected for study on the basis that they had existing legislation and/or that they had already introduced GM crops (or were moving forward in this direction). Some countries not listed have an outright ban on the technology, but have implemented this in the absence of any relevant legislation (e.g., the Democratic Republic of the Congo, Zambia). Vigani and Olper (2013) developed a composite index of GMO standards restrictiveness for 60 countries. However, their scoring did not interrogate the nature of the risk assessment and whether it included consideration of benefits as well as risks; the data in Table 4.2 therefore may be regarded as complementary to their assessment.

The countries in Table 4.2 fall into four broad categories:

- benefits explicitly taken into account and a positive attitude towards GMOs (e.g., the Philippines);
- benefits mentioned in passing, countries somewhat slow to adopt the technology (e.g., Bolivia, Mozambique, Vietnam);
- benefits not explicitly mentioned, but nevertheless GMOs introduced due to a positive political climate (e.g., South Africa, Argentina, Colombia);
- benefits not mentioned and a negative attitude towards GMOs (e.g., Peru, Senegal).

It is of interest to examine further the developing countries that were early adopters of GMOs, such as South Africa and Argentina. In the early days, and before the CPB was introduced, there was less controversy about the technology. Burachik (2010) stresses that political support linked to a recognition of the benefits to farmers, the environment and the economy has influenced the positive adoption of GMOs by Argentina. In South Africa, early field trials were undertaken under pre-existing legislation in the Department of Agriculture, and a positive attitude towards biotechnology existed; specific GMO legislation was introduced at a later date (Morris, 1995), but there was never any intention to block the introduction of GMOs. Yet in recent years there have been several cases where GM applications have been refused approval without valid reasons being presented, and/or where there have been significant delays in decision-making. This is in considerable measure due to the influence of anti-GM activists, who have stirred up negative opinion at the parliamentary level (Morris and Thomson, 2014), while at the same time there is no clear mechanism for the benefits of GMOs to be considered.

In the majority of countries it appears to be political will that determines the overall attitude to GMOs, and in many cases this has also influenced the tone of the legislation. In fact, a disproportionate focus on risk can have a negative impact not only on the

Table 4.2 Summary of benefit consideration in developing country legislation

Country	Reference to benefits?	Document	Wording used	Status
Argentina	No, but there is a requirement to determine the effect on agricultural ecosystems. Benefits have been extensively analysed. Following environmental and food assessments, the commercial advantages and disadvantages of approval are considered.	Resolution 701/2011. Establish procedures for submission and evaluation of applications for experimental release of GM plant organisms that do not have commercial approval.	Evaluation . . . aimed to determine the effects of living modified organisms (LMOs) on the cultivation of agricultural ecosystems under agricultural practice	Third-largest producer of GM crops in the world
Bangladesh	Yes	National Biosafety Framework, 2006	Assessment of the potential risks and benefits. For environmental release of GMOs, risk assessment . . . could include an analysis of the potential benefit of the GMOs.	Bt brinjal released
Bolivia	Yes, but only with reference to economic benefits	Supreme Decree No. 24,676, 1997	The possible economic benefits that the activities with the GMO may yield	Produces GM soybeans, but uncertainty after 'Law of Mother Earth' in 2012 includes a GMO ban
Brazil	No, but is conciliatory in nature. Ministerial committee CNBS has mandate to analyse socio-economic opportunity and the national interest, but is rarely consulted.	Biotechnology Law, 2005 (Mitre and Reis, 2014)	Assessment of the risk	Second-largest producer in the world

Burkina Faso	Yes, authorisation requires that benefits to the country are demonstrated	Biosafety Law, 2013	'L'Agence nationale de biosécurité ne peut délivrer une autorisation que si elle établit que l'importation, l'utilisation en mileu confiné, la dissémination ou la mise sur le marché de l'organisme génétiquement modifié: profite au pays sans causer de risques dommageables pour la santé humaine, animale, la diversité biologique et l'environnement; participe au développement durable; ne nuit pas à l'environnement socio-économique; n'est pas contraire aux rêgles d'éthique'	GM cotton adopted, but was abandoned in 2016 due to poor lint quality
Chile	Not applicable	Regulatory framework not in place		Seed propagation and re-export, import of food products
China	No	Regulations on Safety of Agricultural Genetically Modified Organisms, 2001; Implementation Regulations on Safety Assessment of Genetically Modified Organisms, 2002	Assessment of the danger or potential risk	GM cotton and papaya commercially grown, approval given for other crops but planting not implemented due to negative public opinion
Colombia	No, but Ministry of Agriculture is positive	Decree 4525, 2005	Identify, assess and evaluate potential adverse effects	Produces GM cotton, maize and horticultural crops
Costa Rica	No	Phytosanitary Protection Law, 1997	Evaluation of risks	Seed propagation and re-export, but major public antagonism
Cuba	No	Decree Law 190, 1999	Assessment and management of the risks	Research and extensive field testing of GM maize under 'regulated commercialisation'
Egypt	No	Presidential decrees, draft biosafety law		Ban in place (early 2016) after earlier introduction of GM crops
Ethiopia	No	Biosafety Proclamation No. 65/2009; Biosafety (Amendment) Proclamation No. 896/2015	Preventing the adverse effects of modified organisms or at least managing them down to levels that are of no significance	Preventative legislation recently relaxed to allow the possibility of introduction of GM cotton

(*continued*)

Table 4.2 *(continued)*

Country	Reference to benefits?	Document	Wording used	Status
Ghana	Yes	Biosafety Act, 2011	Assessment of the potential risks and benefits	Field trials
Honduras	No, but positive political environment	Biosecurity Regulation, 1998	Evaluation of risks	Commercial production of GM maize
India	No	Rules, 1989; Guidelines, 1990, 1995 and 1998	Assess and manage risks	Only cotton approved for commercial production
Indonesia	No; religious, ethical, socio-cultural and aesthetic norms also taken into account	Government Regulation No. 21/2005	Biological safety	Local development, moving towards commercial production. Grain imports.
Iran	Yes	National Biosafety Law, 2008	Exploiting the benefits of modern biotechnology	GM rice approved but not commercialised, development of GM cotton but lack of regulatory capacity for further progress
Kenya	No, but makes provision for consideration of socio-economic impact (could be positive or negative)	The Biosafety Act, 2009	Minimise the risks	Release of GM was approved in late 2016
Malaysia	No, but includes a provision that the Minister 'may take into account socio-economic considerations'	Biosafety Act, 2007	Assessment of the risk and adverse effect	Only importation of grain. Trial release of GM mosquitoes and confined field trials of GM papaya.
Mali	No	Biosafety Act, 2008	Evaluation of risks	No GM testing
Mexico	Yes	Biosafety Act, 2005	'Para el análisis de soluciones a problemas particulares se evaluarán caso por caso los beneficios y los posibles riesgos del uso de OGMs'	GM cotton widely grown, but mixed signals regarding acceptance of other GM crops, injunction against GM maize. Support for scientific development of GMOs is embedded in the Act.
Mozambique	Yes, but only with reference to laboratory use and confined field trials. A socio-economic impact report is required for release to the environment.	Decree No. 71/2014	The application forms for contained use and for field trials require the applicant to provide information on the benefits of the GMO compared with conventional methods	No activity with GMOs as yet

Namibia	Yes	Biosafety Act, 2006	Any dealing . . . involves the intentional introduction of a GMO or GMO product into the environment, the application for the permit must include, or be accompanied by, a statement containing information on the potential benefits of such release	Regulations being finalised, law not yet fully operationalised
Nigeria	Yes	National Biosafety Management Act, 2015	'Risk or benefit assessment'	Field trials, general release and marketing of GM cotton approved
Pakistan	No	Pakistan Biosafety Rules, 2005; Biosafety Guidelines, 2005	Minimise the risks	Produces GM cotton, but weak regulatory system hinders further introductions
Paraguay	No, but talks about the demand for biotechnology products from the agriculture and forestry sectors	Decree 9699/2012 (no biosafety law approved as yet)	'Que el sector agropecuario y forestal es el más dinámico en la demanda de liberación de productos de la biotecnologia moderna que luego ingresan en la cadena alimentaria y en el comercio'	Produces GM cotton, maize and soybean, but lags behind neighbouring countries in approvals
Peru	No	Law 29,811 and Supreme Decree 008–2012-MINAM	The Ministry of the Environment is the lead organisation, with its focus being 'to ensure the preservation of centres of origin and biodiversity'	Ten-year moratorium in place from 2011
Philippines	Yes	Administrative Order No. 8, 2002 (nullified by the Supreme Court in 2015 and replaced by Joint Department Circular 2016); Executive Order No. 514, 2006	Executive Order No. 514: safe and responsible use of biotechnology so that the Philippines and its citizens can benefit from its application. Annexes to Joint Department Circular ask applicants to provide a 'summary of potential benefits (describe how the new trait will benefit farming, the farmer, the environment and society as a whole)'.	Leader in commercialisation
St Kitts and Nevis	Refers to impacts (these could be either negative or positive), but the overall tone of the Act is concerned with risks	Biosafety Act, 2012	An assessment of the impacts and risks posed by the GMO or product	Legislation not yet implemented, no GMO activity
Senegal	No	Biosafety Law, 2009	Evaluate the potential adverse effects	No GM testing
South Africa	No, except with reference to socio-economic impacts (could be positive or negative)	Genetically Modified Organisms Act, 1997 and associated regulations	Scientifically based risk assessment; assessment of socio-economic impact	Produces GM maize, soybeans and cotton. No GM crops developed locally have reached the market.

(*continued*)

Table 4.2 *(continued)*

Country	Reference to benefits?	Document	Wording used	Status
Sudan	Yes	National Policy on Biosafety; National Biosafety Law, 2010	Promoting the application of biotechnology as a tool in the sustainable development of the country to benefit the people of Sudan	First commercial release of GM cotton in 2012
Swaziland	Yes, but the overall tone of the Act is negative	Biosafety Act, 2012	An assessment of the potential risks and benefits	Cotton field trials, no commercialisation as yet
Tanzania	Yes, but strict liability clauses have resulted in a prohibitive framework	Environment Management (Biosafety) Regulations, 2009	'determine whether the import, release, contained use or placing on the market of the GMOs or products thereof will benefit the country; and contribute to, and not undermine, sustainable development'	Strict liability clause recently relaxed for research purposes, field trials on drought-tolerant maize started late 2016
Togo	Yes, but only in relation to public communication	Biosafety Law, 2009, being revised	'Every citizen has the right to be informed about the benefits and risks'. Elsewhere in the document there are references only to risk evaluation and prevention.	No GM testing
Uganda	Refers to opportunities from modern biotechnology, but these are not explicitly taken into account	National Biotechnology and Biosafety Bill, 2012	'creates enormous opportunities for modernisation of agriculture, protection of the environment, enhancement of public health and industrialisation'	Bill not yet passed into law. Research and field trials under way, particularly on GM bananas and cassava.
Uruguay	Not directly but refers to the strategic goal of 'Productive Uruguay' with new technologies forming a central component	Decree 353/08	'que el objetivo estratégico de la presente Administración es fomentar el "Uruguay Productivo", porque se considera que innovación, producción con mayor conocimiento y uso de nuevas tecnologías, son componentes centrales de esa estrategia de desarrollo'	Commercialised GM soybean and maize
Vietnam	Not mentioned in the Decree, but subsequent Circulars embed the concept of Exceptional Recognition of Biotechnological Advances for GM crops	Biosafety Decree 69/2010/ND-CP, supplemented by a number of Circulars, including 8/2013/TT-BTNMT, which provides the procedure for granting Certificates of Biosafety	Technical advances of biotechnology are recognised when they meet conditions including economic efficiency, social criteria, environmental protection and food safety (etc.)	GM maize approved for commercial planting

nature of the legislation but also on whether legislation is introduced at all. Politicians in many developing countries, like the general public, tend to have limited understanding of the technology, and are often themselves afraid of it and of the motives of the multinationals which control most of the products. This view is strongly driven by the anti-GM pressure groups, which have been successful at galvanising public opinion against the technology. Politicians' response to this fear may cause them to ban the technology outright (e.g., Zambia), or may cause paralysis of action and the lack of any relevant legislation. It may take a farsighted leader to change the *status quo* and determine that the potential benefits to the country necessitate the introduction of GMOs (e.g., Burkina Faso – although the sentiment there is changing, see Chapter 23). The outcome may be influenced by the relative strength of the 'pro' and 'anti' lobby groups.

Socio-Economic Risks and Benefits

The CPB makes provision for optional inclusion of socio-economic considerations, as described in Chapter 8. Many countries do include socio-economic considerations in their national legislation; some of these are highlighted in Table 4.2. Falck-Zepeda *et al.* (2013) argue against the inclusion of socio-economic impacts in the regulatory process, and point out that they can be used (or abused) as a blanket justification for rejecting the technology. In addition, they can impose an increased regulatory burden, especially where the scope and methodology are not well defined from the start.

Nevertheless, given that many countries do consider socio-economic impacts, the question could be asked whether they are in fact generally beneficial. Fischer *et al.* (2015) conducted a systematic literature review of the socio-economic impacts of GM crops at the farm level. They grouped the impacts into five categories: economic impact; distribution of benefits; access and ownership; wellbeing; and cultural heritage. At this stage there would appear to be insufficient data to provide a comprehensive view of the full range of socio-economic impacts, though from an economic perspective the impact appears to be positive. Brookes and Barfoot (2016) calculated that net economic benefits at the farm level amounted globally to US$150.3 billion over the years 1996–2014, with the benefits divided roughly equally between developed and developing countries.

Marcoux *et al.* (2013) refer to non-safety aspects (i.e., socio-economic aspects) and urge their inclusion in the regulatory decision-making process, stating that 'the refusal to consider nonsafety issues impairs democratic processes by depriving regulators [of the ability] to use relevant and legitimate information when assessing a biotechnology product'. Yet these issues may be perceived very differently by different sectors of the community, with farmers stressing economic benefits while consumers and organic producers may have very different views. The weight given to socio-economic considerations in comparison with the scientifically assessed aspects varies greatly in different jurisdictions. At a country level, the ban on GMOs by some countries has been entirely influenced by socio-economic considerations. For instance, most of the countries in the European Union have decided to ban their farmers from growing GM crops, purely owing to social pressures and against official scientific advice.

Where politicians drive the introduction of GMOs through a consideration of benefits, it is largely in the context of socio-economic issues; improved economic growth and wealthier smallholder farmers are vote-grabbing topics and likely to improve the chances of re-election. This again is reflected in the legislation of a number of developing countries (e.g., Bolivia and Malaysia), where the only reference to benefits is in the context of socio-economic issues. The non-biased term 'impact' (which could be either positive or negative) is most frequently seen in legislation as it relates to socio-economic impact (e.g., South Africa and Brazil).

Yet, even where benefits are mentioned in the legislation, this does not always lead to a positive decision to introduce GMOs. In Malaysia, consideration of socio-economic benefits led to approval for controlled field release of GM mosquitoes (Ramatha and Andrew, 2012). However, this has not progressed further and the government has been reluctant to take further action, fearing outcry from environmental groups. In 2014 a statement was issued saying that the trial will not proceed due to the cost associated with the next phase (USDA, 2015).

Other Risks and Benefits

In contrast to socio-economic aspects, in the context of environmental or health issues the term 'risk' more often predominates and the other aspects of benefit (environmental, health and agricultural) may be neglected. This is particularly seen in the environmental domain; where Ministries of Environment are responsible for the legislation then there is an overwhelming tendency for the legislation to be risk-focused and preventative in nature (e.g., Ethiopia). Vigani and Olper (2015) state that countries with greater demand for restrictive environmental regulation have also more restrictive GMO regulations. This can also be driven by lobbying and political donations from green organisations and associations of organic farmers.

From an environmental perspective, the primary benefits of GM crops relate to their ability to contribute to sustainable agricultural intensification, defined as 'the goal of producing more food with less impact on the environment, intensifying food production while ensuring the natural resource base on which agriculture depends is sustained, and indeed improved, for future generations' (The Montpellier Panel, 2013). While environmental activists focus on 'superweeds', loss of biodiversity and negative effects on non-target organisms, the benefits, as outlined in Table 4.1, include increased yields, reducing the need to bring new or marginal land into agricultural production, as well as reduced use of fertilisers and pesticides.

The lack of explicit consideration of health benefits in many countries may hamper the introduction of the next generation of nutritionally enhanced crops that hold benefits for the consumer. The exceptionally long lead time for the introduction of vitamin A-enhanced 'Golden Rice' is a case in point (see also Chapter 12). Wesseler and Zilberman (2014) compared the perceived economic cost of introducing Golden Rice in India with the health benefits that would be achieved. The perceived costs of introducing the technology in the face of anti-GMO lobby groups (i.e., a socio-economic consideration) appeared to outweigh the potential health benefits from its contribution to combatting the approximately 2.3 million disability-adjusted life years (DALYs) that are lost in India each year through vitamin A deficiency (Stein *et al.*, 2008). Field trials that indicate inconsistencies in yield compared with conventional varieties (IRRI, undated) may of course influence the balance of risk and benefit.

Other developments of GMOs that hold clear health benefits are likely to face similar hurdles unless there is a rethink on how the risks and benefits of these GMOs are assessed and communicated. Examples include the BioCassava Plus project, aimed at providing cassava with increased nutrient levels, improved shelf life, lower content of toxic cyanogens and resistance to viral disease (Sayre *et al.*, 2011) and the African biofortified sorghum project, where, although there was an initial focus on 'benefit messaging', the project later became derailed by anti-GM activism (Kamanga *et al.*, 2014).

Risk and Benefit Analysis

Methodologies for risk analysis (comprising risk assessment, risk management and risk communication) are well established internationally, even though many developing countries still do not have the capacity to carry out effective risk analysis. However, at the current time there are no international standards for benefit analysis, leaving a vacuum both for developed and for developing countries.

Particularly at the level of communication, risk communication predominates. Risk communication has been defined by the Codex Alimentarius Commission (CAC, 2007) as 'an interactive exchange of information and opinions concerning risk among risk assessors, risk managers, consumers and other interested parties' and forms an integral component of risk analysis. The report of a joint expert consultation of the Food and Agriculture Organization (FAO) and the World Health Organization (WHO) in 1998 (FAO/WHO, 1998) states that, among the elements of risk communication, messages could contain information on

The nature of the benefits
The actual or expected benefits associated with each risk.
Who benefits and in what ways.
Where the balance point is between risks and benefits.
The magnitude and importance of the benefits.
The total benefit to all affected populations combined.

From the above it is clear that at international level for many years there has been an acceptance of the need to communicate benefits as well as risks. Yet, while there has been talk about the need for development of effective risk–benefit communication strategies (Frewer *et al.*, 1998) and considerable research has been done on perceptions and communication of food risks and benefits (FoodRisC, 2015), a holistic strategy taking into account and balancing all the dimensions of risk and benefits (environment, agriculture, health/food and socio-economic) does not seem to have been developed. Particularly for developing countries, communication strategies are complex and need to be tailored to specific stakeholder groups (Racovita *et al.*, 2013). Unfortunately, the lack of coherent and well-presented strategies to communicate benefits has left the door open for anti-GM activists, who are expert at communicating only the risk side of the equation, as cited in the African biofortified sorghum example above. The European position, which specifically excludes consideration of benefits, does nothing to help this situation.

The Way Forward

There needs to be more inclusive discussion in international fora regarding strategies and mechanisms for a structured approach to risk–benefit analysis, as advocated by Morris (2011). An *ad hoc* approach leaves regulators, politicians and the general public confused. This is particularly the case with the rise of social media, where mixed messages and plain untruths are widely circulated. Lack of consensus amongst the international community represents a significant challenge, and demands increased collaboration between developed and developing countries.

The next generation of GM crops with real value to the consumer is being developed by the public sector or through public–private partnerships. Such projects should be encouraged and supported as they are not tainted by fear of the multinationals.

Meanwhile, technological developments are proceeding rapidly, as described in Chapter 2, and the concept of a GMO, as embedded in the CPB and the national legislation of many countries, is rapidly becoming outdated. Technologies such as oligo-directed mutagenesis, cis- and intragenesis, grafting on GM rootstocks, reverse genetics, zinc-finger nucleases, synthetic genomics and more recently genome editing using clustered, regularly interspaced, short palindromic repeats (CRISPR) and related systems will necessitate a fresh look at what is actually meant by a GMO. Already these developments have prompted calls for a shift from regulation based on the modification process used to regulation based on the nature of the phenotype developed, an approach adopted from the start by the USA and Canada (see Chapter 5 and Wolt *et al.* (2016)). It is to be hoped that blurring the lines between GM and 'conventional' breeding will result in a less emotive approach to consideration of risks and benefits.

References

Arvidsson, G. (2015). Food security now or wait for research to assess risks? Genetically modified crops and smallholder farmers in Africa. Nordiska Afrikainstitutet Policy Note 3.

Brookes, G. and Barfoot, P. (2016). *GM Crops: Global Socio-economic and Environmental Impacts 1996–2014*. Dorchester: PG Economics Ltd.

Burachik, M. (2010). Experience from use of GMOs in Argentinian agriculture, economy and environment. *New Biotechnology* **27**(5), 588–592.

CAC (2007). Working principles for risk analysis for food safety for application by governments. CAC/GL 62–2007.

Convention on Biological Diversity (2000). Cartagena Protocol on Biosafety to the Convention on Biological Diversity. Text and Annexes. [Online]. Available from www.cbd.int/doc/legal/cartagena-protocol-en.pdf

Convention on Biological Diversity (2012). Guidance on risk assessment of living modified organisms. UNEP/CBD/BS/COP-MOP/6/13/Add.1. [Online]. Available from www.cbd.int/doc/meetings/bs/mop-06/official/mop-06-13-add1-en.pdf

Convention on Biological Diversity (2015). Guidance on risk assessment of living modified organisms and monitoring in the context of risk assessment. UNEP/CBD/BS/RARM/AHTEG/2015/1/4. [Online]. Available from www.cbd.int/doc/meetings/bs/bsrarm-ahteg-2015-01/official/bsrarm-ahteg-2015-01-04-en.pdf

European Food Safety Authority (2006). Guidance document of the scientific panel on genetically modified organisms for the risk assessment of genetically modified plants and derived food and feed. *EFSA Journal* **99**, 1–100.

Falck-Zepeda, J. *et al.* (2013). The current status of the debate on socio-economic regulatory assessments: positions and policies in Canada, the USA, the EU and developing countries. *World Review of Science, Technology and Sustainable Development* **10**(4), 203–207.

FAO/WHO (1998). The application of risk communication to food standards and safety matters. [Online]. Available from www.fao.org/docrep/005/x1271e/X1271E01.htm

Fischer, K. *et al.* (2015). Social impacts of GM crops in agriculture: a systematic literature review. *Sustainability* **7**(7), 8598–8620.

FoodRisC (2015). Website of the project 'Perceptions and communication of food risk/benefits across Europe'. [Online]. Available from www.foodrisc.org

Frewer, L. J. *et al.* (1998). Consumer acceptance of transgenic crops. *Pesticide Science* **52**, 388–393.

Herman, R. A. *et al.* (2013). Bringing policy relevance and scientific discipline to environmental risk assessment for genetically modified crops. *Trends in Biotechnology* **31**(9), 493–496.

IRRI (undated). What is the status of the Golden Rice project coordinated by IRRI? [Online]. Available from http://irri.org/golden-rice/faqs/what-is-the-status-of-the-golden-rice-project-coordinated-by-irri

Kamanga, G. D. *et al.* (2014). Why communication and issues management (CIMS) must occupy a central role in GM projects: case study of the Africa Biofortified Sorghum (ABS) Project. In *Biotechnology in Africa: Emergence, Initiatives and Future*, ed. F. Wambugu and D. Kamanga. Basel: Springer International Publishing Switzerland, pp. 225–241.

Ludlow, K. *et al.* (2013). *Socio-economic Considerations in Biotechnology Regulation*. Berlin: Springer Science and Business Media.

Marcoux, J.-M. *et al.* (2013). The inclusion of nonsafety criteria within the regulatory framework of agricultural biotechnology: exploring factors that are likely to influence policy transfer. *Review of Policy Research* **36**(6), 657–684.

Mitre, M. and Reis, B. P. W. (2015). Science and politics in the regulation of genetically modified organisms in Brazil. *Review of Policy Research* **31**(2), 125–147.

Morris, E. J. (1995). Biosafety regulations in South Africa. *African Crop Science Journal* **3**(3), 303–307.

Morris, E. J. (2011). A semi-quantitative approach to GMO risk–benefit analysis. *Transgenic Research* **20**(5), 1055–1061.

Morris, E. J. and Thomson, J. A. (2014). Genetically modified crops commercialized in South Africa. In *Biotechnology in Africa: Emergence, Initiatives and Future*, ed. F. Wambugu and D. Kamanga. Basel: Springer International Publishing Switzerland, pp. 53–65.

Racovita, M. *et al.* (2013). Experiences in sub-Saharan Africa with GM crop risk communication. *GM Crops & Food* **4**(1), 19–27.

Ramatha, L. and Andrew, J. (2012). Socio-economic aspects in decision-making in the context of the Biosafety Protocol: Malaysia's experience and case studies. *Asian Biotechnology and Development Review* **14**(3), 19–30.

Sayre, R. *et al.* (2011). The BioCassava Plus program: biofortification of cassava for sub-Saharan Africa. *Annual Reviews of Plant Biology* **62**, 251–272.

Siegrist, M. *et al.* (2016). Biased perception about gene technology: how perceived naturalness and affect distort benefit perception. *Appetite* **96**, 509–516.

Stein, A. J. *et al.* (2008). Genetic engineering for the poor: golden rice and public health in India. *World Development* **36**(1), 144–158.

The Montpellier Panel (2013). Sustainable Intensification: A New Paradigm for African Agriculture, London.

USDA (2015). Malaysia Agricultural Biotechnology Annual 2015. Global Agricultural Information Network (GAIN) Report Number MY5011. [Online]. Available from http://gain.fas.usda.gov/Recent%20GAIN%20Publications/Agricultural%20Biotechnology%20Annual_Kuala%20Lumpur_Malaysia_7-10-2015.pdf

Vigani, M. and Olper, A. (2013). GMO standards, endogenous policy and the market for information. *Food Policy* **43**, 32–43.

Vigani, M. and Olper, A. (2015). Patterns and determinants of GMO regulations: an overview of recent evidence. *AgBioForum* **18**(1), 44–54.

Wesseler, J. and Zilberman, D. (2014). The economic power of the Golden Rice opposition. *Environment and Development Economics* **19**(6), 724–742.

Wolt, J. D. *et al.* (2016). The regulatory status of genome-edited crops. *Plant Biotechnology Journal* **14**(2), 510–518.

5 The Canadian Experience with the Creation and Implementation of Regulatory Frameworks for the Environmental Release of GM Crops as a Model for Developing Countries

Phil Macdonald

Introduction

It has been over three decades since the first transgenic plants intended for commercialisation entered into regulatory processes. The application of molecular biological tools has allowed plant breeders to introduce rapidly traits that it would have been difficult or impossible to introduce using more traditional breeding techniques. The application of modern biotechnology broadens the scope of genetic changes that can be introduced into plants to achieve breeding objectives, although it does not inherently result in plants that are less safe than those produced by more conventional techniques (National Academy of Sciences (NAS), 1987). Pre-market risk assessment is a key component of the regulatory approach to biotechnology, and an increasing awareness of environmental issues among citizens prompted the development of regulations on the environmental release of plants altered using biotechnology. The call for environmental risk assessments arose from concerns about the potential risks of the modified plants themselves, as well as risks associated with the transfer of novel traits to related plants. Human exposure in the environment is often also part of the consideration (exposure to allergens through inhalation or contact, accidental consumption), but, for the most part, human health issues are dealt with by food safety evaluations.

The application of biotechnology to plant breeding has engendered concerns that have persisted, posing a difficult dilemma for scientists involved in product development, risk assessment and other regulatory functions and those engaged in public policy. The intersection of policy and science in the sphere of agricultural biotechnology regulation has proven to be challenging and some interesting outcomes have arisen. Canada's experience facing those challenges may provide some useful lessons for countries looking to develop regulatory approaches, and this chapter will demonstrate a science-based, fit-for-purpose example of a regulatory model for the environmental release and commercial cultivation of GM plants.

The Development of a Regulatory Approach to Biotechnology

With the active research and development of new plants using the techniques of biotechnology in the 1980s, companies and industries were looking to governments for guidance on how such plants would be considered and potentially regulated. Canada adopted a broad definition of biotechnology as a direct outcome of the scientific consultations it carried out in the late 1980s.

One of the first of these consultations took place in 1988 at a workshop carried out in conjunction with the Canadian Agricultural Research Council. The scientific perspective that came from the consultations was that plants derived through genetic engineering were not necessarily any riskier than those derived through other breeding techniques such as, for example, chemical mutagenesis, resulting in a regulatory policy that captured 'plants with novel traits' (PNTs) rather than GMOs or GM plants. The transfer of the scientific advice into policy resulted in the 1993 Federal Regulatory Framework for Biotechnology, which described an approach to how biotechnology would be viewed that informed the policy environment. The Federal Regulatory Framework for Biotechnology set out principles that included the use of science-based safety assessments and risk management with the overarching goal of protecting human and animal health, and the environment, but without hindering the innovation that could enhance the prosperity and wellbeing of Canadians. This straightforward guidance provided useful policy guidance on the scope and goals of the risk assessment.

All current regulatory frameworks for transgenic plants incorporate the need for a risk assessment prior to environmental release. The risk assessment will typically involve the identification of any potential hazards, an evaluation of the likelihood of the hazards occurring and a determination of the potential exposure to the hazard in the environment. The risk assessment may also consider benefits as part of the process, depending on the regulatory framework. Along with the risk assessment, risk management strategies and risk communication (which can include summaries of risk assessments or outreach to stakeholder groups) will be incorporated into the risk analysis. Risk management may encompass socio-economic factors such as the cost to growers of implementation or the effects on valued sectors, but the risk assessment component of risk analysis is a science-based process and is applied case by case. The data which support the risk assessment can come from a variety of sources, including experimental data, the published literature or other sources that countries decide are relevant, reliable and meet acceptable standards. The information which can usefully support a risk assessment for a product may include data that do not necessarily achieve the usual standards for a peer-reviewed scientific publication, such as information from grower groups and agriculture extension personnel and grower experiences.

The presence of a small group of trained and skilled regulators can add predictability and consistency to the risk analysis of the environmental release of a GM crop, and Canada has shared that experience through capacity-building activities and by hosting visiting scientists from countries with less developed regulatory systems to allow them to gain hands-on experience. A large team is not necessary when regulatory authorities build networks into the scientific community where expertise can be accessed when needed. Countries often rely on arm's-length scientific advisory groups to provide expertise on specific scientific issues or, in some cases where required by regulations or where resources are insufficient, to actually conduct the risk assessment. Regulators often speak of 'need to know vs. nice to know' information. This statement from regulators articulates the need to determine what data are necessary and sufficient to make a decision about the probability of a risk being realised. Since regulators generally have formal scientific training, it can sometimes be difficult for a regulator to separate the need for information as part of the structured inquiry for risk assessment from the desire for information that fills some knowledge gaps about the product. When external scientific bodies are used, it may be necessary to ensure that these bodies focus on the data required to address safety concerns rather than pursuing a scientific inquiry that satisfies curiosity. A good example of this conflict between seeking information to satisfy scientific inquiry and the identification of sufficient data to address risk often arises around discussions of the mode of action of a new trait introduced into a transgenic plant. A precise understanding of the mode of action can sometimes simplify the risk assessment, but regulators successfully carried out risk assessments of virus-tolerant, or modified, oil crops, before RNA interference was completely understood (Saurabh *et al.*, 2014).

Determining the scope and purpose of the environmental risk assessment of a transgenic crop is a key first step. Guidance in these areas is typically captured in the central objectives that must be addressed by the applicant to assess the environmental safety of a transgenic event. Typically these are described in acts or implementing regulations. It is here that governments will articulate social values and the aspects of the environment that are most valued. Policy frameworks will inform protection goals, and other enabling acts, such as those that describe the protection of species at risk, will be considered. Ideally, the presence of an overarching policy framework guides the process of risk analysis and regulation by informing the key areas of consideration for policy and guidance. When those parameters are clear, the risk assessor and regulator understand the context for regulation and what will constitute acceptable risk. The result is consistency in approach for any applicant approaching with a biotech product that is subject to regulation, clear guidance for the risk assessor and a predictable set of information requirements to inform a risk assessment

Once clear guidance on the goals for regulation of the environmental release of transgenic plants has been articulated, the next step is the crafting of appropriate regulatory oversight that meets the policy objectives of the regulatory framework while addressing the scientific concerns that have been raised over the potential environmental release. Since 1996 in Canada, with the establishment of the relevant regulations, GM plants have been subject to pre-market assessment and approval prior to release into the environment. Environmental release could include a wide range of activities, but the regulations should describe two general categories of release:

- confined release into the environment, where release is conducted under conditions that are intended to minimise the establishment and spread in the environment of the GM plant and facilitate managing the cultivation site after the release; and

- unconfined release into the environment, where release is no longer subject to the same strict conditions and generally cultivation will be no different from that of a conventional crop (although conditions may be applied to the authorisation in order to manage risk).

In Canada and the USA, with some exceptions, environmental release of GM plants will fall into these two broad categories, confined and unconfined, where it will be either regulated or unregulated, respectively. Some countries, such as Mexico, where there are special considerations for some crops such as maize, may have more pre-commercial steps where the scale of the release may become larger and management of the environmental release may become progressively less restrictive in order to enhance public confidence or generate additional data on a larger scale but still under controlled conditions.

Confined Release

Confined release is often used to conduct experiments designed to meet the information requirements necessary for applications for unconfined release. It is also a mechanism to generate biosafety data to support risk assessments and to address questions of fundamental science. For example, in Canada, the Canadian Forestry Service has conducted multi-year research to address biosafety issues related to trees with novel traits, using conifers expressing insect-resistance genes to examine issues such as gene stability and environmental interactions. The confined field trial programme allows small-scale releases under very controlled conditions that are strictly enforced. The trial sites are put under conditions of reproductive isolation, normally by using sufficient distance to prevent outcrossing from related species, but nets, tents, bags and flower removal can also be used. Viable and non-viable material is also controlled, and seed and vegetative materials must be securely stored or destroyed by an approved method. Post-trial conditions that depend on the biology of the species are also imposed, with the goal of preventing the plant from persisting in the environment. In Canada, as in many jurisdictions, there are inspections to ensure that those conducting field trials comply with the terms and conditions.

The development of a regulatory programme that permits small confined field experiments with GM plants is an outcome of the implementation of the principles of the Federal Regulatory Framework, since this programme allows for controlled release of a GM plant into the environment under circumstances of managed risk. Confined field trials are an essential mechanism for generating necessary science outputs that support the risk assessment and regulation of GM crops, but they are also intimately linked to the science inputs that inform the terms and conditions which are imposed on the field trial. For familiar crops such as canola/rapeseed, maize and soybeans, agronomic practices are well understood, the basic biology is well understood and associations that set standards for seed production and identity-preserved production, such as the Canadian Seed Growers Association, publish established isolation distances based on the probability of outcrossing arising from estimated pollen flow. For unfamiliar crops, the information is less apparent and, as countries look to explore niche markets and value-added crops, it is necessary to have the science to inform the terms and conditions of confined field trials with these speciality crops if confidence in the regulatory oversight of this programme is to be maintained.

Confined field trial oversight represents the confluence of science and policy. In this activity, the objectives of all concerned are clear; the policymaker needs assurances that regulated activities can be carried out safely and the scientists understand the need to fill in the knowledge gaps. Both groups have a heavily vested interest since the outcomes from the confined field trials can also provide scientific outputs.

Unconfined Release

Moving beyond small-scale field trials requires a more complete assessment of the GM crop prior to release. Key to the environmental assessment is a thorough knowledge of the crop species that has been subjected to modification by biotechnology to express a new trait. This knowledge is fundamental to conducting a comparative risk assessment. The concept of familiarity is used to identify and evaluate environmental risks that may be associated with the release of a transgenic plant and also to inform management practices that may be needed in order to mitigate recognised risks. In Canada this requirement is satisfied by the creation of individual crop Biology Documents that describe the agricultural, agronomic and environmental behaviour of the crop species specifically in the Canadian environment, including a description of plant growth, reproduction, interactions with related and unrelated species, and management and cultivation practices. The Canadian Biology Documents are similar in focus to the Consensus Documents developed by the OECD's Working Group on the Harmonization of Regulatory Oversight in Biotechnology, but the Canadian document describes management conditions and environmental interactions for the unmodified species that are specific to the Canadian environment. The Biology Document uses the familiarity with the cultivation and management of a species to provide a framework to identify potential hazards during the safety assessment. Biology Documents for the species grown in Canada have been developed in consultation with experts (CFIA, 2016).

Concerns about the Environment

Concerns about protecting biodiversity and endangered species emerged early in the discussions of the application of biotechnology to crop development. For example, concerns have been expressed about the production of 'superweeds' (Union of Concerned Scientists, 2013) or unforeseen environmental consequences arising from unexpected consequences of the genetic modification (Schnell *et al.*, 2015). Potential areas of concern generally fall into the following broad categories, described in Canada as 'the five pillars' (Wozniak and McHughen, 2012; Layton *et al.*, 2015).

- The novel traits increasing the weediness or invasiveness of the plants by conferring greater fitness and allowing the plants to outcompete others in the environment.
- The novel traits will spread through hybridisation of the PNT with wild, sexually compatible relatives, creating new weeds and driving out desirable plant species.

- Some novel traits, such as insect-resistance traits, can inadvertently target unintended species, including beneficial insects or threatened and endangered species, or create unintended consequences from bioaccumulation.
- The novel trait can result in a plant that harbours disease or insect pests.
- The novel trait will result in a plant that threatens biodiversity through direct or indirect measures, including the over-adoption of a PNT variety affecting agricultural biodiversity.

In Canada, the stacking of traits by conventional breeding would only trigger an additional risk assessment if the result of the cross were the introduction of an additional novel trait, for example if the transgenes from the modified plants were designed to interact with each other to produce a unique trait in the offspring or if there were specific stewardship requirements applied to the plant, such as the requirement for an insect-resistance management plan (for insect-tolerant crops) or a herbicide-tolerance management plan (for herbicide-tolerant crops). When stewardship conditions apply to the stacked-trait plant, the review concentrates on any changes that may be required to the stewardship. Frequently, insect-tolerance traits are stacked to pyramid modes of action and reduce refuge requirements. Stacking of herbicide-tolerant traits would raise concerns if a producer's options for weed management were compromised, e.g., if there were no herbicide control options for crop volunteers in subsequent seasons. Otherwise, the outcome of the cross of approved novel traits would trigger oversight no different from that for any of the other breeding steps that would take place for variety development of a crop.

A review of the various regulations and guidance documents produced by countries with well-developed regulatory systems suggests that there are very substantial areas of agreement in the broad descriptions of types of information that have been most commonly considered when countries have performed environmental safety reviews for transgenic plants (García-Alonso *et al.*, 2014). There may, however, often be significant differences between specific reviews, depending upon the relevant circumstances of individual cases and/or the laws and regulations under which the reviews and decisions are being made. The specific data sets reviewed may vary according to the specific characteristics of the plant and the trait introduced as well as the environment of use. Thus, most reviews are still, in actuality, conducted on a case-by-case basis, thereby allowing the consideration of a greater or lesser number of criteria that can be articulated in broad descriptions of the types of information most commonly considered in government-performed environmental risk/safety assessments. The significant overlap in terms of the data sets required and the consistent way in which the concerns are addressed provide ample opportunity for harmonisation, data portability and sharing of risk assessments.

Information Elements for an Environmental Risk Assessment

The characteristics of the transgenic plant under review must be known before conducting an environmental assessment. Information on the host organism, the donor organism, the properties of the new trait and the phenotype of the transgenic plant usually constitutes the basic framework for the environmental risk assessment.

In broad terms, the environmental risk assessment considers the potential effects of the release of the transgenic plant on biodiversity, and this protection goal is applicable both to developed countries and to developing economies. In this regard, highly domesticated crops may be less problematic than, for example, forestry tree species or previously undomesticated plant varieties intended for uses such as biofuels. For the latter types of transgenic plants the possibility of affecting the biodiversity of natural populations either directly through invasion and competition or indirectly through the transfer of undesirable traits may need to be considered. For GM crops the considerations are normally confined to the managed ecosystem, since most domesticated crops, although they may form feral populations, do not compete well outside managed ecosystems (Crawley *et al.*, 2001), and domestication traits often come with a fitness cost (Mercer *et al.*, 2006).

Generally, the environmental risk assessment of transgenic plants identifies and evaluates the risks associated with release into the environment, and the consequences of cultivation of these plants in comparison with a conventional counterpart. The counterpart is usually the most closely related unmodified version of the transgenic plant. The most closely related comparator may in some cases itself be a transgenic plant, such as in the assessment of a subsequent transformation of an already-commercialised transgenic plant. Often, for comparative assessments, other similar varieties of the same species will also be considered as part of the environmental safety assessment to help cover the normal variability and to predict the behaviour of the new transgenic trait or traits in other varieties.

The environmental assessment will normally include information on the molecular characterisation and stability of the genetic modification. Documents such as the OECD consensus document on the Molecular Characterization of a Transgenic Plant provide guidance on the wide variety of elements that countries consider and how they can be used in the Environmental Safety Assessment. Molecular characterisation data can be useful in calculating potential toxicity and characterising potential pathways to harm, but genotypic data are not necessarily predictive of potential risks, given the inherent variability in plant genomes. This fact is especially relevant when considering the extensive data that can be generated from conducting genomic analyses. Applying this type of data is difficult, and, at this time, its relevance to risk analysis of transgenic plants is uncertain without a clear idea of how the data can be linked to a possible harm. Comparative risk assessments will contextualise the environmental release in terms of the cultivation of the unmodified counterpart, i.e., the monoculture of a crop will affect biodiversity, as will the cultivation of the GM crop; a crop produces a variety of chemicals, endogenous toxins and allergens, where does the new trait fit into that continuum? This is the Canadian context for substantial equivalence, but a metabolite-by-metabolite comparison would have little value to a risk assessor without a clear link to causality or harm.

Unexpected Effects from Genetic Modification of a Crop

Unexpected effects can also arise from the introduction of a new trait. Current practice is to rely on a thorough phenotypic evaluation of the modified variety to determine whether there are significant changes in normal agronomic performance. Many jurisdictions rely on post-approval monitoring to evaluate any potential unexpected changes. There is now significant opportunity to modernise the comparative risk assessment and move away from the consideration of each transformation event on a 'case by case' basis. Recent published work by Schnell *et al.* (2015) places the outcomes of inserting a gene into a crop plant in the context of conventional breeding. Although there is a significant focus on the outcomes of inserting genes into a plant, this review illustrates that significant genetic perturbations occur during conventional breeding (a factor long exploited by breeders in search of new varieties) and that there is an opportunity to modernise regulatory approaches, particularly in the case of vegetatively propagated crops such as potatoes where the same genetic construct may be inserted into different potato varieties in the development of the crop. The review illustrated that the insertional effects associated with genetic engineering are similar to the genetic changes that occur in conventionally bred plants. Given the similarity, these effects should present a similar level of risk to genetic changes associated with conventional breeding, so the risk assessment conducted on the original transformant can be applied to subsequent transformants, saving time, effort and resources.

Discussion and Conclusions

Canada has now accumulated more than 20 years of experience in the regulation of GM plants, leading to approvals that have provided growers with access to the products of modern biotechnology. This has resulted in situations where for some crop types, such as canola and corn, the majority of production is GM. In countries where food security is an issue, growers could benefit from access to crops that are tolerant to water stress, or resistant to viruses or other plant diseases. Before countries can access those benefits, citizens need to be assured that there is a system in place to ensure that the release is conducted safely and the system evolves. There have been key factors that have allowed Canada to successfully authorise GM plants while considering environmental safety, and these may be instructive for countries looking to develop similar regulatory oversight. They are as follows.

1. The elaboration of a regulatory framework. Regulatory frameworks provide government oversight on commercial or other activities that potentially raise concerns, whether social, ethical or environmental. The creation of regulatory frameworks to address the potential risks and access the benefits of biotechnology has been, to date, the usual response of governments, although the implementation of regulatory policy has overwhelmingly been focused on addressing potential risks. Irrespective of whether the regulation is process-based, as in Australia or the UK, or truly product-based like the Canadian approach, there needs to be a sound basis in legislation, with regulations derived from the appropriate legal authority.

2. The articulation of a policy framework. Integral to any regulatory development is the application of the concepts of transparency and consultation, and the articulation of a policy framework that describes the goals and social values which underlie the regulations. In Canada, the policy framework clearly states that the regulatory process should allow the release of innovative biotech products that have been thoroughly evaluated for their safety, and the marketplace rather than regulators will choose which products succeed. Developing economies may find that the policy framework could, in some cases, need to be further enhanced to make allowances for higher uncertainty or to incorporate flexibility to accept somewhat greater risk in order to address concerns around food security or to allow emergency response to serious crop failure from a pest or drought to be incorporated into considerations for eventual release or as a consideration for protection goals. Policy oversight may be influenced by a consideration of the sustainability and redundancy in the country's agricultural production. Social licence, consideration of indigenous practices and the ability to impose regulatory controls will often factor into risk management considerations, and these should be guided by the policy framework.
3. Clear description of information required in support of product safety. Equally integral to a regulatory framework is a description of the information required to support the safe deployment of the transgenic crop. There is ample guidance with regard to the issues addressed in a risk assessment and the information that can be applied to address these concerns, and there is substantial overlap of different regulatory regimes. This commonality in information requirements is a strong argument for the portability of data and the possibility of applying risk assessments either in whole or in part to a number of regulatory jurisdictions, leading to, for example, the possibility of mechanisms for regional approvals. For developing economies, the use of existing data and regional approvals can provide a way to speed up access to transgenic crops that growers may want, and minimise the incidence of asynchronous approvals between countries that share trade and/or borders. As countries develop regulatory frameworks, there will be opportunities to take advantage of familiarity and potentially eliminate from consideration some traits or some aspects of the risk assessment of those traits on a case-by-case basis. The information to inform the risk assessment should be fit for purpose, and sufficient to address hypothesis-driven concerns in order to ensure that the risk assessment fulfils its purpose rather than acting as a barrier to innovation.
4. Clear description of roles and responsibilities. The mechanics of the regulatory process may differ with respect to how the functions are conducted inside the regulatory framework and which individuals, whether from government or from private institutions, take responsibility. Frequently, the risk assessment and risk management recommendations are separated from the decision-making part of regulation (e.g., in Canada, Australia and the UK) in order to separate the more science-based aspects of risk assessment from the broader concerns when deciding on the feasibility and economics of risk management measures and the broader concerns that can be applied to decision-making. Whatever the breakdown of roles, it is essential that all involved in delivering the regulations have the skills and necessary information to deliver their

part. In Canada, those involved in the regulation of transgenic crops have expertise in crop production, crop inspection, plant health and plant risk assessment. In countries with limited resources there is a great deal of value in utilising the experience of those involved in the regulation of conventional crops also in the regulation of transgenic crops as part of their duties.

Last Thoughts

The development of transgenic crops and their commercial release has provided advantages to growers worldwide in providing new management tools, and new possibilities for enhancing yields. As with all technologies, there have been both benefits and problems. The adoption of agricultural biotechnology has sparked a larger discussion on the environmental impact of agriculture and how the production of food fits alongside the need to feed burgeoning populations. In Canada, recent reviews have calculated the environmental and economic benefits of the adoption of herbicide-tolerant crops (Smyth *et al.*, 2010). However, along with the reduction of pesticide use, the conservation of soil and the reduced costs to growers there are the potential problems that arise from the loss of very useful herbicides and the evolution of resistant weed biotypes. These are potential problems that can arise at a time when there is very little investment in terms of new herbicide modes of action. In the same vein, although growers have reaped significant economic benefits from the deployment of insect-tolerant crops and health benefits from reduced exposure to sprayed pesticides, there are already reports of resistance evolving to these insect-protected crops, and the alternative may be a return to less sustainable technologies. The implementation of a regulatory framework will not address all of the risks and challenges posed by transgenic crops, but a properly constructed regulatory framework can provide a consistent, adaptable, transparent mechanism to allow the development of transgenic crops and their responsible deployment into the environment.

References

CFIA (2016). Canadian Food Inspection Agency Biology Documents. [Online]. Available from www.inspection.gc.ca/english/plaveg/bio/dir/biodoce.shtml

Crawley, J. *et al.* (2001). Transgenic crops in natural habitats. *Nature* **409**, 682–683.

García-Alonso, M. *et al.* (2014). Transportability of confined field trial data for environmental risk assessment of genetically engineered plants: a conceptual framework. *Transgenic Research* **23**, 1025–1041.

Layton, R. *et al.* (2015). Building better environmental risk assessments. *Frontiers in Bioengineering and Biotechnology* **3**, 110. http://dx.doi.org/10.3389/fbioe.2015.00110

Mercer, K. L. *et al.* (2006). Effects of completion on the fitness of wild and crop-wild hybrid sunflower from a diversity of wild populations and crop lines. *Evolution* **60**(10), 2044–2055.

NAS (1987). *Introduction of Recombinant DNA Engineered Organisms into the Environment: Key Issues*. Washington, D.C.: National Academy Press.

Saurabh, S. *et al.* (2014). RNA interference: concept to reality in crop improvement. *Planta* **239**(3), 543–564.

Schnell, J. *et al.* (2015). A comparative analysis of insertional effects in genetically engineered plants: considerations for pre-market assessments. *Transgenic Research* **24**, 1–17.

Smyth, S. J. *et al.* (2011). Environmental impacts from herbicide tolerant canola production in Western Canada. *Agricultural Systems* **104**(5), 403–410.

Union of Concerned Scientists (2013). 'Superweeds' resulting from Monsanto's products overrun U.S. farm landscape. [Online]. Available from www.ucsusa.org/news/press_release/superweeds-overrun-farmlands-0384.html#.WG0fmX1BkbU

Wozniak, C. and McHughen, A. (2012). *Regulation of Agricultural Biotechnology: The United States and Canada*. Berlin: Springer Science and Business Media.

6 Labelling of Food from GMOs: Options to Consider by Developing Countries

Wilna Jansen van Rijssen and Wynand J. van der Walt

Introduction

A precautionary safety assessment of foods derived from genetically modified organisms (GMOs), according to the guideline of the Codex Alimentarius Commission (CAC), is followed by most countries, by comparing genetically modified (GM) foods with conventional foods with a history of safe use. Why then label GM foods as GM, while standard Codex labelling already applies to all food? The inclusion of labelling of GM food resulted to a large extent within the history of scares like mad cow disease, which resulted in less consumer trust in government decision-making. In addition to the scares, the ruling by the World Trade Organization (WTO) on the dispute between the USA and the European Union on GM foods resulted in the adoption of European Commission (EC) regulation 1829/2003 on food/feed safety assessment and EC regulation 1830/2003 on traceability of GMOs. Certain anti-biotechnology groups and lobbyists, specifically as regards GM crops, exploited public apprehension by raising fears on GM crops and GM food, leading to public distrust in GM technology (Dubock, 2014). The fear-raising has spilled over to other countries. This chapter focuses on issues relating to the labelling of GM food and proposals for developing countries.

Consumers' Rights

Consumers have a number of basic rights in terms of the guidelines set by the United Nations (UN, 2003), including, to name a few, the right to safety (safeguarding against goods that are hazardous to life and property); to information; to choice (between GM and GM-free food); to be heard; to satisfaction of basic needs; to redress, to consumer education; and to a healthy environment. These are to be revised according to the United Nations Conference on Trade and Development (UNCTAD, 2013). The rights are indeed important. Therefore, labelling of GM foods should be seriously considered. The question would be 'how would labelling of GM food impact on these rights?'

Labelling Requirements by International Bodies

The Role of the Codex Alimentarius Commission in Labelling of GM Foods and Harmonisation of Labelling Requirements

The members of the Codex Alimentarius Commission (Codex) have been discussing harmonisation of GM food labelling at many meetings since the early 1990s. An impasse in the discussion resulted from polarised arguments, mainly concerning whether GM foods were sufficiently unique or posed dangers that should require such labelling (Miller and Kershen, 2011). The standards that Codex develops are voluntary, not mandatory. Codex did reach consensus in 2011 to prepare guidelines on general labelling issues (CAC, 2011). The text is a compilation of ten relevant Codex texts. The explanation given by Codex reads as follows:

> Different approaches regarding labelling of foods derived from modern biotechnology are used. Any approach implemented by Codex members should be consistent with already adopted Codex provisions. This document is not intended to suggest or imply that foods derived from modern biotechnology are necessarily different from other foods simply due to their method of production (CAC/GL 76–2011).

The ten Codex guidelines referred to for labelling cover claims, production, pre-packed food, processing and marketing of organically produced foods, use of the term 'Halal' and principles of risk analysis that apply to food in general, as well as to foods derived from modern biotechnology. A specific condition is that the presence in any foods obtained through biotechnology of an allergen shall be declared or the food should not be marketed if adequate information on the presence of the allergen is not provided.

Codex makes specific mention of misleading information, unsubstantiated claims and claims that could give rise to doubt about the safety of similar food or could arouse or exploit fear in the consumer (CAC/GL1–1979). The standards that Codex develops are voluntary, not mandatory.

A general conclusion is that Codex discussions did not contribute to harmonisation of GM labelling requirements other than existing ones for foods in general.

Labelling Requirements of the Cartagena Protocol on Biosafety

Labelling, as a Cartagena Protocol on Biosafety (CPB) requirement (Convention on Biological Diversity, 2000), applies to living modified organisms (LMOs). It should be noted that these labelling requirements are only for export and import of traded shipments and not directly related to consumer use. The labelling requirement was initially worded as 'may contain' for LMOs, but has since then included a second option of 'does contain' and the wording 'contains LMOs'. In the latter case a consignment must comply with an identity preservation system (IPS) (Box 6.1) and an identity (a unique identifier) must be provided for each event. A change in labelling regulations from 'may contain' to 'does contain' will significantly increase the cost of trade and potentially the price of grain. For example, Kenya has several ports of entry to neighbouring countries. Kimani and Gruère (2010) analysed the impact such a change will have on Kenya's border control administration. The 'does contain' option requires sampling and analysing of each import and export consignment, plus the appointment and training of officials to execute all aspects of monitoring. The additional administrative and monitoring requirements will impose costs (both for the government and for consumers) that may (will) increase food prices (Kimani and Gruère, 2010, p. 224).

Bouet *et al.* (2010) evaluated the global trade effects of implementing the 'does contain' rule on the maize sector in all significant trading countries. The authors developed

Box 6.1 Definitions

Segregation refers to maintaining separate handling of GMO crops to avoid commingling with non-GM crops during planting, harvesting, loading and unloading, storage and transport.

Traceability is the ability to maintain credible identification of a product through various steps in the farm-to-retail chain, including production, processing and retailing, as well as specification of its national origin.

Identity preservation is defined as the more stringent handling process requiring strict separation to be maintained at all times. Identity preservation lessens the need for additional testing and lowers the liability and risk for growers and handlers of GMO products (Falck-Zepeda, 2006).

Product labelling. By product labelling is meant that the technology used in the development of the end-product, e.g., mutation breeding, organic agricultural practices or genetic modification, is not indicated on the label. Only products with detectable and quantifiable traces of GM materials or ingredients are required to carry a label. Substantial differences such as high oleic acid content can be monitored with testing diagnostics and can be enforced.

Process labelling. Process labelling requires labelling of food describing the technology from which the product has been derived, or specifying that it contains GMOs. Any product derived from GM crops will have to be labelled, irrespective of whether it contains traces of GM material or not. Maintenance of the identity of a product throughout the links in the manufacturing chain to the end-product and applications included will be necessary.

Labelling ingredients. Individual ingredients can be indicated as GMO-derived or the processed food can be described as such, as in process labelling. A practical problem is determining tolerance levels where the product contains multiple GMOs.

a spatial trade model and used simulated trade scenarios. The results from the study showed that mandatory labelling of LMOs in international trade would cause trade distortion (realignment with trading blocs) and affect welfare for all signatories of the Protocol as well as for non-members that produce GM maize. Consumers in sub-Saharan African countries will probably pay a larger proportional price than consumers in developed countries. A country such as South Africa, which is strongly connected to the world market, may experience even larger losses, both for producers and for consumers. The effects are confirmed in a study for Kenya (Kimani and Gruère, 2010).

The African Union Model Law on Biosafety

The African Union Model Law on Biosafety is largely based on the Convention on Biological Diversity (CBD). The labelling requirements specify detailed information so as to guarantee traceability from the country of origin to the destination for each trait.

Specificities such as characteristics, allergenicity and risk are to be provided (African Union, 2014). The model law has been criticised because of its very strict requirements, such as 'does contain' labelling, which has cost implications.

The Requirements of the WTO: Technical Barriers to Trade

The Sanitary and Phytosanitary (SPS) Agreement of the WTO (1995a) specifically refers to safety regulations, whereas labelling is a consumer privilege for more information and not a safety issue. The WTO (2011) cites Codex's food safety standards, guidelines and recommendations for facilitating international trade and protecting public health. Should Codex standards be incorporated into domestic laws it is assumed that food complies with WTO scientific safety requirements. As the food from GMOs currently on the market is considered safe, labelling of foods from GMOs, if not harmonised, could result in technical barriers to trade in conflict with the Technical Barriers to Trade Agreement (WTO, 1995b). Furthermore, the WTO SPS Agreement clearly states that the SPS applies only to 'likelihood of entry, establishment or spread of a disease in the territory of the importing country and of biological or economic consequences or adverse effects on human or animal health from additives, contaminants, toxins, or disease-causing organisms' (abbreviated). The rule in international trade is to comply with the country of importation's legislation. However, as GM labels are not safety warnings, this may be an argument for not labelling under the WTO's SPS requirements.

The Reality of Food Labelling

Global Problems in Understanding Food Labelling

There are various problems with general food labelling that occur in practice. For instance, not all consumers read labels, and the literacy rate of consumers could be an obstacle to providing label information. Consumers are not sufficiently educated to interpret label messages, and activism adresses emotions. One of the issues is the number of official languages in some countries, making labelling difficult. Africa has an estimated 2,000 languages (Chimhundu, 2002). This highlights the practical problems in meeting Codex Alimentarius guidelines for clear, understandable label wording.

Challenges in Africa to Legislation on Labelling

Labelling regulations in Africa are still at the level of discussion by most countries as biosafety legislation in this region is progressing very slowly. However, research in a number of GM crops is in the pipeline in different African countries. According to Morris (2014, pp. 65–67) the majority of 52 African countries drafted national biosafety frameworks and policies. As these countries are signatories to the CPB, it is implied that regulatory requirements for labelling of transboundary movement of GMOs are in the pipeline or have been captured in policies, decrees or legislation. However, fewer than

Box 6.2 South African GMO Label Legislation

A specific example of conflicting labelling requirements is the South African labelling of GM foods, feeds and other consumer products. This has been a point of contention (Jaftha, 2014, pp. 194–195). Labelling regulations were made by the Department of Health after extensive consultation with stakeholders. It was approached from a safety and nutrition perspective according to the requirements of the food safety law (South Africa, 2004). However, the Department of Trade and Industry then entered the food arena with mandatory labelling, defending their stance by stating that labelling is a form of consumer information (South Africa, 2011). The two regulations contradict one another as the former is voluntary and the latter mandatory.

half the African countries have enacted legislation (or a decree) and even fewer have specified requirements for labelling. Namibian law provides for labelling regulations (Namibia, 2006), Kenya has mandatory labelling with 1% tolerance (Kenya, 2012) but has been criticised as current GM regulation on labelling is very prohibitive and may result in additional cost on labelling and cause shortage of food (Oh and Ezezika, 2014), and South Africa has two labelling regulations (Box 6.2).

Various issues are hampering the implementation of GM legislation. A survey of the views and positions of a number of stakeholders conducted by Adenle *et al.* (2013) listed these issues as follows: politicians who were still negative towards GMOs; lack of scientific expertise; limited capacity for risk analysis of GMOs; and inadequate consultation with local farmers in some countries. Other factors at the international level influence the adoption of GMOs in Africa. The lack of a common position at United Nations level has been questioned (Adenle *et al.*, 2013), while other authors have identified the difficulty of interpreting the precautionary principle on which the CPB is based, resulting in requests for more information, and the requirement for socio-economic information, mostly not empirically substantiated, which could lead to rejection of requests for approval of permits (Jansen van Rijssen *et al.*, 2015).

GM Crops Produced by Smallholders and Marginal Farmers and Sold by Street Vendors

Many Africans live below the breadline, many sell their produce on the streets and many are still illiterate. Strong initiatives from governments do exist to improve these situations. However, in spite of this good will, improvement may be gradual. Therefore, the rights of people will have to be accorded a high priority. The commercialisation of the GM potato, Spunta, was recently rejected by the South African regulatory authorities, one of the reasons being the economic implications for these kinds of producers of GM crops, including the cost of segregation and labelling of the GM product by these famers. But should these categories of products be exempted from labelling requirements? This case illustrates some of the practical problems of GM labelling.

The Plethora of Conflicting Labelling Regimes

The current situation with GM labelling of food from crops adds to the confusion in more than one respect. Globally, there exist different policies and approaches to labelling of GM foods: no labelling; labelling on specific products only; and voluntary or mandatory labelling. Product and process labelling are defined in Box 6.1.

In 2007 a labelling survey, conducted in 17 countries on what and how GM labelling is regulated, showed that the European Union (EU), Brazil and China have mandatory food labelling (Gruère *et al.*, 2009). India, one of the leading countries in the development and use of GM crops, introduced mandatory labelling that was, according to Gruère and Rao (2007), 'among the most stringent globally'. The EU is the forerunner in mandatory labelling. In 2003, the EU extended product labelling to process labelling and traceability, and also covered GM additives and flavourants. The EU approves GM food and feed, but has no labelling for meat and dairy products from animals raised on GM feed (GMO-Compass, 2007). Some countries have only mandatory labelling; others have mandatory and voluntary labelling, depending on what has to, or may, be labelled (Australia, New Zealand, Japan, South Korea and Taiwan), with Australia requiring labelling only on contents in processed foods. Several countries have lists that dictate which foodstuffs, food products or ingredients must be labelled (Gallani, 2015). A number of countries have labelling stating 'derived from GM', which is process labelling rather than product labelling. Process labelling will be near-impossible to verify or enforce without a strict IPS. Variations on the specific policy are plentiful.

At the same time, thresholds for GMO presence vary, being 0.9% in the EU, 1% in Brazil, 1% in Kenya and zero in China. The lower the threshold concentration, the greater the risk of false-positive or false-negative detection results. This could affect the cost of compliance, causes disputes and eventually has an impact on the food system in general. Sample taking and laboratory analyses pose challenges. An important issue is monitoring for compliance with legislation. This is often not sufficient and needs much more attention (Branquinho *et al.*, 2010). Apart from the costs of setting up a registered, accredited GMO testing laboratory with proficient personnel, using recognised diagnostic and automated equipment, there will be additional costs associated with using qualified parties having inspectors employing prescribed sampling protocols to visit grain storage silos (elevators) and food manufacturing plants. One remaining problem issue is that there will always be an element of uncertainty associated with the outcome of GMO detection. The European Commission, Joint Research Centre (EC-JRC) has addressed some of these issues (EC-JRC, 2015; Trapman *et al.*, 2009). JRC GMO-Amplicons is the first attempt at making sequences specifically related to GMOs publicly available (Petrillo *et al.*, 2015). The ultimate requirement is that GMO testing laboratories receive international accreditation (OECD, 1992).

Cost of Labelling

The cost of labelling systems depends on a number of issues: the threshold level; the capacity of the industry to comply; and the capacity of the authority to enforce labelling regulations. A number of authors (Plastina and Giannakas, 2007; Bansal and

Gruère, 2015; Huffman and McCluskey, 2014) analysed the effects of GM labelling on the market and welfare effects in small open economies. A general conclusion was that the effects of the introduction of GM crops in small countries that have been exporters of conventional products have been shown to be case-specific and dependent on (a) the labelling regime(s) in the world market, (b) the labelling regime in the domestic market, (c) segregation costs and the marketing margins under the different labelling scenarios, (d) attitudes of domestic consumers towards GM products, (e) the price premium enjoyed by the non-GM crops in the world market, (f) the relative cost effectiveness of GM crops under local production conditions and (g) the market power of GM seed suppliers. A general conclusion is that the more a country produces and uses products that contain GM food, the more costly a mandatory labelling regime will be. It is not possible to give a single scenario that would apply to each developing country as they are at different stages of developing and implementing GM legislation.

Studies were conducted to estimate the cost of various scenarios for Canada, Australia, the UK, the Philippines, the US state of Oregon and the Canadian province of Quebec. The main cost estimates range from US$1 to $10 or more per capita per year (Gruère and Rao, 2007). A study on mandatory labelling for the Philippines (De Leon *et al.*, 2004), a developing country, showed that it was associated with a 10% increase in consumer prices for certain products. Oh and Ezezika (2014) raised some doubt about the real cost increase to consumers and warned against overemphasising of the aspect of cost by the GM food industry.

A comprehensive socio-economic study on the costs of mandatory labelling of maize and soybean-based food products was undertaken for the Philippine Bureau of Food and Drugs (De Leon *et al.*, 2004). The requirements of mandatory labelling inevitably required an IPS from farm level to the phases of production and marketing in the food chain. The authors estimated on the basis of this study an increase in cost of 11%–12% for industry. The food industries in the Philippines have limited capacity to absorb additional costs. Consequently, the increase in cost could be passed on to consumers in terms of higher retail prices and/or to farmers in terms of lower farm-gate prices (De Leon *et al.*, 2004, p. 97).

Industry has ways to deal with some of these challenges. Manufacturers or importers may specialise exclusively in GM-free or GM products. That may reduce the cost of an IPS by simplifying silo capacity, transportation, storage and duplication of manufacturing equipment (Mann, 2015; Varacca *et al.*, 2014).

Discussion

The analysis of GM labelling policies revealed a complex multifaceted phenomenon perhaps not hitherto experienced in the world of food and feed security. The primary concern in all areas relating to labelling is that a proliferation of standards at international, regional and national levels will create barriers to trade and ultimately confuse consumers and other end-users (Glowka, 2003, p. 44). The ideal situation would have been harmonisation, but Codex did not manage to reach this goal.

An analysis of the complexity of GM labelling is revealing of patterns in political–economic trends. Gruère *et al.* (2009) explained the difference in mandatory labelling policies with a political–economic model that accounted for products, consumers/voters and civil groups. Their conclusions were as follows: countries that label were usually more developed and less dependent on agriculture, whereas those which do not label were signatories of the CPB and do not have active non-governmental groups; countries producing crops and exports are more pragmatic and have less stringent labelling policies; low consumer acceptance resulted in mandating of labelling; and imitation amongst countries occurred as a result of regional and trade relationships. Huffman and McCluskey (2014) evaluated the different GM labelling policies. The policies depend on the size of the GM-free markets. Voluntary GM labelling works well for a small GM-free market and may work for a large GM-free market; however, an enforcement mechanism will be required. Mandatory GM labelling will impose costs on all for the benefit of a few in a small market of GM-free food. In a large market it could work and is not different from voluntary labelling. A study by Bansal and Gruère (2012) showed that mandatory labelling in India of domestic GM-derived cottonseed oil and imported soybean oil would probably have market effects but would be unlikely to change access to information and choice. A subsequent study anticipating the effect on GM brinjal and GM rice showed that the effect depends on the type of product, the latter two having direct effects on the marketing systems and posing many challenges (Bansal and Gruère, 2015).

GM foods approved by governments for sale following extensive risk assessments meet consumers' right to safety. Another right is to have information and to be able to make choices. It is questionable whether the abbreviation 'GMO' or 'GM' provides much information on a highly technical subject such as biotechnology. Unless the consumer is well versed in biotechnology, it may result in stigmatisation of GM food (Oh and Ezezika, 2014). An alternative approach would be voluntary or mandatory labelling of foods that are 'GM-free' or 'organic'.

A burning question is what effect would any kind of GM labelling have on the livelihood of smallholder farmers, upcoming farmers and street vendors who make a living by selling their products? Of all the consumer rights, which will receive priority consideration? Surely it must be the right to have enough affordable and safe food on the table! These are basic needs. By approving GMOs for sale the message conveyed by government to consumers is that GM food is considered safe. Proof of this statement is that after more than 15 years no incidence of unsafe GM foods has been reported globally. Concern for GM food safety by those that have limited knowledge and information should be guided by trust in government decisions.

The cost of GM labelling and food security remains one of the most critical issues. Although some argue (Oh and Ezezika, 2014) that no study has as yet been conducted to evaluate the real cost and whether the industry would absorb costs, it cannot be ignored that implementation of a new system will affect law-enforcement capacity and the capacity of the food industry. Consumers are inclined to consider only the price of food, especially when they are living below the breadline, as is the case in many African countries.

Greater consumer choice may not be achieved, as experience has shown that retailers have removed GM products from the shelves due to perceived consumer aversion to GM products (Carter *et al.*, 2012). Mandatory labelling has affected the European markets in such a way that GM food has almost disappeared from the food shops (Carter and Gruère, 2003). This has been a drastic result that needs to be prevented. With voluntary labelling there is the opportunity for the food industry to seek niche markets as a choice for consumers who prefer GM-free food.

Governments should conduct thorough research into the matters mentioned in this review of GM food labelling. Matters such as consumer majority requirements for labelling, the effect on consumer choice and contribution to consumer information should be assessed. It is of great importance that the cost of labelling be investigated and compatibility with the country's general economic goals and its international obligations considered. There are no acceptable pro-label options that would not affect consumers and their wellbeing. The onus is on governments to take the right course of action.

Proposals

1. That GM regulatory frameworks specifying any labelling requirements be urgently put into place.
2. That GM food intended for consumption prepared and sold by small vendors be exempted from labelling. The term 'small vendors' here includes smallholder farmers, marginal farmers and street vendors.
3. That those developing countries which are experiencing relatively low levels of consumer aversion accept voluntary GM labelling. They will be better off, as some of the GM labelling analysis has concluded, with voluntary labelling than with mandatory labelling.
4. That the less restrictive tolerance of 5% be recommended for voluntary labelling.

References

Adenle, A. A. *et al.* (2013). Status of development, regulation and adoption of GM agriculture in Africa: views and positions of stakeholder groups. *Food Policy* **43**, 159–166.

African Union (2014). African Union Model Law on Biosafety, Art. 14: Identification and Labelling. Addis Ababa: African Union Commission, Department of Human Resources, Science and Technology.

Bansal, S. and Gruère, G. P. (2012). Implication of mandatory labelling of GM food in India. Evidence from the supply side. *Food Policy* **37**(4), 467–472.

Bansal, S. and Gruère, G. P. (2015). Labelling GM food in India: anticipating the effects of GM-brinjal and rice marketing chains. *AgBioForum* **18**(2), 156–167.

Bouet, A. *et al.* (2010). From 'may contain' to 'does contain': the price and trade effects of strict information requirements for GM maize under the Cartagena Protocol on Biosafety. Selected paper prepared for presentation at the Agricultural & Applied Economics Association's 2010 AAEA, CAES & WAEA Joint Annual Meeting, 25–27 July, Denver, Colorado.

Branquinho, M. R. *et al.* (2010). Survey of compliance with labelling legislation in food containing GMOs in Brazil. *Journal of Food Composition and Analysis* **23**(3), 220–225.

Carter, C. A. and Gruère, G. P. (2003). Mandatory labelling of genetically modified foods: does it really provide consumer choice? *AgBioForum* **6**(1&2), 68–70.

Carter, C. A. *et al.* (2012). California's Proposition 37: effects of mandatory labeling of GM food. *ARE Update* **15**(6), 3–8 (University of California Giannini Foundation of Agricultural Economics).

CAC (2011). Codex Alimentarius Commission compilation of Codex texts relevant to labelling of foods derived from modern biotechnology, CAC/GL 76–2011. [Online]. Available from www.codexalimentarius.org/standards/list-of-standards/en/

Chimhundu, H. (2002). *Language Policies in Africa.* Final report of the Intergovernmental Conference on Language Policies in Africa, revised version. UNESCO.

Convention on Biological Diversity (2000). Cartagena Protocol on Biosafety to the Convention on Biological Diversity. Text and Annexes. [Online]. Available from www.cbd.int/doc/legal/cartagena-protocol-en.pdf

De Leon, A. *et al.* (2004). The cost implications of GM food labelling in the Philippines. *Crop Biotech Brief* **4**, 2. [Online]. Available from www.isaaa.org/kc/Publications/pdfs/briefs/Brief4-2.pdf

Dubock, A. (2014). The politics of Golden Rice. *GM Crops & Food* **5**(3), 210–222.

EC-JRC (2015). New guidelines to enhance GMO testing across the European Union. [Online]. Available from https://ec.europa.eu/jrc/en/news/new-guidelines-gmo-testing?search

Falck-Zepeda, J. B. (2006). Coexistence, genetically modified biotechnologies and biosafety: implication for developing countries. *American Journal of Agriculture Economics* **88**(5), 1200–1208.

Gallani, B. (2015). Labelling of genetically modified (GM) ingredients in foods and beverages. In *Advances in Food and Beverages Labelling: Information and Regulations*, ed. P. Berryman. Amsterdam: Elsevier, pp. 177–189.

Glowka, L. (2003). *Law and Modern Biotechnology: Selected Issues of Relevance to Food and Agriculture.* Rome: Food and Agriculture Organization of the United Nations, p. 44.

GMO-Compass (2007). Labelling of foodstuffs produced from animals – the discussion continues. [Online]. Available from www.gmo-compass.org/eng/news/stories/286.gmo_labelling_animal_products_discussion_continues.html

Gruère, G. P. *et al.* (2009). Explaining international differences in genetically modified food labelling policies. *Review of International Economics* **17**(3), 393–408.

Gruère, G. P. and Rao, S. R. (2007). A review of international labelling policies of genetically modified food to evaluate India's proposed rule. *AgBioForum* **10**(1), 51–64.

Huffman, W. E. and McCluskey, J. J. (2014). The economics of labelling of GM foods. *AgBioForum* **17**(2), 156–160.

Jaftha, J. (2014). Guidelines on the role and functions of a biosafety committee (NBC) within a regulatory system – a case study from South Africa. In *Biosafety in Africa: Experiences and Best Practices*, ed. D. P. Keetch *et al.* East Lansing, MI: Michigan State University Press.

Jansen van Rijssen, F. W. *et al.* (2015). The precautionary principle: making managerial decisions on GMOs is difficult. *South African Journal of Science* **111**(3, 4), 1–9.

Kenya (2012). The Biosafety Act (No. 2 of 2009), The Biosafety (Labelling) Regulations, 2012, Arrangements of Legislation, Legal Notice No. 40. [Online]. Available from www.biosafetykenya.go.ke/Docs/labelling.pdf

Kimani, V. and Gruère, G. (2010). Implication of import regulations and information requirements under the Cartagena Protocol on Biosafety for GM commodities in Kenya. *AgBioForum* **13**(3), 222–241.

Mann, S. (2015). Is 'GMO free' an additional 'organic'? On the economics of chain segregation. *AgBioForum* **18**(1), 26–33.

Miller, H. I. and Kershen, D. I. (2011). A label we don't need. *Nature Biotechnology* **29**(11), 971–972.

Morris, E. J. (2014). Biosafety regulatory systems in Africa. In *Biosafety in Africa: Experiences and Best Practices*, ed. D. P. Keetch *et al.* East Lansing, MI: Michigan State University.

Namibia (2006). Biosafety Act (No. 7 of 2006), Government notice no. 3736.

OECD (1992). OECD Principles on Good Laboratory Practice, revised. [Online]. Available from www.oecd.org/chemicalsafety/testing/goodlaboratorypracticeglp.htm

Oh, J. and Ezezika, O. C. (2014). To label or not to label: balancing the risks, benefits and costs of mandatory labelling of GM food in Africa. *Agricultural and Food Security* **3**(8), 1–8.

Petrillo, M. *et al.* (2015). JRC GMO-Amplicons: a collection of nucleic acid sequences related to genetically modified organisms. [Online]. Available from www.ncbi.nlm.nih.gov/pmc/articles/PMC4589694/

Plastina, A. and Giannakas, K. (2007). Market and welfare effects of GMO introduction in small open economies. *AgBioForum* **10**(2), 104–123.

South Africa (2004). Foodstuffs, Cosmetics and Disinfectants Act (Act No. 54 of 1972). Regulations relating to labelling of foodstuffs obtained through certain techniques of genetic modification, Government Gazette No. 25,908, 16 January 2004.

South Africa (2011). Consumer Protection Act (Act No. 68 of 2008). Government Gazette 32186 (2009); 526; Regulations (R293, 1 April 2011), Government Gazette 34180 (2011); 550. [Online]. Available from www.thedti.gov.za/business_regulation/acts/consumer_protection.PDF

Trapman, S. M. *et al.* (2009). Guidance document on measurement uncertainty for GMO testing laboratories. European Commission Joint Research Centre. [Online]. Available from http://publications.jrc.ec.europa.eu/repository/bitstream/JRC50052/101-2009_as%20published.pdf

UN (2003). United Nations Guidelines on Consumer Protection, Department of Economic and Social Affairs. [Online]. Available from http://unctad.org/en/PublicationsLibrary/UN-DESA_GCP1999_en.pdf

UNCTAD (2013). United Nations Conference on Trade and Development, United Nations Guidelines on Consumer Protection. [Online]. Available from http://unctad.org/en/Pages/DITC/CompetitionLaw/UN-Guidelines-on-Consumer-Protection.aspx

Varacca, A. *et al.* (2014). Economic aspects of segregation between GM and non-GM crops in Italy. *AgBioForum* **17**(2), 123–132.

WTO (1995a). World Trade Organization, Uruguay Round Agreement: Agreement on the Application of Sanitary and Phytosanitary Measures. [Online]. Available from www.wto.org/english/docs_e/legal_e/15sps_01_e.htm

WTO (1995b). World Trade Organization, Uruguay Round Agreement: Agreement on the Application of Technical Barriers to Trade. [Online]. Available from www.wto.org/english/docs_e/legal_e/17-tbt_e.htm

WTO (2011). The WTO and the FAO/WHO Codex Alimentarius. [Online]. Available from www.wto.org/english/thewto_e/coher_e/wto_codex_e.htm

7 Building Human Capacity and Skills in Biosafety: Lessons Learned and Emerging Best Practices

John Komen and Muffy Koch

Introduction

An expanding body of literature confirms the farm-level and societal benefits of growing genetically modified (GM) crops, as summarised and analysed by Brookes and Barfoot (2014) and, for Africa, by Falck-Zepeda *et al.* (2013). A recent study by Klümper and Qaim (2014), which presents a meta-analysis approach on 147 published studies conducted during 1995–2014 worldwide, found that on average GM crops have reduced chemical pesticide use by 37%, increased crop yields by 22% and increased farmers' profits by 68%.

In the next decade, as the emerging bioeconomy expands to include second-generation biofuels and biomaterial production, biotechnology and GM crops are expected to play an even greater economic role, particularly in developing countries. It is generally accepted that GM crops must be assessed for food, feed and environmental safety before they can be released in farmers' fields (McLean *et al.*, 2002). These principles are reflected in an internationally binding agreement, the Cartagena Protocol on Biosafety, which has been ratified by 170 countries (as of December 2015). However, only in countries with a functional biosafety policy environment that enables the safe use of biotechnology products will people and national economies be able to take full advantage of the bioeconomy.

It is often overlooked that much of the required capacity and skills can already be found in existing environmental management agencies, food safety agencies and agricultural organisations such as phytosanitary inspection services and agricultural research institutes. This would suggest that tailored, hands-on capacity development initiatives are needed in order to hone existing skills. At the same time, over the past two decades we have seen substantial investment in rather generic biosafety capacity and skills development, supported by a range of multilateral and bilateral donor agencies.

From our review of capacity development programmes and recent reviews of biosafety capacity-building and training initiatives, a number of lessons were learned and emerging best practices identified. These include the following:

(1) development of need-driven, customised and strategic training and capacity-building approaches;
(2) implementation of sub-regional approaches to biosafety capacity development;

(3) development of clear and workable systems for monitoring and evaluation (M&E) of capacity-development programmes;
(4) encouragement of donor coordination in biosafety capacity development to minimise duplication and obtain wider impact for the available funding.

Recent Biosafety Progress in Developing Countries – Focus on Sub-Saharan Africa

Recognising the important benefits of the bioeconomy, governments across Africa have taken steps to establish an enabling regulatory framework to deal with GM crops and derived products. Progress for the year 2015 included, but was not limited to, the following developments.

- In February 2015, **Ghana** established its National Biosafety Authority (NBA) as the primary mechanism to implement the Biosafety Act (2011). This 13-member scientific advisory board now monitors progress with confined field trials (CFTs) of GM cowpea (insect-resistant), rice (salt-tolerant, water-efficient, nitrogen-use-efficient) and cotton (insect-resistant, herbicide-tolerant). The NBA drafted implementing regulations under the Biosafety Act, which were approved by Cabinet and await review by Parliament.
- **Nigeria**'s President signed the National Biosafety Management Agency Act (2015). This Act established the National Biosafety Management Authority (NBMA), which currently oversees CFTs for GM cowpea (insect-resistant), cassava (biofortified, virus-resistant) and sorghum (biofortified). In February 2016, the NBMA approved CFTs for GM maize (insect-resistant, herbicide-tolerant), rice (salt-tolerant, water-efficient, nitrogen-use-efficient) and cotton (insect-resistant, herbicide-tolerant).
- In **Kenya**, two applications were submitted in 2015 for general, unconfined release. These were for GM maize (insect-resistant) and cotton (insect-resistant, herbicide-tolerant). Kenya's National Biosafety Authority (NBA) determined, in January 2016, that GM maize can proceed to national performance trials prior to variety registration. A decision approving general release of GM cotton was made in September 2016. The CFT portfolio includes cassava (virus-resistant), sorghum (biofortified) and sweet-potato (insect-resistant). It is hoped that technological progress will help overturn a November 2012 decision by Kenya's Cabinet to ban imports of GM commodities, following unsubstantiated claims that GM maize would cause cancer.
- **Uganda** has broadened the support base for the National Biotechnology & Biosafety Bill (2012), which has been approved by Cabinet and undergone Parliamentary review. Following general elections in May 2016, Uganda's Parliament is expected to decide on the bill in March 2017. This will strengthen a steadily expanding portfolio of CFTs managed by the National Agricultural Research Organisation in a range of international collaborative programmes (see Chapter 24 in this volume). CFTs include banana, cassava, cotton, maize, rice, sweet potato and potato with locally relevant traits.

- **Tanzania**'s Environment Division at the Vice-President's Office adopted revised biosafety regulations in March 2015, which provide more enabling conditions to conduct research experiments and CFTs. Previously, strict liability clauses deterred research and development (R&D) and CFTs. A CFT application was submitted in early 2016 for drought-tolerant maize as part of the Water Efficient Maize for Africa (WEMA) project.
- **Malawi** continues implementation of its national biosafety framework. Following CFTs with insect-resistant cotton in 2011, the National Biosafety Regulatory Committee has reviewed an application for the general release of GM cotton submitted in early 2015. The Environmental Affairs Department issued a decision approving GM cotton's general release in April 2016. In September 2016 the Department of Agriculture authorised variety-registration trials, which commenced in December 2016. In January 2016 CFTs with insect-resistant cowpea were carried out, followed by bunchytop virus-resistant banana trials in April 2016.
- Reflecting progress in a number of member states, the Common Market for Eastern and Southern Africa (**COMESA**) started implementation of its Regional Policy on Biotechnology and Biosafety, which was adopted in February 2014. An implementation plan, developed in 2015, was endorsed by member states. When implemented, the policy will provide member states with a mechanism for regional risk assessment of GMOs intended for commercial planting, regional trade and food aid, backed by regional scientific advisory panels and capacity-building efforts.

While a few African countries stand out as champions, many more are caught up in different stages of developing functional frameworks. Many African countries face political opposition to GM crops and/or have high-profile anti-GM activism that has derailed or slowed down the development of biosafety policies and legislation, and prevented access to GM crops. Success requires African governments to establish and actively support national regulatory frameworks. The absence of this support has resulted in only three countries in Africa (Burkina Faso, South Africa and Sudan) approving release of GM crops to farmers. For a continent that could benefit greatly from improved planting material including GM crops, this progress is slow.

In industrialised countries national biosafety frameworks were often developed sequentially, beginning with voluntary guidelines and standards developed cooperatively by academia, industry and government. These were incorporated over time into legal instruments, either under existing legislation covering environmental management, food and agricultural products, or under new legislation dealing specifically with biosafety. The need for biosafety decisions drove this process. In contrast, attempting to develop regulations in the absence of a clearly defined need – such as an application for field trials involving GM crops – and without experience has resulted in a large number of unworkable and unused national biosafety systems. This is evidenced by the long list of 'draft' National Biosafety Frameworks (NBFs) developed with support from the Global Environment Facility (GEF) but not yet formally adopted by developing-country governments. In many cases, these drafts include highly precautionary language that has effectively discouraged access to, and testing of, relevant technologies. A regularly cited case in point involves the long lead time taken before drought-tolerant maize lines can be tested in Tanzania, as part of the WEMA project. In 2014, the WEMA project manager stated that

> Although scientists in Tanzania have been given a go ahead to continue with research to determine the safety of GM crops, the strict liability clause in the Biosafety Regulations is a setback. For example, it has not been possible for the WEMA project to test GM drought-tolerant and insect-pest protected maize in Tanzania. These GM maize hybrids are being tested in other WEMA countries such as South Africa, Kenya and Uganda. (Kitabu, 2014)

The evolution of biosafety policy and its implementation is ongoing, reflecting the diversity of new technology and products. It is not unusual to have a mix of voluntary and statutory mechanisms, even in those countries with long-established frameworks. In Egypt, the Philippines, South Africa, Kenya and Uganda, among others, significant experience was gained with contained experiments and CFTs under existing legislation or amended acts, well before the development of more comprehensive biosafety laws and regulations.

Continuing Challenges for Biosafety Capacity Development

Despite significant effort and resources having been devoted to biosafety capacity development over the past two decades, good progress in some countries notwithstanding, many countries still do not have adequate capacity to design and implement biosafety regulations (Convention on Biological Diversity (CBD), 2010). This remains a significant barrier to the testing and adoption of new crop varieties that would offer farmers a means to grow more food, enhance incomes and reduce the environmental impact of agriculture. Furthermore, an uncertain regulatory environment discourages private- and public-sector investment in development of the pro-poor crops and traits that farmers need the most.

Many donor-funded support programmes have attempted to build national capacity for the regulation of GM crops, with mixed success. Generally, countries with existing capacity for biotechnology R&D that already receive applications for activities with GMOs, and have high-level political support for biotechnology and biosafety capacity building, have made most advances and have benefited most from technical assistance. A comprehensive review of biotechnology and biosafety capacity-building programmes conducted by the United Nations University (Johnston *et al.*, 2008) is one of the few attempts to look at the inherent challenges with these programmes. This assessment found that most of the developing countries, including most countries of Africa, Central Asia, Oceania and the Caribbean, were not yet able to manage modern biotechnology and implement their national biosafety frameworks. The study noted that capacity deficiencies are

> [. . .] so pervasive and broad that there is no effective international system of biosafety at the moment. In addition, the volume of resources available to address these needs in the coming years appears insufficient to provide the necessary support for countries to implement their basic obligations under the Cartagena Protocol on Biosafety. (Johnston *et al.*, 2008, p. 11)

Regarding the support for human skills and expertise, the study concludes that

> More innovative, varied, yet targeted approaches are required for the design and delivery of training and capacity building programmes, including an enhanced emphasis on learning-by-doing and iterative approaches. Projects need to move away from the heavy reliance on workshops as the delivery mechanism. (Johnston *et al.*, 2008, p. 14)

At the core of biosafety assessment is the need for skilled people. These are lacking in many countries, particularly where university systems are weak or private-sector firms are minimally involved in the life sciences. Where there is a need for biosafety decision-making, a sustained effort to develop these skills is needed. For many countries, biosafety skills are new and not immediately applicable, unlike other professional skills. Many donor-supported programmes equate capacity development with training and workshops, or with an increase of physical infrastructure. It is often assumed that developing individual capacities will automatically lead to improved organisational capacity and performance, e.g., in terms of timely biosafety review processes and science-based decision-making. Following two decades of investment in biosafety capacity development, this is still not a common outcome.

People Skills and Expertise Underpin Biosafety Decisions

The people working in biosafety systems are the public- and private-sector scientists seeking to test GMOs and members of the review committees who assess risk and decide on proposed releases. The applicant and members of the biosafety review committees have different roles, but are engaged in a cooperative effort with a shared goal – the safe deployment of GMOs. All those involved need to be familiar with the environmental issues associated with biotechnology products and must have a working knowledge of the biosafety review process. They need to recognise what constitutes a potential risk and what risk management strategies may be applicable. Because advances come quickly and biosafety concepts are in constant evolution, all these people need ready access to the latest scientific, regulatory and biosafety information.

Given the limited experience with GMO releases in developing countries, the need for introductory, awareness-raising training will continue. This is different from the longer-term, in-depth technical skills development for countries that are actively reviewing applications and taking biosafety decisions. In response to identified needs, biosafety support programmes and donor organisations are supporting new methods in capacity development, some of which are presented below.

Biosafety Development Support: Achievements and Challenges

Technical assistance and financing for biosafety capacity building received a boost following the adoption of the Cartagena Protocol in 2000, and the designation of the Global Environment Facility (GEF) as the financial mechanism for supporting the implementation of the Protocol. GEF support was allocated according to a strategy adopted in 2000 to assist countries to prepare for the entry into force of the Cartagena Protocol. While a detailed analysis of GEF support to biosafety is beyond the scope of this chapter, the type of projects supported and overall results have been analysed in reports prepared by the main implementing agency, the United Nations Environment Programme (UNEP, 2006; 2008), and are summarised in Table 7.1.

GEF-supported projects make up most of the biosafety capacity-building projects in developing countries. They have been complemented by a range of projects supported by other UN agencies, international organisations, bilateral donor agencies and private foundations. A non-exhaustive overview for these projects is provided in Table 7.1. More comprehensive analysis can be found in Maredia *et al.* (2011). These initiatives have resulted in some important achievements, including those in the following list.

(1) Many countries have developed, or are in the process of developing, biosafety policies and laws, implementing regulations and guidelines largely through donor-funded projects and programmes, although progress has been uneven across countries.
(2) Support programmes have produced valuable resource materials, e.g., guidelines and manuals for environmental risk assessment, which can be adapted and used by many countries as they define and implement their biosafety systems.
(3) The number of countries with experience in GMOs and biosafety decision-making has grown and opportunities for tailored, hands-on training have expanded. For example, some countries that have approved CFTs with GM crops have hosted prospective field trial managers from other countries at local hands-on training courses.
(4) Regional political and economic organisations have taken an interest in biosafety, partially as a result of capacity-development support projects. These include the African Union, the Common Market for Eastern and Southern Africa (COMESA), the West African Economic and Monetary Union (WAEMU), the Caribbean Community (CARICOM), the Organization of American States (OAS) and the Association of Southeast Asian Nations (ASEAN), which have active programmes on biotechnology and biosafety. Such regional and sub-regional initiatives could become a conduit for promoting collaboration among countries, sharing resources and biosafety information, and providing financial support for continued biosafety capacity development.
(5) Some smaller countries, such as Lesotho and Guinea-Bissau, have reviewed risk assessment processes proposed by regional bodies or used in larger, neighbouring countries and determined that these are sufficient for informing their own national decisions on specific GMOs.

Nonetheless, the effective implementation of national biosafety systems continues to be hampered by the lack of capacity in many developing countries. As the CBD reported in its Action Plan for Capacity Building (CBD, 2013), 114 of the 143 Parties to the Protocol recorded a lack of capacity in various biosafety management areas. The key question therefore is why, after more than 15 years of technical assistance and donor support, have so many countries not developed adequate capacity to fully implement a functional national biosafety system? Clearly, there is a need to rethink and improve

Table 7.1 Overview of biosafety capacity-development programmes, 2000–2016

Project/programme	Implementing agency	Country focus	Funding (excluding country co-financing)	Goal/objectives
Global Environment Facility (GEF)-supported projects				
Development of National Biosafety Frameworks (NBFs) (2001–2007)	United Nations Environment Programme (UNEP)	123 countries in Africa, Asia, Pacific, Latin America, Caribbean, Central and Eastern Europe	GEF: US$34 million	Defining national biosafety frameworks (NBFs); regional and sub-regional collaboration
Building Capacity for the Effective Participation of Parties in the BCH (2004–2008)	UNEP	109 countries in Africa, Asia, Pacific, Latin America, Caribbean, Central and Eastern Europe	GEF: US$13.5 million	Supporting countries in fulfilling BCH obligations
Implementation of NBFs (2001–2006)	UNEP (8 countries) United Nations Development Program (UNDP) (2) World Bank (2)	12 countries: Bulgaria, Cameroon, China, Colombia, Cuba, India, Kenya, Malaysia, Mexico, Namibia, Poland and Uganda	GEF: US$9.2 million	Establish a regulatory regime, in line with national priorities and international obligations
Ditto (2006–2010)	UNEP	11 countries: Cambodia, Czech Republic, Egypt, Estonia, Lithuania, Mauritius, Moldova, Slovakia, Tanzania, Tunisia and Vietnam	GEF: US$7.4 million	As above
Ditto (2010–2014)	UNEP	Albania, Bhutan, Costa Rica, Ecuador, El Salvador, Guatemala, Lao PDR and Madagascar	GEF: US$5.1 million	As above
Ditto (ongoing)	UNEP	Bangladesh, Ethiopia, Ghana, Jordan, Lesotho, Malaysia, Mauretania, Macedonia, Mozambique, Nigeria, Rwanda, Sri Lanka, Syria, Swaziland, Turkmenistan, Turkey and Venezuela	GEF: US$15.7 million	As above
'Phase II' NBF implementation (2011–, ongoing)	UNEP	India	GEF: US$2.7 million	Strengthen the biosafety management system

(cont.)

Table 7.1 (*cont.*)

Project/programme	Implementing agency	Country focus	Funding (excluding country co-financing)	Goal/objectives
Multi-Country Capacity-Building for Compliance with the CPB (2008–2011)	World Bank CIAT	Brazil, Colombia, Costa Rica and Peru	GEF: US$5 million	Strengthening technical capacity in risk assessment and management, decision-making and public awareness
West Africa Regional Biosafety Programme (2007–2010)	World Bank WEAMU	Benin, Burkina Faso, Côte d'Ivoire, Guinea-Bissau, Mali, Niger, Senegal and Togo	GEF: US$5.4 million	Establish and implement a shared biosafety regulatory framework
Regional Project for Implementing NBFs in the Caribbean Sub-region (2011–2015)	UNEP CARICOM	Antigua & Barbuda, Bahamas, Barbados, Belize, Dominica, Grenada, Guyana, St Kitts & Nevis, St Lucia, Surinam, St Vincent & Grenadines and Trinidad & Tobago	GEF: US$6.1 million	Implementing NBFs in participating countries; establishing regional cooperation in biosafety
Other UN agencies and international organisations				
FAO Building Biosafety Capacities (ongoing)	FAO	• 15 country projects • 4 sub-regional projects • 2 global projects	FAO: US$7.5 million	Assist countries and regions in building strong technical, institutional and information-sharing capacities; ensure safe use of modern biotechnologies; enhance sustainable agriculture and food production
e-Biosafety Training Programme (ongoing)	UNIDO	Global	Fee-based courses	Strengthen professional expertise in biosafety management
ICGEB Biosafety Unit (ongoing)	ICGEB	• Global • Africa (through Gates Foundation grant)	• Fee-based courses • Gates Foundation: US$3 million	Capacity building/training; information dissemination; international collaboration
Programmes supported by bilateral donors and foundations				
South Asia Biosafety Programme (ongoing)	• ILSI Centre for Environmental Risk Assessment • IFPRI	Bangladesh and India	USAID: US$2 million	Developing and applying sound science to the environmental risk assessment of agricultural biotechnologies
BioSafeTrain (2004–2010)	University of Copenhagen	Kenya, Tanzania and Uganda	DANIDA: US$1.4 million	Build regional capacity in biosafety and ecological risk assessment; enhance regional collaborations in biosafety; promote North–South collaboration

African Union Biosafety Project (2006–2011)	African Union (AU) GTZ, Germany	AU member states	BMZ: US$2.6 million	Provide capacities and instruments to support AU member states in implementing the CPB and the African Union Model Law on Biosafety
Norwegian Centre for Biosafety (ongoing)	Genøk, Norway	• Global (training programme) • South Africa • Zambia	NORAD: • US$2.9 million (training) • US$0.9 million (South Africa) • US$0.5 million (Zambia)	Provide countries with scientific and technical capacity for biosafety assessment and regulation; build capacities to conduct scientific risk assessment and risk management, address other technical biosafety regulation needs
African Biosafety Network of Expertise (ongoing)	ABNE	Africa	Gates Foundation: US$10 million	Technical assistance, biosafety-related tools and resources, training workshops for regulators
Capacity Strengthening for the Safe Management of Biotechnology in Sub-Sahara Africa (2009–2015)	FARA	Burkina Faso, Ghana, Nigeria, Kenya, Malawi and Uganda	Syngenta Foundation: US$1.3 million	Training in biotechnology stewardship; train stewardship leaders
Southern Africa Biosafety and Environment Programme (2003–2012)	RAEIN-Africa	SADC countries	The Netherlands (no funding)	Development and implementation of biosafety systems; promote and foster networking and collaboration
ASEAN Centre for Biodiversity	ACB	ASEAN countries	European Union (no funding)	Strengthen biosafety management in ASEAN countries; regional biosafety guidelines
Hemispheric Biotechnology and Biosafety Programme – biosafety component (ongoing)	IICA	Latin America and the Caribbean	No funding	Development, implementation and communication of transparent, science-based policies and regulatory frameworks; facilitate regional harmonisation
Assuring Agricultural and Food Safety of Genetically Modified Organisms (GMOs) in Southern Africa, GMASSURE (2013–2016)	University of Pretoria	Southern African Development Community (SADC) countries	EU: €900,000	Improve knowledge about, and increase capacity of, agricultural biosafety and biotechnology; and the safe use of genetically modified (GM) agricultural crops

the general approach to biosafety training and capacity development. To date, the dominant approach has been short-term technical training to support drafting a set of legal and technical requirements for implementation at national level, without adequate consideration of resources required for implementation by national governments. Such an approach has succeeded in more advanced countries, with functioning institutional governance and R&D infrastructure, but has had limited success in low-income and fragile states.

Innovative Approaches to Capacity and Skills Development

In response to these challenges, biosafety support organisations are developing innovative approaches to capacity development, particularly in terms of novel approaches to technical training and skills development. The key will be to establish a robust cadre of in-country scientists and regulators who apply available risk assessment data in actual reviews and decision-making. This group would be supported by the vast and growing array of regulatory data and published literature that supports biosafety and food/feed safety assessments. Considering that the generic workshop training approach has limitations, innovative approaches are being pursued, often with encouraging results. Illustrative examples include the following.

(a) In order to balance the need for international exchange of biosafety expertise with the practical limitations associated with overseas training, the United Nations Industrial Development Organization (UNIDO) launched a distance-learning e-biosafety programme in 2006. Since then this 1-year, academically accredited programme has been held annually, managed by universities in Belgium and Italy. The course is a combination of distance-learning modules focused on risk assessment of GM plants with a short on-campus period in Europe for practical training and case-study analysis (Pertry *et al.*, 2014). Pertry *et al.* (2014) acknowledge the limitations of distance learning in an essentially hands-on field like biosafety and the fact that internet connectivity represents a logistical challenge to developing-country course participants.

(b) Michigan State University in the USA, and Aberystwyth University in Wales in collaboration with the University of Adelaide in Australia, have run short biosafety courses and a Masters course in biosafety, respectively. After many years of refining these curricula to meet the needs of developing-country regulators and biosafety advisors, these universities have been working to transfer the course materials and methodology to African universities. Closer proximity is expected to make the course more accessible to a larger group of regulators and students, and to continue to hone the curricula to make them more appropriate for the region. Collaboration started in 2014 with the University of Burkina Faso and in 2015 with the University of Nairobi and University of Ghana to establish these courses locally.

(c) Experience in the Philippines, one of the countries early to adopt GM crops, showed that posting experienced regulators in developing-country regulatory offices for relatively short, three-week periods over several years can have a positive impact on the rate of implementation of new regulatory systems. Experienced regulators are able to guide implementation steps, help with legal and administrative requirements, and provide options and input for addressing needs as they arise. This process is being introduced in Africa by the International Centre for Genetic Engineering and Biotechnology (ICGEB), where regulators from countries such as Australia and Canada are working in national biosafety offices as advisors to assist in implementation of national regulatory systems. These visits are complemented by visits of senior regulators from Africa to Australia and Canada, in order to learn from advanced biosafety offices where many applications and permits of diverse nature are processed.

(d) The Program for Biosafety Systems (PBS), in its regulatory training efforts, has moved to small-group (sometimes individual), targeted training interventions supporting a clearly defined goal, and addressing a concrete biosafety challenge. A case from the Philippines illustrates how policy development went hand in hand with technical training. During the period 2008–2011, the PBS collaborated with the Philippines Bureau of Plant Industry (BPI) in providing technical assistance and training to build insect- and weed-resistance management policies for GM maize varieties that are commercially available in the country. PBS-supported internship programmes in the USA and Canada for key BPI personnel aimed at drafting and updating insect-resistance management (IRM) guidelines and training materials, providing on-the-job training in IRM that targeted crop-protection officers from regions in the Philippines. Knowledge gained in IRM schemes is currently being applied in developing IRM guidelines for the planned introduction of insect-resistant aubergine. South–South collaboration and exchanges are encouraged, including the organisation of internship programmes for Vietnamese regulatory officials in the Philippines, focusing on environmental risk assessment and GMO decision-making.

(e) One of the best drivers for implementation of a regulatory system and hands-on skills development is a clearly identified need. When applications for permission to carry out activities with GMOs are received, new national regulatory bodies must implement national biosafety systems and apply their training to provide efficient, effective and transparent regulatory decisions. At this stage, providing experienced regulatory advisors to work alongside national regulators as they administer and review applications has proved effective. Ultimately the decisions are made by national regulators, but reaching the decision-making point is facilitated by access to an experienced regulatory advisor. The PBS and the African Biosafety Network of Expertise (ABNE) have successfully supported this type of capacity-building support aimed at a direct transfer of expertise.

Recommendations

Donor organisations and national, regional and international implementing agencies need to invest in training initiatives as an ongoing and major component of biosafety capacity development. This should be balanced by commitment from governments for contributions to the programmes and vocal support for national biosafety systems. Countries require a steady supply of trained scientists able to implement risk assessment and make science-based safety recommendations. In addition, emphasis must be placed on customisation and tailored solutions for particular needs and challenges in different countries. Whereas large agricultural producing and exporting countries will need comprehensive, functioning NBF structures, those countries that primarily deal with agricultural commodity imports could opt for more streamlined solutions. These include handling biosafety decision-making through existing agricultural, environmental and food safety institutions, and applying available safety data generated for other countries. In these countries the focus should be on ensuring capacity for effective decision-making using regional risk assessment recommendations.

These options have implications for organisations supporting biosafety capacity development.

Firstly, training plans should be practical and directly linked to a country's strategic objectives, or government strategy for biotechnology and biosafety, and should contribute directly to efficient administration and effective decision-making. Secondly, organisations supporting biosafety training and capacity development – such as those described in Table 7.1 – could work towards a rational division of labour with local universities focusing on regular, introductory and advanced biosafety training, while international projects focus on supporting strategic long-term, hands-on skills development of key individuals responsible for biosafety decision-making. Thirdly, improved collaboration and coordination among biosafety capacity-development projects, training programmes and supporting donor organisations is essential. This would encourage more strategic training approaches at country level, particularly in countries where multiple programmes are active, and there is benefit from a division of labour between local and international partners. Collaboration and coordination would lower duplication of efforts as training events frequently target the same organisations and individuals, even though they are conducted by different international programmes. A formal capacity-building coordination mechanism was established by the CBD Secretariat in 2004, involving donor organisations and capacity-development projects and programmes. From this, a group of biosafety service providers focusing on sub-Saharan Africa meets annually to review priorities and work plans, and to determine areas for collaboration.

Systems for monitoring and evaluation (M&E) of training results are equally important. Currently, the M&E systems used to gauge the performance of biosafety capacity development appear underdeveloped. Beyond internally commissioned donor evaluations, independent evaluations at the project or country level are not common.

Finally, collaboration on biosafety training and skills development should expand among countries in the same (sub-)region. Regional collaboration would help to ensure standards of safety information and risk assessment methodology. Regional collaboration would lead to countries sharing experts and expertise, safety information and safety assessments. There are currently some initiatives at the regional level that encourage technical and political collaboration among countries on biosafety regulatory issues.

While they are still evolving, they could promote biosafety collaboration among countries, share resources and provide financial support for regional biosafety capacity development.

Conclusion

This chapter highlights the key components of risk analysis, particularly where capacity development plays an important role in developing functional biosafety systems in developing countries. The old model of training workshops and courses is being replaced with more effective capacity-building models that enable regulators to apply their training and skills to biosafety activities at in-country regulatory agencies. This helps ensure that training is timely, appropriate and focused on delivering effective biosafety decision-making. Partnering new, national biosafety bodies with experienced regulatory consultants and with regulators from functioning biosafety systems is beginning to prove effective at streamlining and driving implementation of new national biosafety frameworks. While workshops will continue to train regulatory officers in risk assessment, risk management, risk communication and biosafety administration principles, this training should be linked to hands-on decision-making in order to improve the implementation and efficiency of developing-country regulatory structures.

References

Brookes, G. and Barfoot, P. (2014). Economic impact of GM crops: The global income and production effects 1996–2012. *GM Crops & Food* **5**(1), 65–75.

CBD (2010). Assessment and Review under Article 35 of the Cartagena Protocol on Biosafety: Discussion Paper on a Proposed Framework for the Second Assessment and Review. Report UNEP/CBD/BS/COP-MOP/5/15, 22 January 2010.

CBD (2013). Framework and Action Plan for Capacity-Building for the Effective Implementation of the Cartagena Protocol on Biosafety (2012–2020).

Falck-Zepeda, J. *et al*., eds. (2013). *Genetically Modified Crops in Africa: Economic and Policy Lessons from Countries South of the Sahara*. Washington, D.C.: International Food Policy Research Institute.

Johnston, S. *et al*. (2008). *Internationally Funded Training in Biotechnology and Biosafety: Is It Bridging the Biotech Divide*? Yokohama: UNU Institute of Advanced Studies.

Kitabu, G. (2014). Here comes drought-tolerant hybrid maize varieties. *The Guardian*, 17 January 2014. [Online]. Available from http://wema.aatf-africa.org/files/guardian-oikeh-interview.pdf

Klümper, W. and Qaim, M. (2014). A meta-analysis of the impacts of genetically modified crops. *PLoS ONE* **9**(11), e111629.

Maredia, K. *et al*. (2011). Capacity building in biosafety. In *Environmental Safety of Genetically Engineered Crops*, ed. R. Grumet *et al*. East Lansing, MI: Michigan State University Press.

McLean, M. A. *et al*. (2002). A Conceptual Framework for Implementing Biosafety: Linking Policy, Capacity, and Regulation. Briefing Paper No. 47. The Hague: International Service for National Agricultural Research.

Pertry, I. *et al.* (2014). Biosafety capacity building: experiences and challenges from a distance learning approach. *New Biotechnology* **31**(1), 64–68.

UNEP (2006). *A Comparative Analysis of Experiences and Lessons from the UNEP-GEF Biosafety Projects*. Geneva: UNEP-GEF Biosafety Unit.

UNEP (2008). *Guidance towards Implementation of National Biosafety Frameworks: Lessons Learned from the UNEP Demonstration Projects*. Geneva: UNEP-GEF Biosafety Unit.

Part II

Diversification of Expertise

8 Regulation of GMOs in Developing Countries: Why Socio-Economic Considerations Matter for Decision-Making

José Falck-Zepeda and Marnus Gouse

Introduction

Increased use of genetically modified crops, especially in developing countries, has contributed to an increase in the number of proposals for the inclusion of socio-economic considerations (SEC) in biosafety decision-making. In the negotiations leading to a major international regulatory agreement, the Cartagena Protocol on Biosafety (CPB) focusing on the trans-boundary movement of genetically modified (GM) organisms, there was considerable disagreement amongst parties and non-parties to the CPB on whether SEC should be included in its text. In what is considered to be a 'compromise text', in order to addresses the positions of disagreeing parties during the CPB negotiations, Article 26.1 provides for the inclusion of SEC in decision-making. However, after more than a decade of meetings, discussion fora and the formation of an Ad Hoc Technical Expert Group, there is still no clarity or agreement on what Article 26.1 allows for, or what is actually implied, and there is a lack of conceptual clarity that would empower countries to move forward starting from an agreed definition of what SEC are (Ludlow *et al.*, 2014).

The objective of this chapter is not to solve incongruities in Article 26.1. This chapter's objective is to highlight under what conditions and for what purpose carrying out a socio-economic study may be a good idea when considering the commercial release of a GM commodity and to describe the challenges and limitations associated with the inclusion of SEC in the making of such decisions.

We provide narratives to illustrate some real-life examples of why socio-economic impact assessments (SEIAs) may be necessary and important – not as barriers to trade or to the application of technology, but to facilitate adoption and sustainable use. A SEIA may be applicable especially in developing countries, where the absence of a strong and functioning 'invisible hand' and/or functional legal frameworks can often lead to market failures. This discussion will also include a set of institutional issues that limit potential gains from GM crop adoption and insights that may be gained from conducting socio-economic research. Moreover, we focus our discussion on real-life limitations and methodological issues that practitioners have faced in conducting such research.

Background

The CPB, an implementing protocol of the Convention on Biological Diversity (CBD), was adopted in 2000 and entered into force in 2003. Up to 2015, the Protocol had been ratified by 170 countries, but noticeably not by large exporters of GM crops such as the USA, Canada and Argentina.

In Article 26.1 of the CPB, the Protocol allows for the inclusion of socio-economic considerations in biosafety decision-making.

There are numerous and divergent interpretations of Article 26.1. The countries which oppose the inclusion of SEC in biosafety decision-making emphasise the fact that inclusion is optional, that the scope of the socio-economic assessment allowed for in the Protocol is limited to the scientifically assessed potential impact on biological diversity of living modified organisms (LMOs) and that any socio-economic concern that is raised should be justifiable within the framework of the country's other international agreements and responsibilities. This statement prominently refers to, but is not limited to, those obligations a country has taken on as a party to the World Trade Organization and its Sanitary and Phytosanitary Agreement.

Countries that tend to oppose GM crops view the scope of Article 26 as basically all agricultural and related activities and argue for the inclusion of a broad scope of SEC, including potential cultural, traditional, religious and ethical issues. Countries that support this view tend to focus on the potential negative impacts of GM crops, including in some instances hypothetical and even speculative risks that may be associated with the use of GMOs (Falck-Zepeda *et al.*, 2013; Falck-Zepeda, 2014).

The countries which oppose the wider scope of socio-economic assessment tend to be the innovators, and large-scale adopters of GM crops. These countries focus on the scientific risk assessment role of the Cartagena Protocol, ensuring public and environmental health, and argue that socio-economic issues should be considered on a national level rather than under an international biosafety agreement. These countries fear that inclusion of a seemingly endless list of direct and indirect SEC in biosafety decision-making would result in unworkable regulatory systems and delays and rejection of technologies on the basis of unscientific deductions.

The countries that support the wider scope of socio-economic assessment tend to be net importers, have limited investment in biotechnology, and generally have limitations in terms of science, technology and innovation capacity, including the capacity to implement biosafety assessments (Falck-Zepeda, 2014). Though some of these countries see the inclusion of a wider range of SEC as a necessity in the development of what they perceive to be a responsible regulatory system for their specific countries, other countries see SEC as a tool to keep GM crops out of their countries.

Binimelis and Myhr (2016) report that, up to 2015, 34 countries, mainly from Africa and Latin America, have included SEC in their national biosafety legal frameworks. Seven of these countries have approved commercial production of GM crops but, with the exception of Argentina, where only economic (trade) criteria rather than broader socio-economic issues are considered, the manner and degree in which SEC were actually included in biosafety decision-making is not clear. More than half of the 34 countries reported that they have no experience in SEC assessment.

Study of Socio-Economic Considerations

Society can gain knowledge and information on technology impacts from the examination of socio-economic considerations, especially as a result of rigorously well-done assessment studies. Rigorous assessments can help identify valuable technologies and discard those which are unacceptable for society.

Many socio-economic impact assessment studies have been conducted, for various technologies, leading to relevant gains to society. Detailed meta-analyses and reviews of the socio-economic impact assessment literature, including Klümper and Qaim (2014), Areal *et al.* (2013) and Smale *et al.* (2009), conclude that GM crops have shown overall positive benefits, but there is considerable variability across farmers, households, regions and countries. An overall positive benefit does not mean that there are no negative outcomes and no losers from adopting the technology; it just means that the positive impacts are larger than the negative.

Here we need to clearly differentiate between studies carried out before deliberate release or commercialisation (the *ex-ante* studies) and those done post-release for technology evaluation, monitoring and other purposes (the *ex-post* studies). This is an important distinction as by definition there will not be any data on adoption or on farm-level and wider impacts in an *ex-ante* study, which must thus rely on confined field trials, performance studies and trials, and primary and secondary data collected from existing producers. This limits the type and complexity of potential methodologies that may be applied in the socio-economic study (Smale *et al.*, 2009; Falck-Zepeda and Zambrano, 2011). In addition, while it is possible, to a certain extent and on the basis of prevailing institutional issues and arrangements, to predict how a value chain would react to the introduction of a new technology, some *ex-post* studies have shown that other unpredictable factors often have a considerable impact on adoption and sustainable production in general. These factors are difficult or impossible to evaluate in an *ex-ante* study.

Identifying Potential Release Issues

Assessment studies can help stakeholders take advantage of the benefits from such technologies by identifying limiting issues relevant to technology deployment and adoption. For example, an *ex-ante* study done on insect-resistant and herbicide-tolerant (Bt/RR) cotton in Uganda highlighted the need to address all aspects of cotton productivity in the country (Horna *et al.*, 2013). Even though the benefits were relatively small and positive, they were particularly sensitive to prices and other production variables, especially existing low cotton yields. From the standpoint of those stakeholders interested in developing the cotton sector in Uganda, there is a need to carefully examine not just existing productivity but also market coordination and integration mechanisms, especially for a commodity-type crop such as cotton. This implies that there is a need to address pricing mechanisms carefully in developing countries as well as those factors affecting seed systems.

The importance of a functioning seed system is echoed by Gouse (2012), who reported that in South Africa Monsanto held farmer days in nine areas across four provinces in 2001 to introduce Bt maize to approximately 3,000 small-scale farmers. In the following season, in 2002, impressed by the performance of the small sample of Bt seed they had been given free of charge, a considerable number of these farmers wanted to purchase Bt maize seed. However, due to seed multiplication problems and increased demand by large-scale farmers, there was not enough seed available for smallholders. The enthusiasm and anticipation created through the farmer days and one season of on-farm trials were lost due to failure of the seed system.

An *ex-ante* study of a GM banana by Kikulwe *et al.* (2011a, 2011b, 2014a, 2014b) highlighted the need for the developer, the National Agricultural Research Organization (NARO) in Uganda, to increase its communication efforts to address negative perceptions about GM technology in the country in order to reduce the possibility of reduced adoption levels due to uncertainty. Furthermore, efforts need to emphasise consumers in urban areas, who tend to be more uncertain about the technology. In terms of the impact of the R&D and regulatory processes on innovation, the study shows that, even though social gains – even when considering irreversible costs and benefits – can be high for a food security crop such as bananas in a country like Uganda, there are limits to the total cost of development beyond which the investment is not viable. Using reasonable assumptions about discount and interest rates, the estimated total cost of development must not surpass US$108 million for the fungus-resistant banana. This number is important, as it will require an evaluation of the regulatory and decision-making process so that it does not become a binding hurdle to innovation and feasible proposals.

Improved Institutional Context Understanding

An *ex-post* study on Bt cotton in South Africa highlighted the impact that institutions can have on the type and level of benefits that technology may bring to farmers. Gouse *et al.* (2005) and Gouse (2009) documented how the successful introduction and adoption of Bt cotton by small-scale cotton farmers on the Makhathini Flats in South Africa largely came to a halt due to institutional failure. One cotton company was supplying input credit and extension services on the Flats and acted as the only buyer of cotton. This system is rather common in Africa, where communal land ownership does not allow for secured credit, and failure of credit and input markets tends to lead to interlocked transactions and regionally isolated firms provide inputs on credit with the resulting output serving as the collateral. This system worked reasonably well for most of the 1990s, but in 2002 a new cotton gin started sourcing cotton on the Flats and set in motion a chain of events whereby farmers borrowed from the first company but delivered to the second. As a result, the first company suffered a huge financial loss, as a result of which it stopped all support and extension activities on the Flats, and farmer numbers decreased tenfold. While farmers enjoyed yield increases and a reduction in insecticide use from Bt cotton relative to conventional varieties, adverse selection and institutional failure resulted in a collapse of the system.

The above-mentioned studies also pointed out that, even though production technologies can increase productivity, markets matter. Despite approving and releasing nine GM cotton events since 1997, the South African cotton sector has seen a drastic decline in production since the early 1990s due to a number of factors, including the deregulation and abolition of the Cotton Board in 1997, the depressed world cotton price caused by increased production (especially in India), the collapse of the South African textile industry and considerably higher profit margins for competing crops such as maize, soybeans and sunflower seed. While the increase in profitability induced by GM technology was substantial for the South African cotton industry, in the real world market, it was not enough.

The adoption of insect-resistant and herbicide-tolerant (Bt/RR) maize in the Philippines was examined in two *ex-post* studies (Mutuc *et al.*, 2013; Yorobe and Smale, 2012). Adoption has increased to an estimated 700,000 hectares, mostly located in northern Luzon and in the southern part of Mindanao. These studies have shown the positive and significant benefits accruing to farmers, but there are questions as to the role private financiers are playing in terms of defining adoption and what the impact of an increased dependence on private financing is. In addition, these studies showed that adopting farmers tend to be better educated, have higher income and have a more positive attitude towards the technology. This is not an unexpected result, and the effect may be amplified in the case of GM technologies as they tend to be more expensive than conventional technologies, and thus there may be a credit constraint in place.

An *ex-post* study on Bt/RR maize in Honduras (Falck-Zepeda *et al.*, 2012; Falck-Zepeda *et al.*, 2015) describes the example of a small country with very low levels of R&D investment, but one that was able to set in place an enabling policy and regulatory environment by which the country was able to tap into existing GM technologies in maize. The adoption of Bt/RR maize in Honduras was derived from concrete steps taken by its government to facilitate the deployment of the technology. On the other hand, and in spite of the significant and positive benefits obtained by adopting farmers, adoption has levelled off at roughly 30,000 hectares, which represents approximately 7%–10% of the total maize-growing area. There are a several seed market and institutional issues that may help explain the apparent paradox of a high-return crop with relatively low adoption.

Inclusion of SEC in Decision-Making – Is It Necessary and Important?

Costs of compliance with biosafety regulations will increase if SEC is included as conducting an *ex-ante* socio-economic assessment or even a basic review will unequivocally increase implementation costs. The more complex and broader the portfolio of issues included in an SEC assessment, the more expensive the activities needed to address those questions. Implementing an economic study focused on a very specific topic (i.e., the impact on exports/trade or on small-scale producers) certainly does not involve the same level of complexity as performing one that may examine impacts on biodiversity and long-term sociological and anthropological issues.

Regulatory cost may be especially important in those situations where compliance costs are high enough that they become an insurmountable hurdle to a developer. This scenario refers especially to the public sector in developing countries that are likely to develop technologies of a public-good nature. The introduction of additional regulatory hurdles can impact national and international public investments in R&D, which in turn can have an important impact on technology flows by causing a reduction in the number of potential technologies available to producers and society. This may be a result of additional regulatory complexity, cost implications and/or uncertainty (Smyth *et al.*, 2014).

However, the cost of compliance might not be the most important factor; rather, potential regulatory delays can have a major impact on benefits. A paper by Bayer *et al.* (2010) showed that delays can have a big impact on the net benefits from four GM crop applications in the Philippines. This paper showed that even small delays of three years compared with the baseline can significantly decrease net benefits to producers, while increases in the cost of compliance had only a small effect on net benefits (see Figure 8.1). If inclusion of socio-economic studies in the biosafety regulatory process is not done in a careful manner, that is, in close coordination with other assessment processes, it may also increase the amount of time necessary to complete the assessment process, with the aforementioned potential investment and financial impacts.

Inclusion of SEC may potentially introduce more uncertainty into the regulatory and decision-making process. A workable system can be partially defined as one where all elements of society are able to define, describe and trust the process and its outcome.

For example, if developers (be they in the public or private sector) face a predictable regulatory system, they can assign a value to the outcome by projecting potential gains by producers and consumers, and incorporate the respective probabilities of success into their decision-making process. If developers cannot assign such probabilities because of unpredictability or uncertainty in the regulatory system, the likelihood is that they will not make investments in such jurisdictions. The potential impact of uncertainty on the public sector will be of relevance to many developing countries that have already invested in developing products of their own.

In essence, inclusion of SEC may lead to an unworkable system, especially if the rules and standards for decision-making are not known by all of the stakeholders. A regulatory system must clearly define what will be required, how studies will be evaluated, the required quality standard, how a decision will be made and how the socio-economic study will relate to the environmental and food safety assessment, amongst other issues. Having an unworkable system, or one that cannot render a decision in a timely, cost-efficient manner and whose decisions are not robust, protective and accepted by society, is undesirable and will result in a questionable use of scarce societal resources.

Finally, there exists the possibility that the inclusion of socio-economic considerations can be manipulated consciously (or unconsciously) to support a specific position. Elements of best practice need to be applied in order to secure the best possible assessments, following accepted practice that is based on evidence and well-defined

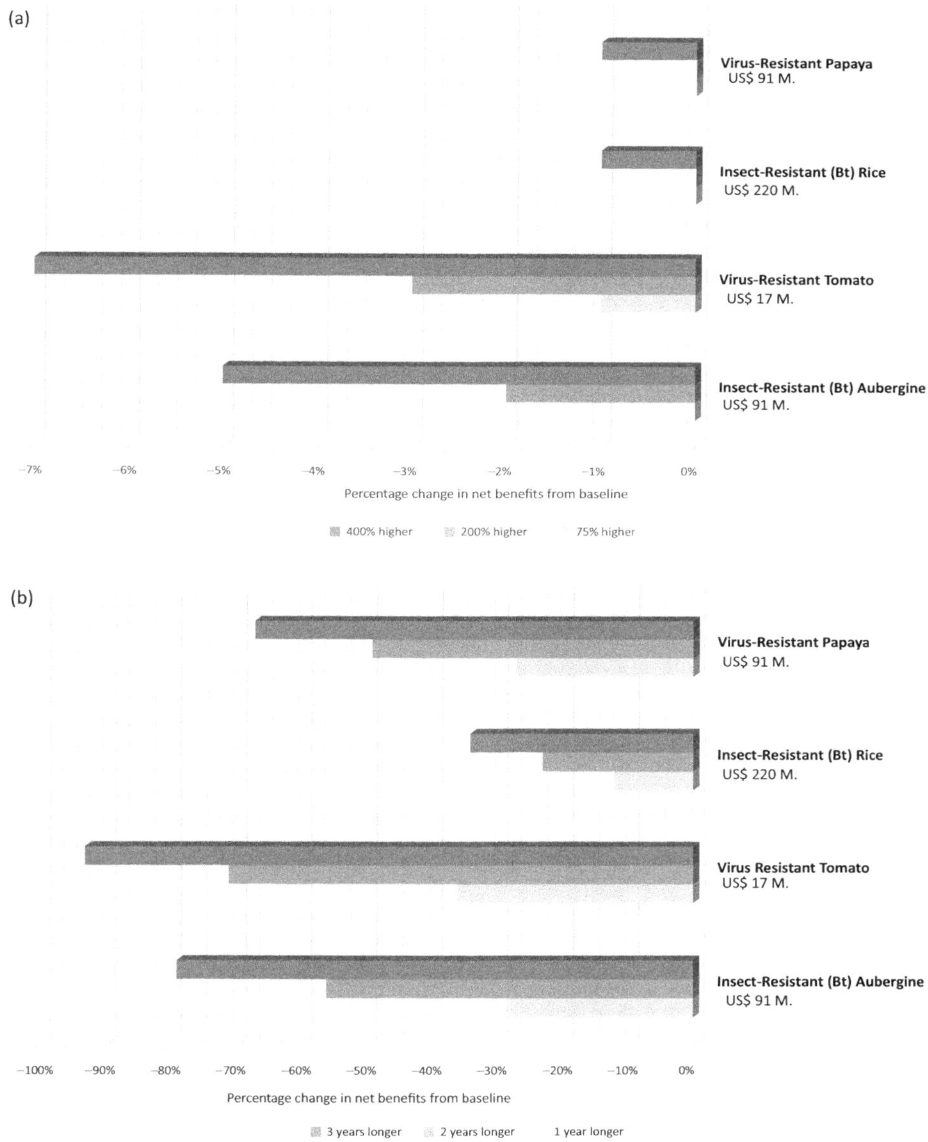

Figure 8.1. Impacts of increases in cost and regulatory delays for four technologies in the Philippines. Source: Estimated from Bayer *et al.* (2010).

decision-making standards. However, defining a decision-making standard on which all stakeholders agree will be difficult if not impossible, as there are multiple views and standards according to which stakeholders examine issues (e.g., economic versus ecological approaches). This may imply that the inclusion of socio-economic criteria will in the end be a political decision, not a decision taken by those affected by the use of a specific GM technology. In an ideal situation, those affected would decide by evaluating and adopting those technologies which are applicable and valuable in their own specific situation.

If a decision has been taken, policymakers will need to examine all of the issues related to options and alternatives for potential inclusion of socio-economic criteria in decision-making and thus there is a need to examine all choices in order to ensure that in the end a feasible and functional biosafety system is achieved. Most risk assessment experts with actual field experience agree that the best approach is to have a science-based regulatory process focused on accepted biosafety regulatory practices. Thus the approach pursued by Brazil and Mexico, which is to allow the science-based risk assessment to proceed and then, if the product is deemed safe or risk 'manageable', if, and only if, any socio-economic issue is identified, commission a formal socio-economic study, is a feasible option. After a general process flow has been established, there is a need to further decide on the multiple procedural issues that may enter into a guideline and/or standard operating procedures, as discussed by Falck-Zepeda and Zambrano (2011).

Experiences from the Socio-Economic Assessment Practitioner Trenches

In this section we review some of the challenges and limitations derived from the experiences we have encountered while conducting mainly *ex-post* socio-economic assessments in practice. Most of these issues are also applicable to *ex-ante* studies, with the added constraint that actual adoption information and data do not exist. We do not imply that this is an exhaustive discussion, as there are many more challenges, including logistical issues and funding availability, which are even more binding constraints in a developing-country setting. Our objective is to illustrate those issues which we have considered necessary to address in order to ensure a continued evolution of methods and approaches to socio-economic assessments.

Challenges and Limitations

1. Availability of accurate data on adoption and distribution by type/size of farmers (large-scale, small-scale, commercial or subsistence) and the lack of accurate records on seed sales and users of traditional seeds, open-pollinated varieties (OPVs), conventional hybrids and GM seed. Without accurate data, the possibility of drafting sampling protocols and understanding the statistical significance of results becomes compromised.
2. Issues with maintaining long-term adoption, farm-level, production and consumption panel data.
    (a) Crop failures due to erratic weather and damage from animals
    (b) Entry/exit of farmers due to extemporaneous conditions – off-farm employment, discontinuity of government programmes, availability of seed
    (c) Interactions between crop choice and climatic conditions

3. Availability of socio-economic assessment methods to examine particular farmers' motivations for production in a developing-country setting, including for profit, subsistence or safety first, in which downside risk may be a more pressing issue as ensuring survival is a priority. Current methods tend to emphasise one motivation at a time. In practice, households in developing countries may be pursuing a diversified portfolio of objectives, in some cases in situations where there is not a pecuniary market that sets prices.
4. Absolute and relative input and output measurements.
   (a) Manual harvest units
   (b) Sizes and shapes of plots and seeding rates
   (c) Lack of official produce measurement for subsistence farmers and consumption throughout season
   (d) Focus on partial productivity versus total factor productivity
   (e) Measurement of effectiveness and efficacy of farmer damage abatement activities
   (f) Accuracy of farmer survey recall data
   (g) Farmer illiteracy
5. Small samples linked to small adopting populations and limited seed availability.
6. Loss of a counterfactual group after a number of seasons of adoption.
7. Adopters in government programmes or programmes initiated by seed companies are not 'real adopters' as the adoption decision is not theirs.
8. Timely availability of support services and complementary inputs in government-supported programmes.
9. Climatic conditions and production practices of individual farmers influence 'problem pest' pressure and hence the 'problem pest' control benefit of the new technology.
10. Single-season studies have limited usefulness because of varying weather, insect pressure, adoption and adaptation by farmers, identification and development of improved research methodologies and focus questions.
11. Choice of a comparable alternative (counterfactual) versus a baseline (actual) approach. This is an old issue, which probably will never be resolved. Sociologists and other social scientists tend to prefer a baseline without the intervention, whereas agricultural economists tend to prefer a counterfactual approach, especially in those situations where a long time has elapsed since the determination of the baseline. Other factors that are unrelated to the technology or intervention may influence the outcome compared with the baseline.
12. Statistical biases, including self-selection, simultaneity, sampling and other early-adoption biases. Self-selection arises when adopters choose the technology by themselves; they are not assigned it randomly. Simultaneity arises from making input decisions that may be related (for example GM adoption and fertiliser use). Sampling bias arises especially during the early-adoption stage, when, in spite of pursuing random sampling strategies, one may capture a large number of farmers with better resource endowment, who may be the better-off farmers. In the latter case, the measured difference in outcome, such as yield or profit, may be

overestimated (or underestimated) as it may be a reflection of the better farmers rather than the technology.

Countries' Policy Responses on Socio-Economics

Expansion to Broader Considerations

As mentioned previously, several countries have expanded their national legislation or biosafety frameworks to incorporate broader considerations beyond impacts on biodiversity and decisions concerning imports. Where broader socio-economic concerns have been included in legislation, few countries actually have clarity about what this requirement means and no guidelines on the definition of socio-economic issues, methods, rules and standards are available from the regulators.

Few Real-World Experiences

Argentina (a non-party to the CPB) has a mandatory process that assesses economic consequences on exports and competitiveness before commercialisation. Brazil and Mexico have a system whereby they allow the risk assessment to proceed, that is, the environmental and food/feed safety assessment, and then, if any socio-economic issue is raised, they commission a study from a third party. India and China do not have a formal requirement in their legislation/policy, but both have conducted socio-economic studies, with different degrees of impact (Pray *et al.*, 2006).

The implication of this state of affairs is that there is a need to draw from other regulatory processes to gain knowledge about what to do and what the limitations of such inclusion are. Potential examples include the National Environmental Policy Act (NEPA) in the USA and the newer (and still developing) approaches being developed in the European Union.

Conclusions and Policy Recommendations

- Inclusion of SEC into biosafety decision-making can have both positive and negative impacts, partially depending on how inclusion is implemented in a specific country.
- In some situations inclusion is a policy decision seeking to address market imperfections or failures and to address a weak institutional and regulatory setting, especially in developing countries.
- If policymakers make the decision to include SEC, they will need to carefully examine all of the issues related to options and alternatives for potential inclusion. These discussions will require appropriate choices in order to ensure that a feasible and functional biosafety system is achieved.
- There is some agreement that the best approach is to have a science-based risk assessment process focused on accepted biosafety regulatory practices and then commission a formal socio-economic study only if the product is deemed 'safe' and a socio-economic issue has been identified.

- Experience in some countries has shown that allowing a process that may consider SEC for the approval of specific GM technology may provide some positive net benefits to society. This is the result of purposely focusing on having a functional system and avoiding any step that eventually becomes an insurmountable hurdle.

References

Areal, F. J. *et al.* (2013). Economic and agronomic impact of commercialized GM crops: a meta-analysis. *The Journal of Agricultural Science* **151**(1), 7–33.

Bayer, J. C. *et al.* (2010). Cost of compliance with biotechnology regulation in the Philippines: implications for developing countries. *AgBioForum* **13**(1), 53–62.

Binimelis, R. and Myhr, A. I. (2016). Inclusion and implementation of socio-economic considerations in GMO regulations: needs and recommendations. *Sustainability* **8**(1), 62.

Falck-Zepeda, J. B. (2014). Non-adopters of GM crops in Latin America. In *Edward Elgar Handbook on Agriculture, Biotechnology and Development*, ed. S. J. Smyth *et al.* Cheltenham: Edward Elgar.

Falck-Zepeda, J. B. and Zambrano, P. (2011). Socio-economic considerations in biosafety and biotechnology decision making: the Cartagena Protocol and national biosafety frameworks. *Review of Policy Research* **28**(2), 171–195.

Falck-Zepeda, J. *et al.* (2012). Caught between Scylla and Charybdis: impact estimation issues from the early adoption of GM maize in Honduras. *AgBioForum* **15**(2), 138–151.

Falck-Zepeda, J. *et al.* (2013). The current status of the debate on socio-economic regulatory assessments. *World Review of Science, Technology and Sustainable Development* **10**(4), 203–227.

Falck-Zepeda, J. B. *et al.* (2015). In *Analyses: Africa's Future . . . Can Biosciences Contribute?*, ed. P. Mitton and D. Bennett. Cambridge: Banson Publishers and Biosciences for Farming in Africa (B4FA).

Gouse, M. (2009). Ten years of Bt cotton in South Africa: putting the smallholder experience into context. In *Biotechnology and Agricultural Development: Transgenic Cotton, Rural Institutions and Resource-Poor Farmers*, ed. R. Tripp. London: Routledge.

Gouse, M. (2012). GM maize as subsistence crop: the South African smallholder experience. *AgBioForum* **15**(2), 163–174.

Gouse, M. *et al.* (2005). Bt cotton in KwaZulu Natal: technological triumph but institutional failure. *AgBiotechNet* **7**(134), 1–7.

Horna, D. *et al.*, eds. (2013). *Socioeconomic Considerations in Biosafety Decision Making – Methods and Implementation.* [Online]. IFPRI Research Monographs. Available from www.ifpri.org/publication/socioeconomic-considerations-biosafety-decisionmaking

Kikulwe, E. M. *et al.* (2011a). A latent class approach to investigating demand for genetically modified banana in Uganda. *Agricultural Economics* **42**(5), 547–560.

Kikulwe, E. M. *et al.* (2011b). Attitudes, perceptions, and trust. Insights from a consumer survey regarding genetically modified banana in Uganda. *Appetite* **57**(2), 401–413.

Kikulwe, E. M. *et al.* (2014a). Incremental benefits of genetically modified bananas in Uganda. In *Edward Elgar Handbook on Agriculture, Biotechnology and Development*, ed. S. J. Smyth. Cheltenham: Edward Elgar.

Kikulwe, E. M. *et al.* (2014b). If labels for GM food were present, would consumers trust them? Insights from a consumer survey in Uganda. *Environment and Development Economics* **19**(6), 786–805.

Klümper, W. and Qaim, M. (2014). A meta-analysis of the impacts of genetically modified crops. *PLoS ONE* **9**(11), e111629.

Ludlow, K. *et al.* (2014). Assessing the SEC landscape and moving forward. In *Socio-economic Considerations in Biotechnology Regulation*, ed. K. Ludlow *et al.* New York, NY: Springer, pp. 295–305.

Mutuc, M. E. *et al.* (2013). Which farmers benefit the most from Bt corn adoption in the Philippines? Estimating heterogeneity effects. *Agricultural Economics* **44**(2), 231–239.

Pray, C. E. *et al.* (2006). Costs and enforcement of biosafety regulations in India and China. *International Journal of Technology and Globalisation* **2**(1–2), 137–157.

Smale, M. *et al.* (2009). *Measuring the Economic Impacts of Transgenic Crops in Developing Agriculture during the First Decade: Approaches, Findings, and Future Directions. Food Policy Review 10*. [Online]. Washington, D.C.: International Food Policy Research Institute. Available from http://cdml5738.contentdm.oclc.org/utils/getfile/collection/pl5738coll2/id/26575/filename/26576.pdf

Smyth, S. J. *et al.* (2014). Investment, regulation, and uncertainty: managing new plant breeding techniques. *GM Crops and Food: Biotechnology in Agriculture and the Food Chain* **5**(1), 44–57.

Yorobe, J. M. and Smale, M. (2012). Impacts of Bt maize on smallholder income in the Philippines. *AgBioForum* **15**(2), 152–162.

9 Food Safety Assessment of Genetically Modified Crops in Developing Countries: The Experience in Africa

Godwin Lemgo, Narender Nehra and Hector Quemada

Introduction

Sustained food security is fast becoming a global challenge, with impacts that are likely to be more pronounced in the developing world, especially in sub-Saharan Africa (Wedding and Tuttle, 2013; Anderson *et al.*, 2016). Genetically modified (GM) crops can contribute to the efforts of developing countries to address the agricultural challenges which they face. Traits such as disease and pest resistance, as well as improved nutritional value, are particularly useful in these countries. In order to benefit from these crops, assessments of both food safety and environmental risk must be conducted by the appropriate regulatory authorities to ensure that the crops being imported or developed by scientists in-country are safe for cultivation and consumption. This chapter deals with food safety assessments, with a focus on Africa in order to illustrate the regulatory and legal challenges, as well as the potential for benefits, that may be common for developing countries.

International Standards for Food Safety Assessment

Guidance on the safety assessment of foods derived from GM crops has been developed and published by several international organisations. The World Health Organization (WHO) was one of the first to do so, defining issues to be considered for the safety assessment of foods derived from genetically engineered microbes, plants and animals (WHO, 1991). With regard to crops, this guidance identifies the issues of toxicological and nutritional importance in assessing the safety of foods derived from them.

Toxicological changes are defined as follows:

> (a) the presence or increased content of natural toxicants; (b) the presence of new expressed toxic materials resulting from genetic modifications (e.g., biopesticides); (c) development of allergenicity; (d) accumulation of toxicants or microbial contaminants derived from the environment; (e) changes in the availability of toxins as a result of processing.

Nutritional changes are defined as follows:

> (a) modification of major nutrients, micronutrients or anti-nutrients in the food; (b) changes in the bioavailability of macro- and/or micronutrients; (c) changes in nutritional components as a result of processing.

In order to address these issues, information in four main areas is deemed necessary:

> (a) characterisation of the host, donor/vector and modified organisms; (b) characterisation of the precision of the process and possibility of pleiotropic or secondary effects; (c) characterisation of novel expressed material; and (d) characterisation of the modified food for wholesomeness.

This guidance elaborates a paradigm for food safety assessment that relies first on molecular, biochemical and chemical data, in order to determine whether animal toxicological testing is warranted. Subsequent guidance from other organisations retains the same overarching approach, with increased levels of detail regarding the nature of those characterisations (Codex Alimentarius Commission, 2003; European Commission, 1997, 2003, 2004; European Food Safety Authority, 2006, 2008; Food and Agriculture Organization (FAO), 1996; Jonas *et al.*, 1996; Howlett *et al.*, 2003; Chassy *et al.*, 2004; Organisation for Economic Co-operation and Development (OECD), 1993, 1998; US Food and Drug Administration (USFDA), 1992; WHO, 1995, 2000).

A foundational idea first introduced in the WHO's 1991 guidance is the concept that genetically engineered varieties should be evaluated in comparison with traditionally bred varieties as the standard of safety (recognising that these traditional varieties were not absolutely safe themselves). This concept was more formally elaborated by the OECD (1993):

> For foods and food components from organisms developed by the application of modern biotechnology, the most practical approach to the determination of safety is to consider whether they are substantially equivalent to analogous conventional food product(s), if such exist. [. . .] The concept of substantial equivalence embodies the idea that existing organisms used as food, or as a source of food, can be used as the basis for comparison when assessing the safety of human consumption of a food or food component that has been modified or is new.

Codex Alimentarius Guidance

The various guidance documents written throughout the world support a consensus approach to the assessment of safety of new crop varieties and the foods derived from them. This consensus is best articulated in the guidelines published by the Codex Alimentarius Commission (Codex Alimentarius Commission, 2003), hereafter referred to as Codex, which reaffirms the concept of substantial equivalence as key to food safety assessment. However, Codex clarifies that substantial equivalence is the starting point for the assessment, enabling the identification of similarities and differences between the GMO and the comparator, thereby allowing the determination of issues that would be of safety concern. The Codex guidelines provide a common basis for evaluating the safety of foods derived from GM crops.

Capacity for GM Food Safety Assessment in Africa

While international guidelines for food safety assessment have been in place for 25 years, experience in implementing these guidelines remains limited in most developing countries. This is the case on the African continent, where only a handful of the 56 countries have reviewed any GM crops for commercial cultivation. A closer look at this region illustrates the challenges to building GM food safety experience and capacity in developing countries.

Legislative and Management Capacity for Safety Assessment of Foods Derived from Genetically Engineered Crops

Many developing countries have preferred to enact new legislation instead of revising existing food safety legislation to accommodate foods derived from genetically engineered crops (Nang'ayo *et al.*, 2014). Thus, under the United Nations Environment Programme-Global Environment Facility (UNEP-GEF) biosafety capacity-building project, 42 African countries (of the 51 which ratified the Cartagena Protocol on Biosafety) received assistance to develop their biosafety frameworks (Global Environment Facility, 2006). However, only 20 have enacted their biosafety laws (or decrees) as of 2015. Like any law, these biosafety laws require accompanying regulations and guidelines to make them enforceable (Morris, 2014).

Operating manuals and country guidelines for regulators are key to effective implementation of legislation. Considering the potential for transboundary movement of GMOs, and noting that most African countries are members of the World Trade Organization (WTO), it is essential that such manuals and guidelines for food safety assessment are in alignment with internationally accepted standards such as Codex (Codex Alimentarius Commission, 2003; WTO, 1998), in order to facilitate trade within the region. Generally, such manuals and guidelines for food safety assessment and scientific risk assessment are lacking (Obonyo *et al.*, 2011), or, where available, they are mostly out of date or inconsistent with Codex guidelines. Only a few countries, such as South Africa, have guidelines that explicitly recognise Codex as the basis for food safety assessment (Department of Agriculture, South Africa, 2004).

Institutional responsibilities for food safety are often shared among various ministries, departments and agencies in many African countries, sometimes with overlapping roles (Mwamakamba *et al.*, 2012; Juma and Serageldin, 2007; Food and Agriculture Organization (FAO)/WHO, 2005). For example, in Ghana, the responsibility for food safety is shared among the ministries of Health and Agriculture, local governments and the National Bureau of Standards, which is under the Ministry of Trade (FAO/WHO, 2006). With the enactment of biosafety legislations in many African countries comes the establishment of new national competent authorities commonly domiciled in ministries which hitherto had limited or no oversight on food safety matters. This was the case in Uganda, where the biosafety competent authority, the Uganda National Council for

Science and Technology (UNCST), was until recently domiciled in the Ministry of Finance. Overlapping roles may lead to disagreements regarding which ministry should play a lead role, leading to blockages in the deployment of GM crops, as occurred in Egypt (Adenle *et al.*, 2013).

Currently, almost all the countries with biosafety laws have established their national competent authorities for administrative management of the system (e.g., Ghana, Nigeria, Kenya, Malawi, Namibia, Zimbabwe and Zambia). Others without biosafety laws do have temporary administrative systems within the Cartagena Protocol on Biosafety national focal point institutions (e.g., Uganda) (ABNE, 2016). A coordinated approach whereby the biosafety competent authority serves as the coordinating body for the safety of GM foods using existing institutional arrangements will ensure that applications are handled efficiently. This approach is used by Kenya, where dossiers are generally forwarded by the National Biosafety Authority to the Kenyan Bureau of Standards for a scientific opinion on food safety. This approach provides opportunity for synergies amongst various ministries, departments and agencies on issues of GM foods, in order to avoid possible duplications, mandate overlaps and/or institutional conflicts. However, only a few African countries – South Africa and, to some extent, Egypt, Kenya (food aid), Burkina Faso and Sudan (cotton cake for feed) – have this experience.

Technical Capacity for Scientific Risk Assessment (Review Food Safety Data)

In general, technical capacity for food safety assessment in Africa is limited, as evidenced by requests for capacity building made by many African countries to UNEP and biosafety service providers such as the African Biosafety Network of Expertise (ABNE) (Adenle *et al.*, 2013; Obonyo *et al.*, 2011; Biosafety Clearing House, 2015). Many factors may have contributed to this limited capacity.

First, most biotechnology-related activities in Africa remain at the research level – either in containment or in confined field trials, where food safety considerations are not relevant. Thus, current capacity-building efforts for biosafety in most African countries have generally focused on environmental issues, particularly on reviewing and managing confined and multi-locational field trials (Ezezika *et al.*, 2012). In addition, very little consideration has been given to the possibility of trading GM commodities or processed foods, which may also require food safety assessment. This notwithstanding, efforts have been and are being made to include aspects of food safety in training sessions for African regulators by several biosafety capacity-building service providers active in sub-Saharan Africa, including, but not limited to, (i) the Centre for Safety Assessment of Food and Feed (CSAFF); (ii) the African Biosafety Network of Expertise (ABNE), an Africa-owned, Africa-led initiative of the NEPAD Agency of the African Union; (iii) the Program for Biosafety Systems (PBS) of the International Food Policy Research Institute (IFPRI); (iv) the International Centre for Genetic Engineering and Biotechnology (ICGEB); and (v) GMASSURE, an African, Caribbean and Pacific

(ACP) Science and Technology Programme. A good example is a joint effort by the CSAFF, ICGEB, PBS and ABNE to organise a food safety workshop (ABNE, 2014).

The composition of the various technical sub-committees responsible for scientific risk assessment is another challenge. In many jurisdictions in Africa, membership on food safety assessment bodies is by institutional representation as prescribed in the biosafety legislations, and is often broad, in an effort to accommodate a wide range of activities and interest groups. Some of these areas of expertise have little relevance to food safety assessment. For instance, the scientific advisory committee in some countries includes representatives from such institutions as a council of non-governmental organisations and from government ministries such as those for Labour, Justice, Forestry, Women and Child Affairs, National Parks and Wildlife, and the Police. These are important stakeholders to consider as part of the decision-making process, but their participation in a food safety assessment body is inappropriate.

Another challenge has been the high turnover rate of trained regulators. It is possible that many of the regulators who have had some training in food safety assessment may have moved on to different careers. Others may not have had opportunities to maintain current training in new advances, or, sadly, lost most of the knowledge they had acquired, due to many years of inactivity and lack of practical experience. However, considering that many of the African countries active in biotechnology research and development are approaching advanced stages in their research and commercial release is on the horizon, now is a good time to re-focus capacity-building efforts in the area of food safety assessments.

Capacity to Generate Food Safety Data

Biotechnology research and development activities in Africa are typically conducted by African public research institutions in collaboration with private-sector or developed-country researchers, with support from charitable donors and private industry. Research and development has mostly focused on food security crops for the 'public good', such as banana, cowpea, cassava and sorghum, which have not been commercialised elsewhere in the world. However, some of the work uses genes that have been expressed in products that have already been approved. The information and experience gained with the deployment of a gene in other crops provide important precedents upon which African countries can rely. For example, the protein (Cry1Ab) expressed in an insect-resistant cowpea currently under development is the same as that which has long been commercially available in GM crops like MON 810 maize (AATF, 2015). Thus, even though *Maruca*-resistant cowpea has not been commercialised anywhere else, there are considerable safety data and a history of safe use of the Cry1Ab protein.

In contrast, some proteins will require a new set of regulatory data for an adequate safety assessment. Obtaining *de novo* regulatory data in line with the Codex 2003 guideline is a resource-intensive undertaking and may involve financial resources that could be out of reach for a typical public research institution in Africa. Few facilities capable of generating *de novo* food safety data exist in Africa, and the expertise to establish and staff such laboratories is likewise lacking. Capacity at existing regional hubs and centres of excellence like Biosciences Eastern and Central Africa–International Livestock Research Institute (BecA–ILRI) in Kenya and the Southern Africa Network for Biosciences (SANBio) hub at the Council for Scientific and Industrial Research (CSIR) in South Africa could require human and infrastructural capacity upgrades for the purpose of generating food safety data.

Harmonisation, Joint Reviews and Mutual Recognition of Food Safety Data

Food safety data sets are usually universal in nature and do not differ significantly from one country to another for any particular GM food. This fact makes it possible to explore opportunities to harmonise food safety reviews and allow data portability with necessary adjustments to accommodate differing food cultures, processing practices and consumption levels. It also presents a unique opportunity for African countries with limited capacities to base their food safety decisions on safety assessments carried out elsewhere where these foods have already been commercialised.

Regional or continent-wide harmonisation will serve to prevent unnecessary regulatory burdens on public research institutions working on GM crops for the ‘public good’ and also reduce lengthy regulatory timelines (Atkinson *et al.*, 2015). It will greatly facilitate the dissemination of ‘public good’ GM crops within Africa, particularly to countries where they are needed the most. For example, cassava is a very important staple in West and East Africa. With adequate harmonisation, data generated for nutritionally enhanced or virus-resistant cassava produced in Nigeria could accelerate the regulatory pathway to making these crops available for food and feed uses in other countries by eliminating the need to repeat food safety testing and review.

Possible Approaches to Harmonisation of Food Safety Regulatory Procedures

Centralised Risk Assessment Mechanism to Support National Decision-Making

Several harmonisation mechanisms have been proposed for Africa, but the most practical in terms of cost of institutional structures and respect for national sovereignty is a regional regulatory system that conducts risk assessment and makes safety recommendations that can be used to support national decision-making (Timpo, 2011).

A centralised food safety authority for Africa is being discussed by the African Union (AU) (USDA-FAS, 2015). This proposed body could also serve as the continental body responsible for coordinating and providing scientific and technical guidance on safety of GM foods. The report and recommendations of this continental body would then feed into a national decision-making process. In the absence of this continental food safety authority, harmonisation can still be pursued at the continental level. A small coordinating team can be established at the African Union (AU) level to coordinate a pool of food safety experts to provide scientific opinions on the safety of specific transgenic crops to AU member countries.

At the sub-regional level, various efforts to harmonise biosafety policies (including food safety) within regional blocs are already under way, with notable progress in the Common Market for Eastern and Southern Africa (COMESA), which has already adopted a harmonised policy (Waithaka *et al.*, 2015), and the Economic Community of West African States (ECOWAS), which is at an advanced stage. COMESA's harmonised policy is based on a centralised risk assessment mechanism that feeds into a national decision-making process, while the initial draft ECOWAS policy sought to establish a regional body that will make binding decisions on GMOs on behalf of member states (Chambers *et al.*, 2014).

Performing Joint Reviews

Countries could establish agreements to harmonise their food safety data requirements and review procedures, and to perform joint reviews of applications that have been submitted to each country. The outcome would be recognised by all participating countries. This helps to prevent asynchrony in the approvals, considering that many of Africa's borders are porous, with weak or non-existent monitoring mechanisms. A good opportunity to test this approach would be to start with transgenic crops of mutual interest currently being tested in multiple African countries, such as transgenic cowpea in Ghana, Nigeria, Malawi and Burkina Faso. These countries could pool their expertise and resources to generate joint food safety data and perform joint food safety reviews such that outcomes are recognised in all participating countries.

Mutual Recognition of Food Safety Assessment Reports

For almost two decades, biotech crops have been commercialised in a growing number of countries – 28 countries as of 2013 (James, 2015). There is thus a considerable body of food safety data for these currently commercialised GM foods, publicly available on websites of various regulatory agencies, such as Food Standards Australia–New Zealand (FSANZ), and inter-governmental platforms such as the OECD Bio Track, the FAO GM Foods platform, the Biosafety Clearing House and a database maintained by the International Life Sciences Institute–Centre for Environmental Risk Assessment. There is also a considerable history of use of the current GM foods that have received regulatory approvals across the world and are considered safe by the WHO (WHO, 2015). African countries could choose to consider such existing data in their national decision-making, particularly regarding importation of GMOs meant for food, feed or processing.

Since many of the 28 countries where these crops have been approved are also major trading partners of African countries, the latter could synchronise their approvals with their trading partners and the occurrence of unauthorised GMOs meant for food, feed or processing could be minimised. Clauses in some African biosafety laws, such as those of Ghana and Kenya, permit this policy.

Case Study: Virus Resistant Cassava for Africa (VIRCA)

Among the crops being grown by smallholder farmers in sub-Saharan Africa, cassava (*Manihot esculenta* Crantz), a root crop, is an important staple. It is estimated to be a major source of food and income for over a quarter of a billion people in Africa and nearly a billion people globally (OECD, 2009; Gbadegesin *et al.*, 2013; Anderson *et al.*, 2016). Despite substantial efforts being made by agronomists and breeders on crop improvement, the yield of cassava in Africa remains very low per hectare compared with other parts of the world. This low yield is primarily due to the inability of smallholder farmers to afford inputs, as well as several biotic and abiotic factors imposing severe limitations on the yield and productivity of cassava (OECD, 2009; Wedding and Tuttle, 2013).

Viral Diseases – A Major Threat to Cassava Production

Cassava Mosaic Disease (CMD) and Cassava Brown Streak Disease (CBSD) are two viral diseases that cause major yield losses (Alicai *et al.*, 2007; Taylor *et al.*, 2012). CMD is a foliar disease that has resulted in devastating yield losses in sub-Saharan Africa. Recently, several tolerant varieties have been developed through conventional breeding. However, no conventional resistance to CBSD is available. The disease is spreading in East Africa, and is considered to be a major threat to cassava production in this region (Alicai *et al.*, 2007). CBSD has been recognised as one of the seven most dangerous crop diseases in the world (Pennisi, 2010).

VIRCA – An Effective Technology to Combat Viral Diseases

The Virus Resistant Cassava for Africa (VIRCA) project uses genes expressing small-interfering RNAs (siRNA) to combat CBSD. This approach is based on RNA interference (RNAi), a naturally occurring phenomenon that mediates resistance to pathogens in many plant species. Using the siRNA technology, researchers at the National Crops Resources Research Institute in Uganda, the Kenyan Agricultural and Livestock Research Organisation in Kenya and the Danforth Centre in the USA are developing improved virus-resistant varieties of cassava.

Studies to Generate Food Safety Data for VIRCA Varieties

The VIRCA project is developing a robust science-based regulatory dossier that will meet national requirements for release of virus-resistant cassava varieties in Kenya and Uganda, as well as international standards and guidelines (OECD, 1993; Craig *et al.*, 2008; Codex Alimentarius Commission, 2009). For developing such a science-based approach to regulatory approval, the project team will emphasise that the RNAi-based technology has been successfully used for development of several products such as fatty-acid-modified soybean, amylopectin potatoes and virus-resistant squash, potato, papaya, plum and common bean. These products have been approved for cultivation and food use by a number of regulatory agencies around the world, and have a history of safe use. Although issues surrounding the regulation of RNAi-based GM plants with particular focus on environmental risk assessment of non-target organisms are still being discussed by the research community (Roberts *et al.*, 2015), the existing international frameworks for evaluation of GM products are sufficient for food safety assessment of RNAi-expressing GM products.

Following the Codex guidelines, the VIRCA team is working with national regulatory authorities in Kenya and Uganda to define the specific data requirements for food and feed safety evaluation of VIRCA varieties for eventual cultivation and food/feed use. A consensus document prepared by the OECD on compositional analysis of cassava (OECD, 2009) will be a key starting point for determining key constituents such as nutrients, anti-nutrients and toxicants that are relevant to consumers of cassava in Kenya and Uganda, taking into account recommendations of others, such as Jansen van Rijssen *et al.* (2013). Furthermore, because no viral protein will be produced by the RNAi-producing genes, toxicity and allergenicity studies would not be applicable in their case. While the kanamycin-resistant marker gene (*nptII*) used in the transformation of these constructs does produce a protein, published toxicity and allergenicity studies exist, and demonstration of sufficient identity to the protein used in those studies should allow the project to benefit from those data.

Conclusion

The promise of benefits to be gained by the deployment of crops such as virus-resistant cassava depends upon a robust and capable regulatory system that can assess the safety of foods derived from genetically engineered crops, in accord with internationally accepted standards. While the development of appropriate legislative and regulatory mechanisms is a prerequisite for establishing this system, the capacity to implement these regulations is also required. It is often a challenge for developing countries, especially those in Africa, to achieve both goals. Regional cooperation and reliance on food safety data generated in other countries can alleviate the lack of capacity in individual countries and the legal complications arising from the high probability of movement across borders.

References

AATF (2015). *Maruca* resistant cowpea – frequently asked questions (FAQs). [Online]. Available from www.aatf-africa.org/userfiles/CowpeaFAQ.pdf

ABNE (2014). Workshop on food/feed safety assessment of GM crop (news). [Online]. Available from www.nepadbiosafety.net/workshop-on-foodfeed-safety-assessment-of-gm-crop

ABNE (2016). ABNE in Africa – towards building functional biosafety systems in Africa. [Online]. Available from http://nepad-abne.net/wp-content/uploads/2016/01/ABNE-in-Africa-2016.pdf

Adenle, A. A. *et al.* (2013). Status of development, regulation and adoption of GM agriculture in Africa: views and positions of stakeholder groups. *Food Policy* **43**, 159–166.

Alicai, T. *et al.* (2007). Re-emergence of cassava brown streak disease in Uganda. *Plant Disease* **91**, 24–29.

Anderson, J. A. *et al.* (2016). Emerging agricultural biotechnologies for sustainable agriculture and food security. *Journal of Agricultural and Food Chemistry* **64**(2), 383–393.

Atkinson, J. H. *et al.* (2015). Africa needs streamlined regulation to support the deployment of GM crops. *Trends in Biotechnology* **33**, 433–435.

Biosafety Clearing House (2015). National Reports on Biosafety. [Online]. Available from https://bch.cbd.int/protocol/cpb_natreports.shtml

Chambers, J. A. *et al.* (2014). GM agricultural technologies for Africa – a state of affairs. Report of a study commissioned by the African Development Bank. [Online]. Available from www.ifpri.org/publication/gm-agricultural-technologies-africa-state-affairs

Chassy, B. *et al.* (2004). Nutritional and safety assessments of foods and feeds nutritionally improved through biotechnology. *Comprehensive Reviews in Food Science and Food Safety* **3**, 38–104.

Codex Alimentarius Commission (2003). Guideline for the conduct of food safety assessments of foods derived from recombinant-DNA plants. CAC/GL 45–2003. Joint FAO/WHO Food Standards Programme.

Codex Alimentarius Commission (2009). *Ad Hoc* Inter-Governmental Task Force on Foods Derived from Modern Biotechnology in 1999–2003 and 2004–2009. International Food Standards. [Online]. Available from www.codexalimentarius.org/standards/en/

Craig, W. *et al.* (2008). An overview of general features of risk assessments of genetically modified crops. *Euphytica* **164**, 853–880.

Department of Agriculture, South Africa (2004). Guideline document for work with genetically modified organisms. [Online]. Available from www.nda.agric.za/doaDev/sideMenu/biosafety/doc/GUIDELINE4WORKwithGOM.pdf

European Commission (1997). Commission recommendation of 29 July 1997 concerning the scientific aspects and the presentation of information necessary to support applications for the placing on the market of novel foods and novel food ingredients and the preparation of initial assessment reports under Regulation (EC) No. 258/97 of the European Parliament and of the Council (97/618/EC). *Official Journal of the European Union* **L253**, 1–36.

European Commission (2003). Guidance document for the risk assessment of genetically modified plants and derived food and feed. The Joint Working Group on Novel Foods and GMOs. Brussels: European Commission.

European Commission (2004). Genetically modified crops in the EU: food safety assessment, regulation, and public concerns – overarching report. [Online]. Available from https://orbilu.uni.lu/bitstream/10993/12409/1/K%C3%B6nig%20A.%20et%20All%20Genetically%20modified%20crops%20in%20the%20EU%20Food%20safety%20assessment,%20regulation,%20and%20public%20assessment.pdf

European Food Safety Authority (2006). Risk assessment of genetically modified plants and derived food and feed. Guidance document of the scientific panel on genetically modified organisms of the European Food Safety Authority. *EFSA Journal* **99**, 1–100.

European Food Safety Authority (2008). Safety and nutritional assessment of GM plants and derived food and feed: the role of animal feeding trials. Report of the EFSA GMO Panel Working Group on Animal Feeding Trials. *Food and Chemical Toxicology* **46**, S2–S70.

Ezezika, O. C. *et al.* (2012). Factors influencing agbiotech adoption and development in sub-Saharan Africa. *Nature Biotechnology* **30**, 38–40.

FAO (1996). Report of a joint FAO/WHO consultation, 30 September–4 October, 1996. FAO Food and Nutrition Paper 61. Rome: Food and Agriculture Organisation of the United Nations.

FAO/WHO (2005). National food safety systems in Africa – a situation analysis, CAF 05/2. FAO/WHO Regional Conference on Food Safety for Africa, Harare, Zimbabwe, 3–6 October 2005.

FAO/WHO (2006). Food safety and nutrition food laws and guidelines. [Online]. Available from www.afro.who.int/en/clusters-a-programmes/hpr/food-safety-and-nutrition-fan/fan-publications.html

Gbadegesin, M. A. *et al.* (2013). African cassava: biotechnology and molecular breeding to the rescue. *British Biotechnology Journal* **3**, 305–317.

Global Environment Facility (2006). Evaluation of GEF support for biosafety. [Online]. Available from https://openknowledge.worldbank.org/handle/10986/7034

Howlett, J. *et al.* (2003). The safety assessment of novel foods and concepts to determine their safety in use. *International Journal of Food Sciences and Nutrition* **54**, S1–S32.

James, C. 2015. Global status of commercialised biotech/GM crops: 2015. ISAAA Brief No. 51. Ithaca, NY: The International Service for the Acquisition of Agri-biotech Applications.

Jansen van Rijssen, F. W. *et al.* (2013). Food safety: importance of composition for assessing genetically modified cassava (*Manihot esculenta* Crantz). *Journal of Agricultural and Food Chemistry* **61**, 8333–8339.

Jonas, D. A. *et al.* (1996). The safety assessment of novel foods. Guidelines prepared by the ILSI Europe Novel Food Task Force. *Food and Chemical Toxicology* **34**, 931–940.

Juma, C. and Serageldin, I. (2007). Freedom to innovate: biotechnology in Africa's development. (Report of the High-Level African Panel on Modern Biotechnology). [Online]. Available from http://belfercenterre.ksg.harvard.edu/files/freedom_innovate_au-nepad_aug2007.pdf

Morris, E. J. (2014). Biosafety regulatory systems in Africa. In *Biosafety in Africa: Experiences and Best Practices*, ed. D. P. Keetch *et al.* East Lansing, MI: Michigan State University Press, pp. 64–77.

Mwamakamba, L. *et al.* (2012). Developing and maintaining national food safety control systems: experiences from the WHO African Region. *African Journal of Food, Agriculture, Nutrition and Development* **12**, 6291–6304.

Nang'ayo, F. *et al.* (2014). Regulatory challenges for GM crops in developing economies: the African experience. *Transgenic Research* **23**, 1049–1055.

Obonyo, D. N. *et al.* (2011). Identified gaps in biosafety knowledge and expertise in Sub-Saharan Africa. *AgBioForum* **14**, 71–82.

OECD (1993). *Safety Evaluation of Foods Derived by Modern Biotechnology: Concepts and Principles*. Paris: OECD.

OECD (1998). OECD documents: report of the OECD workshop on the toxicological and nutritional testing of novel foods, SG/ICGB(1998)1. Paris: OECD.

OECD (2009). Consensus document on compositional considerations for new varieties of cassava (*Manihot esculenta* Crantz): key food and feed nutrients, anti-nutrients, toxicants and allergens. Series on the Safety of Novel Foods and Feeds, No. 18. Paris: OECD.

Pennisi, E. (2010). Armed and dangerous. *Science* **327**, 804–805.

Roberts, A. F. *et al.* (2015). Biosafety research for non-target organism risk assessment of RNAi-based GE plants. *Frontiers in Plant Science* **6**, 958; doi: 10.3389/fpls.2015.00958.

Taylor, N. J. *et al.* (2012). The VIRCA Project: virus resistant cassava for Africa. *GM Crops and Food* **3**(2), 1–11.

Timpo, S. E. (2011). Harmonizing biosafety regulations in Africa: surmounting the hurdles. Socio-economics Policy Brief No. 2. [Online]. Available from http://nepad-abne.net/wp-content/uploads/2015/08/Policy-Brief-No-2_Harmonisation-of-Biosafety-Regulations-_Finalized.pdf

USDA-FAS (2015). Africa contemplates establishing continental food safety body. GAIN Report No. ET1502. [Online]. Available from https://gain.fas.usda.gov/Recent%20GAIN%20Publications/Africa%20Contemplates%20Establishing%20Continental%20Food%20Safety%20Body_Addis%20Ababa_Ethiopia_2-26-2015.pdf

USFDA (1992). Statement of policy – foods derived from new plant varieties. *Federal Register* **57**, 22984 (29 May).

Waithaka, M. *et al.* (2015). Progress and challenges for implementation of the Common Market for Eastern and Southern Africa policy on biotechnology and biosafety. *Frontiers in Bioengineering and Biotechnology* **3**, 109; doi: 10.3389/fbioe.2015.00109.

Wedding, K. and Tuttle, J. N. (2013). *Pathway to Productivity: The Role of GMOs for Food Security in Kenya, Tanzania and Uganda. A Report of the CSIS Global Food Security Project.* Lantham, MD: Rowman and Littlefield.

WHO (1991). *Strategies for Assessing the Safety of Foods Produced by Biotechnology.* Geneva: WHO.

WHO (1995). Application of the principles of substantial equivalence to the safety evaluation of foods and food components from plants derived from modern biotechnology. Report of WHO Workshop WHO/FNU/FOS/95.1. Geneva: WHO.

WHO (2000). Safety aspects of genetically modified foods of plant origin. Report of a Joint FAO/WHO Expert Consultation on Foods Derived from Biotechnology, May 29–June 2, 1996, Rome, Italy. Geneva: WHO.

WHO (2015). Frequently asked questions on GM foods. [Online]. Available from www.who.int/foodsafety/areas_work/food-technology/faq-genetically-modified-food/en/

WTO (1998). Understanding the WTO Agreement on Sanitary and Phytosanitary Measures. [Online]. Available from https://www.wto.org/english/tratop_e/sps_e/spsund_e.htm

10 Being Scientific about Socio-Economics in GMO Decision-Making in Developing Countries

Monica Racovita

Introduction

In the GMO decision-making area, the jury is still out on socio-economic issues. There is no commonly accepted definition. Any understanding of socio-economic considerations (SEC) is highly contextual. Developing countries seem to have a particular interest in SEC, as can be seen from the history of the negotiations for the Cartagena Protocol on Biosafety (CPB) (Convention on Biological Diversity, 2003), yet the inclusion of SEC in national regulations can prove more challenging for developing countries than for developed ones (Falck-Zepeda *et al.*, 2013). Special attention needs to be paid to how SEC inclusion would affect the relationship with international agreements, in particular CPB and World Trade Organization (WTO) agreements (Tung, 2014). At the national regulatory level, the lack of specificity and consensus at the international level largely translated into a lack of specific provisions in developing countries or emergence of various regulatory frameworks in developed countries (emphasising sustainable development, participation in decision-making, scientific method or enlarging the scope of SEC).

A developing country interested in including SEC in GMO decision-making would need to balance the various arguments pro and contra. Fragile social and economic conditions might prompt decision-makers to be overly cautious when adopting new technologies (Chaturvedi *et al.*, 2007). Yet assessments of SEC require considerable institutional support, additional funding and experts. SEC are also in higher danger of being politicised than health or environmental impacts, which, combined with volatile political situations, can lead to decisions that end up benefiting no one (Falck-Zepeda, 2009).

Under these conditions, a regulatory framework for GMOs including SEC for developing countries would greatly benefit from a return to the scientific method and legitimate decisions. Such a framework would need to be flexible and inclusive, and should allow the identification of and focus on key issues as well as using information available elsewhere. This chapter proposes a logical framework with these characteristics that is based on the concepts of evidence-based policy and problem formulation.

Current Status of Socio-Economics in GMO Decision-Making

Within GMO impact analysis, SEC are by far the most controversial concept. The discussions do not focus only on whether socio-economics should or should not be a part of GMO analysis or GMO decision-making, but also on what socio-economic issues are.

Some countries or organisations chose to define SEC either more narrowly or more broadly. For example, Argentina focuses on the impact on trade that a future commercial GMO release might have (Pinski, 2012). The CPB, the international agreement on the biosafety of GMOs, refers to SEC in Article 26 as issues that might harm 'the conservation and sustainable use of biological diversity', with special regard to 'the value of biological diversity to indigenous and local communities' (Convention on Biological Diversity, 2000). Some countries tend towards a broader understanding of socio-economics, including cultural, ethical or religious issues (Council of the European Union, 2011). The concept can be stretched even beyond cultural, ethical or religious issues to 'human health related or ecological dimensions that are not addressed in risk assessment' (Ad Hoc Technical Expert Group on Socioeconomic Considerations (AHTEGSEC), 2014).

Currently, there is no commonly accepted definition for GMO SEC. For any practitioner working in the field of GMOs this is hardly a surprise. Even if they are used only in agriculture, for example, GMOs are in a position to touch, and therefore influence, complex relationships between nature, food, health and society. Furthermore, such relationships change depending on context. What would be an important issue in one region could be marginal or even insignificant in another. Thus, it becomes necessary for each country to set its own policy objectives regarding SEC and hence even provide contextualised definitions. Harmonisation, which is important for international trade, could arguably be better achieved through pursuing similar decision-making frameworks.

Yet definitions are not the most difficult part of SEC. Intense debates surround the relevance of SEC inclusion in GMO decision-making, in particular in the developing world (Convention on Biological Diversity, 2015). Judging by the history of negotiations for the CPB, policymakers in the developing world have generally considered SEC as central for their countries and for the Protocol from the very beginning (Convention on Biological Diversity, 2003). Although the participating countries formed different interest groups, not necessarily according to the developed-versus-developing dichotomy, developing countries were generally the main proponents for mainstreaming SEC in the Protocol, while the developed countries mainly opposed its inclusion in the text of the Protocol for various reasons. Negotiations on SEC brought a compromise in the form of Article 26.

SEC remained relevant for developing countries as they continued to place SEC on the agenda for discussions throughout successive Conferences of the Parties serving as the Meeting of the Parties to the Cartagena Protocol on Biosafety (COP-MOP) (Perron-Welch, 2012). The interest in further clarifying Article 26 started to become more concrete at COP-MOP 5 with the results of a survey conducted by the United Nations Environment Programme (UNEP)–Global Environment Facility (GEF) Coordination with funding from the UK Department for International Development. The survey sought to explore SEC experiences in biosafety decision-making at the grassroots level for various types of stakeholders in 154 countries. The results showed, among other things, that over 85% of respondents found the inclusion of SEC in GMO decision-making very or somewhat important and that the majority favoured a specific, identifiable mechanism for inclusion. Although the results were not analysed separately for developed and developing countries, the high level of agreement on the importance of SEC makes it safe to assume the importance of SEC for the various stakeholders in the developing world (UNEP, 2010).

Even if it is relevant, the inclusion of SEC in GMO decision-making in developing countries is problematic, as this could further complicate an already cumbersome process. Along the same lines, it could be argued that for countries with already volatile political situations the extra regulatory burden brought by SEC could further cripple regulatory systems. SEC would also increase the assessment costs both for governments and for GMO developers, and it could also be argued that it might delay the release of technologies that could help farmers mitigate, among other things, the effects of climate change, growing population and poverty (Falck-Zepeda *et al.*, 2013). At the international level, measures taken on the basis of SEC might be considered as barriers to trade, and countries passing such measures risk being brought in front of the Dispute Settlement Body of the World Trade Organization (WTO) (Tung, 2014).

Concerning the last aspect, debates on potential conflicts between the WTO and the CPB due to SEC have been renewed with recent interest in clarifying Article 26. The CPB and WTO have different scopes for GMOs regarding trade and uncertainty: while the CPB places the burden on exporters, the WTO focuses on countries imposing restrictions; while under the CPB countries can use the ‘precautionary principle’ to manage risks, under the WTO countries can analyse risks only by employing objective and accurate scientific data (Eckersley, 2004). Scholars predict that an increase in the level of detail in the rules and implementations of the CPB would increase the potential for conflict with the WTO (Kelly, 2006; Komen, 2012).

From a different point of view, however, the developing world has characteristics that might strongly advocate the inclusion of SEC, such as the presence of many crop and livestock centres of origin or diversity, the cultivation of non-GM plants as an important source of revenue and the fact that the wellbeing of a majority of the population depends upon agriculture (Chaturvedi *et al.*, 2007). It has also been noted that assessing the social and economic impacts of a technology often takes years, by which time they have already become irreversible (Dano, 2007). Higher costs would therefore be incurred to mitigate any negative long-term effects connected with the introduction of a GMO. In the case of the developing world this reason might justify having an evaluation step on the potential socio-economic impacts before introducing a GMO product. Furthermore, paragraph 1 of Article 26 of the CPB states that countries could undertake SEC as long as they respect previous international obligations, which would avoid the possibility of tensions between the WTO and the CPB. In the case of countries with difficult political situations, where it might be argued that SEC could not be addressed for political reasons, the results from the UNEP survey seem to contradict this assumption. Respondents from countries that did not have SEC in their decision-making indicated as reasons for the lack of SEC, in order of importance, 'mechanism not in place', 'lack of capacity', 'lack of institutional coordination', 'regulation does not require it' and only in last place a 'political reason' (UNEP, 2010).

Regardless of the pros and cons, the interest of developing countries in SEC is growing, and, even though their experience with socio-economics is limited, the body of evidence on SEC impacts of GMOs in the developing world is slowly expanding (Smale *et al.*, 2008). Yet there might be a significant gap between interest and inclusion, with inclusion possibly representing a complex set of measures as the following sections will show.

Functioning and Proposed Frameworks for Inclusion of Socio-Economics in GMO Decision-Making

It has been estimated that 56% of the existing National Biosafety Frameworks at least mention SEC. Yet most of them do not have any specific provisions for SEC or are marred by methodological issues (Chaturvedi *et al.*, 2012). When decision-making has taken into consideration SEC, the decisions focused in the majority of cases on imports of GM seeds, but regarded equally their use for food, feed or processing, or introduction into the environment (UNEP, 2010). Socio-economic provisions are included in many national regulations for GMOs throughout the developing world. Countries such as Brazil, Honduras, Indonesia, Kenya, Nigeria and Uganda have more general provisions, while Argentina, India, the Philippines and South Africa include more specific regulations (Falck-Zepeda, 2009).

If provisions do exist for the inclusion of SEC, they can be either mandatory (Argentina) or voluntary (India). Furthermore, in relation to risk assessment, most of the countries concerned (including Brazil and India), when they include SEC in GMO decision-making, are conducting SEC assessments separately from the environmental and health assessments, done by separate commissions, albeit working at the same time (Perron-Welch, 2012). In Argentina, however, the economic assessment is done after the completion of the environmental and health assessments, in a 'sequential' manner (Falck-Zepeda and Zambrano, 2011) (see Chapter 21 for details).

In Norway the assessment of GMOs has three steps: determining effects on health and the environment; applying the precautionary principle; and following sustainable development principles. It must be emphasised that the sustainable development principles apply even to health and environment impacts (Spök, 2010). GMO applications are evaluated by an independent body, the Norwegian Biotechnology Advisory Board (NBAB), which is composed of natural scientists but also representatives of various stakeholder groups (Binimelis and Myhr, 2016).

A framework that places a particular emphasis on participation in impact assessment for SEC has been developed by France. The Haut Conseil des Biotechnologies (HCB) provides recommendations on the health and environmental impacts of GMOs (one scientific committee), as well as on economic, ethical and social impacts (a different committee formed of representatives of many stakeholder groups as well as various experts). Both committees provide recommendations to the European Food Safety Authority (EFSA) (Remondet, 2011).

The Ad Hoc Technical Group mandated by the Executive Secretary of the sixth COP-MOP to search for ways of achieving conceptual clarity for Article 26 proposed a working framework to address SEC. Notable elements of this framework were as follows: the inclusion of human rights agreements as international agreements with which countries need to comply; the need to consider 'local circumstances, priorities and needs'; measures need to be taken in a 'clear, transparent, and non-discriminatory' manner; human health impacts related to the conservation and sustainable use of biological diversity should be taken into account (e.g., impact on nutritional status) provided that they are not already addressed in the risk assessment; it allows the option of SEC being considered in a complementary process to the decision-making process if desired; and it makes it mandatory to have provisions for public participation and consultation as part of the process of taking SEC into account. It also mentions types of methodological approaches that could be considered for SEC as well as factors that could affect such approaches (AHTEGSEC, 2014).

Another proposed framework comes from the European Union. The European GMO Socio-Economic Bureau (ESEB) released a proposed framework for inclusion of SEC in GMO decision-making. It focuses on indicators and methodologies that could be used in an assessment of socio-economic impacts. A noticeable characteristic of the ESEB framework is the emphasis on a 'science-based' approach. The main drawback indicated by the framework is the lack of specific data necessary for some methodologies. The document further recommends a strategy for choosing indicators to measure socio-economic impacts: indicators should be amenable to assessment either quantitatively or qualitatively; the causal mechanism connecting the use of a GMO and its impacts needs to be 'plausible'; and the method chosen to assess the impact needs to be scientifically proven (Kathage *et al.*, 2015).

Evidence-Based Policy and its Application to Socio-Economic Considerations

One concept outlining a scientific approach to public policy is that of evidence-based policy. This approach has been promoted since the 1990s by the UK government (the Blair administration) in order to introduce objective, rigorous and systematic evidence at the core of public policy decisions, as opposed to ideologically driven policies and opinion-based decisions. The approach gained strong support in the developed world, but registered successes in the developing world as well (e.g., the Government of Tanzania achieved a reduction of child mortality by 40% in two pilot districts) (Sutcliffe and Court, 2005). Evidence of 'what works' in public interventions can be used at different levels, depending on human and institutional capacity. Moreover, both objective and subjective sources of information should be considered, since the latter can bring important insights from practice and citizens' views. Regardless of the hierarchy of evidence, data need to be of high quality, credible and relevant, but also cost-efficient. In addition, although acknowledging that policymaking is a political process, placing more emphasis on evidence can render it more efficient and accountable (Sutcliffe and Court, 2005).

A report by the Overseas Development Institute (ODI) analysed the relevance of the concept for developing countries. The recommendations were based on a literature review, a workshop and a series of interviews with representatives from government, academia and think tanks. The report acknowledges that the application of evidence-based policy in a developing-country context confronts more challenges than it does in a developed country: there is higher political volatility or complexity; centralised policy formulation is prevalent; and implementation is troubled by poor accountability and participation, corruption and lack of incentives and capacity (both human and monetary) to draw on evidence (Sutcliffe and Court, 2005). However, the same report indicated that these conditions might not apply to all countries in the developing world or at all levels of decision-making, or on all issues. When conditions allow it, the authors further state, evidence-based policy can and should be used.

Taking it one step further, this chapter argues that, in spite of the challenges, evidence-based decision-making, with its emphasis on reducing wasteful spending, increased use of innovative programmes and an increase in accountability (Pew-MacArthur, 2014), can be particularly relevant for developing countries, with full acknowledgement of the differences between developed and developing contexts.

In developed countries good evidence is often necessary and available at every step in policymaking, from budget development to the evaluation of policy impact (Pew-MacArthur, 2014). In developing countries necessary data are frequently missing, incomplete and/or of low quality. Furthermore, there may not be enough time, money or political will to base each step on evidence (Sutcliffe and Court, 2005). Nonetheless, available evidence of good quality can still exist elsewhere, and policymakers in developing countries could make use of it through systematic reviews (SRs). This methodology aims to identify, synthesise and critically analyse peer-reviewed studies but also 'grey literature' studies (e.g., working papers, academic theses and ongoing studies). Such a review starts with a specific research question/hypothesis and proceeds according to a peer-reviewed protocol, which indicates the details of the reviewing process, often including interested stakeholders, especially in the first steps of the review (Collaboration for Environmental Evidence (CEE), 2013). Even if SRs can be cumbersome and resource-intensive, they can be a less costly alternative to generating empirical data, without compromising on data quality.

Another issue is participation in evidence-based decision-making. In developed countries, it is deemed essential for a democratic process (Sutcliffe and Court, 2005), but in the developing world the first experiences with participatory decision-making came about as a result of obligations assumed under international agreements (Kimera and Mboyah, 2007). Yet the development of communication technologies, together with the emergence of powerful international and transnational non-governmental organisations (NGOs), makes participation a necessity rather than a luxury in the developing world.

Further challenges might appear, depending on the type of evidence needed. In developed countries empirical research, which is of high quality, credible, relevant, accessible and conducted by experts, is the pinnacle of evidence used in decision-making, with 'hard' data (e.g., statistics) being considered more valuable than 'soft' data (e.g., interviews) (Sutcliffe and Court, 2005). In developing countries such data can be almost impossible to deliver consistently to decision-makers. Yet again SRs can be a valuable alternative, as newer SR guidelines emphasise the need to consider evidence of a 'lesser' quality for the unique perspectives this can bring, without compromising the quality of the conclusion (CEE, 2013). Developing countries' policymakers can also learn from their peers in neighbouring countries, countries with similar situations or richer experience.

Yet developed countries do not have an ideal situation either. Politics is a major impediment for evidence-based policy. Time and electoral pressure, self-interest, ideologies, lobbyists and pressure groups, amongst other things, can significantly affect and even derail decision-making. Evidence-based policy is still a goal rather than an achievement in many developed nations (Pew-MacArthur, 2014). Both developed and developing countries could pursue evidence-based policies from monetary (cost reductions) and value-related (increase accountability) perspectives.

When applying evidence-based decision-making to SEC, developing countries would need to do more than just show interest. There is widespread disagreement on what the place of SEC should be in decision-making on GMOs: should SEC be part of the risk assessment or complementary to it; how should one select the issues that should be considered; and how should impacts be evaluated? But there is also hope: many practitioners believe that there is a way to approach SEC in a systematic and rigorous way, with initiatives to construct guidelines on methodologies to be used in decision-making being under way (Kathage *et al.*, 2015; Convention on Biological Diversity, 2015; Binimelis and Myhr, 2016). Agreement seems to have been achieved on using quantitative, mainly economic data, both linear and non-linear pathways, to describe causality. Qualitative data, alongside social, ethical and cultural issues, are still under discussion (Convention on Biological Diversity, 2015; Binimelis and Myhr, 2016).

Employing Problem Formulation in SEC Decision-Making

The US Environmental Protection Agency (EPA) presents problem formulation (PF) as one of the three phases of the ecological risk assessment framework, together with analysis and risk characterisation (EPA, 1992). In risk analysis PF is used as an incipient step to identify and formulate risks in terms of hypotheses that can later be broken down in a chain of events linked by causality.

However, for SEC, PF cannot always be applied in this form because some concerns cannot be expressed as a hypothesis or put in a causal chain form. Furthermore, SEC are often analysed in terms of both risks and benefits. The use of PF in risk analysis sidesteps benefits. Yet this chapter argues that PF as a concept does not necessarily reject benefits or complex causality. Hypotheses can be constructed in terms of risks and benefits, and complex causality can be broken down into smaller linear steps. The

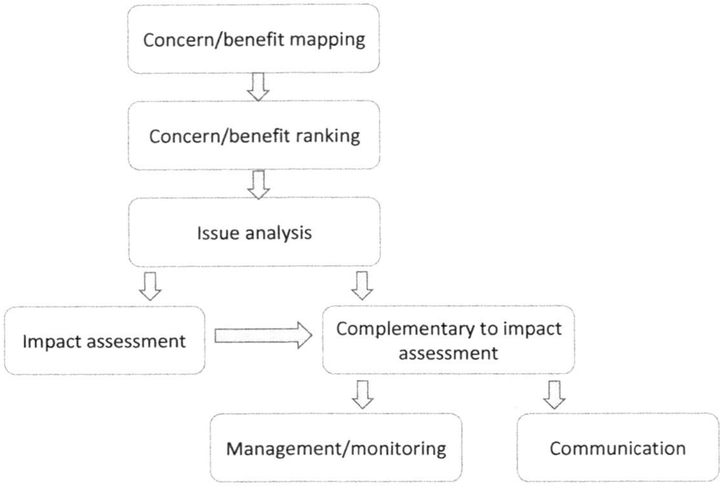

Figure 10.1. A logical framework for SEC.

value of PF as a focus point for a simple yet flexible logical framework is explored in Figure 10.1.

A Framework for SEC

Policymakers in developing countries need to decide first whether SEC should form part of decision-making, either on a permanent basis or in a particular case. If the answer is affirmative, the particular SEC to be considered either can be selected because they have already been included in policy goals or can present themselves on a case-by-case basis. As shown in Figure 10.1, a first step is represented by concern/benefit mapping, or collecting possible issues of concern or indicated benefits. For a better understanding of the logical framework, Figure 10.2 presents a hypothetical case of a fungi-resistant papaya and Figure 10.3 follows through a single issue. Concerns/benefits need to be sufficiently precise to allow in-depth analysis. A step of concern/benefit ranking can be done to order the issues according to urgency and thus focus scarce resources on the most stringent issues. This can be done by consultations among decision-makers, with participation by experts and the general public. The reason for separating risks and benefits is that they can affect different groups disproportionately.

The further analysis of the issue identified could proceed along the lines of an impact assessment. Various methodologies for impact assessment have been proposed (Morris, 2011; Falck-Zepeda *et al.*, 2013; Kathage *et al.*, 2015). The choice of the methodologies used could be based on the data at hand, resources and the conditions pertaining in a particular country. A benchmark against which to measure changes to selected indicators for the concerns and benefits evaluated will need to be established as part of the impact assessment. Data can be obtained from statistics offices of various

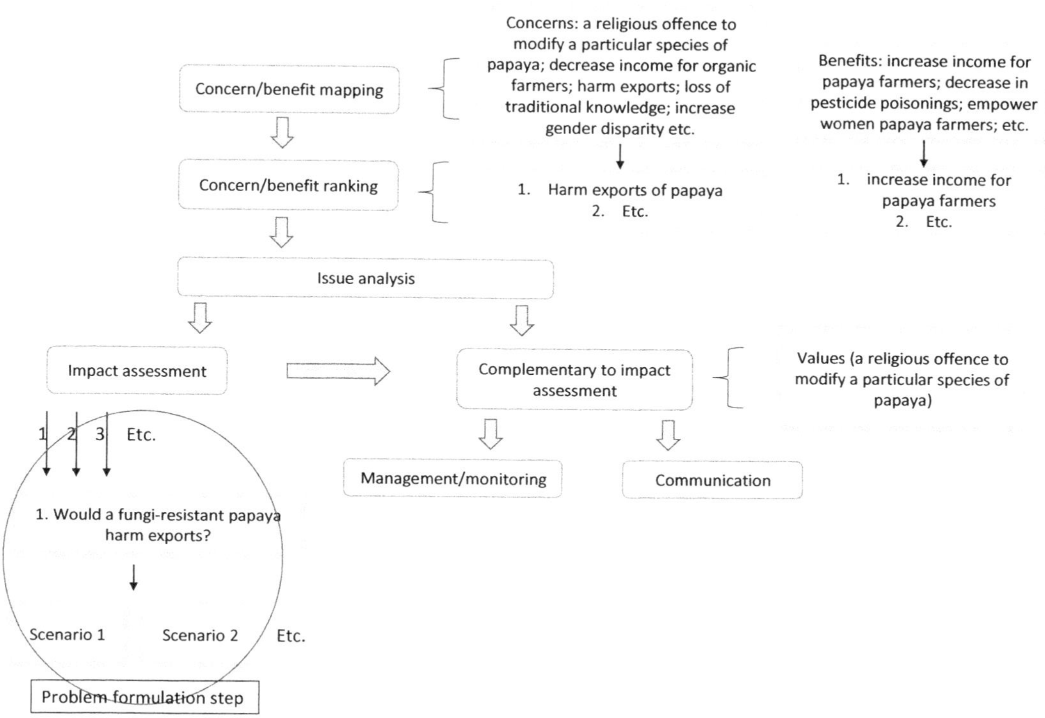

Figure 10.2. Fungi-resistant papaya example.

governmental agencies, using information gained in similar experiences in the country or elsewhere. If necessary, an SR can be done to assess available evidence on a particular subject, or an *ex-ante* study can be performed. In the absence of any evidence, countries in a region could pool resources to investigate possible critically important issues for decision-making.

The analysis can start with a PF approach in constructing a list of hypotheses or research questions for the most important concerns and benefits which can be analysed along the lines of a logical causal chain or tree. When it reaches unknown yet vital information, the assessors would look for the information or make arrangements to generate it. The usefulness of this approach resides in the possibility of discarding issues to analyse solely on the basis of the possible chain of events, without the need to perform an assessment, thus saving both money and time. For those issues which cannot be discarded, impact assessments could be undertaken. The result should be a list of scenarios with different degrees of likelihood of occurring both for concerns and for benefits. Such scenarios should then be evaluated and decisions to manage or monitor should be taken.

Issues that cannot be subjected to concern or benefit analysis (e.g., cultural and moral issues) could be collected in the concern/benefit step and then examined in a step that is complementary to impact assessment. In this step it would be possible to manage or

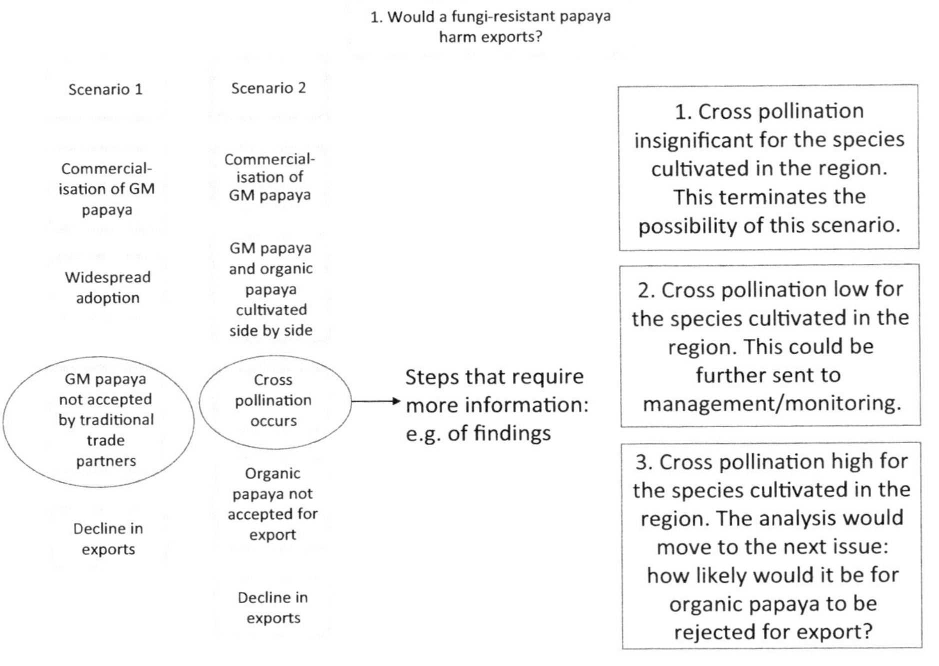

Figure 10.3. Problem formulation step example.

monitor issues following risk analysis decisions, take decisions on value-based issues and/or conduct communication activities in further informing the public and receiving input from it.

The benefits brought by the PF and evidence-based approach would thus include the following: a clarification of local protection goals; a better focus of already scarce resources on appropriate data gathering and analysis methodologies, especially through the use of SRs; and an emphasis on scientifically based assessments. Additionally, after the impact assessment step, communication can be improved by including stakeholders in the identification of concerns and benefits, providing clearer evidence and a simpler process focused on specific and appropriate mitigation measures (Racovita *et al.*, 2013).

The framework has the inherent limitations of the evidence-based policy and problem formulation. It requires a high degree of precision and coordination in addition to political will, all of which are potentially problematic in the developing world. Yet, at the same time, the framework is a pragmatic tool that allows the identification and analysis of main concerns and benefits and makes it possible to devise conclusions and mitigation measures while being inclusive of other stakeholders. It further needs to be stated that overconfidence in evidence should be avoided. As previous experience with evidence-based policy has shown (Blankesteijn *et al.*, 2014), participation of interested stakeholders from the first steps of decision-making is key in avoiding public controversies. Furthermore, there are no zero-risk solutions.

Conclusion

Whether SEC should or should not be included in GMO decision-making is a decision each developing country needs to take depending on its particular mix of reasons pro and contra. If a country has resolved to include SEC, governance of SEC should preferably be based on evidence for monetary (cost reductions) and value-related (increased accountability) reasons. Yet, due to the complexity of issues included under the SEC umbrella, it is difficult to come up with a one-size-fits-all approach. In addition, including SEC can be problematic even for countries with a stable political climate and plentiful human and financial resources. Furthermore, although working or proposed frameworks are available, more research on best frameworks and best practices would be needed on individual country/region level. It is hoped that the proposed logical framework will be a pragmatic and easy-to-use tool for developing-country decision-makers wishing to address SEC.

Acknowledgements

The author would like to thank Dr Armin Spök and the editors of this book (in particular Jane Morris and Ademola Adenle) for comments that greatly improved the manuscript.

References

AHTEGSEC (2014). Report of the Ad Hoc Technical Expert Group on Socioeconomic Considerations. UNEP/CBD/BS/AHTEG-SEC/1/3, Seoul Convention on Biological Diversity.

Binimelis, R. and Myhr, A. I. (2016). Inclusion and implementation of socio-economic considerations in GMO regulations: needs and recommendations. *Sustainability* **8**(1), 62.

Blankesteijn, M. *et al.* (2014). Contested science – public controversies about science and policy. The Hague, Rathenau Instituut. [Online]. Available from https://www.rathenau.nl/en/file/173/download?token=wsT8MmBW

CEE (2013). Guidelines for systematic review and evidence synthesis in environmental management. Version 4.2. Environmental Evidence. [Online]. Available from www.environmentalevidence.org/Documents/Guidelines/Guidelines4.2.pdf

Chaturvedi, S. *et al.* (2007). Environmental risk assessment, socio-economic considerations and decision-making support for LMOs in India. Technical report. New Delhi: ICGEB and RIS.

Chaturvedi, S. *et al.* (2012). Approval of GM crops: socio-economic considerations in developing countries. *Economic & Political Weekly* **47**(23), 53–61.

Convention on Biological Diversity (2000). Cartagena Protocol on Biosafety to the Convention on Biological Diversity. Text and Annexes. [Online]. Available from https://www.cbd.int/doc/legal/cartagena-protocol-en.pdf.

Convention on Biological Diversity (2003). The Cartagena Protocol on Biosafety: a record of the negotiations. [Online]. Available from https://bch.cbd.int/database/attachment/?id=10886

Convention on Biological Diversity (2015). Portal on socioeconomic considerations: online discussion 2015 – Discussion Groups. [Online]. Available from http://bch.cbd.int/onlineconferences/portal_art26/discussion_groups/?threadid=6621

Council of the European Union (2011). Complementary considerations on legal issues on GMO cultivation raised in the opinions of the legal service of the Council of the European Union of 5 November 2010 and of the legal service of the European Parliament of 17 November 2010 – WTO Compatibility. Commission staff working paper. Brussels: European Commission.

Dano, E. (2007). *Potential Socio-economic, Cultural and Ethical Impacts of GMOs: Prospects for Socio-economic Impact Assessment*. TWN Biotechnology & Biosafety Series 8. Penang: Third World Network.

Eckersley, R. (2004). The Big Chill: the WTO and multilateral environmental agreements. *Global Environmental Politics* **4**(2), 24–50.

EPA (1992). Framework for Ecological Risk Assessment. Risk Assessment Forum. EPA/630/R-92/001.

Falck-Zepeda, J. (2009). Socio-economic considerations, article 26.1 of the Cartagena Protocol on Biosafety: what are the issues and what is at stake? *AgBioForum* **12**(1), 90–107.

Falck-Zepeda, J. B. and Zambrano, P. (2011). Socio-economic considerations in biosafety and biotechnology decision making: the Cartagena Protocol and National Biosafety Frameworks. *Review of Policy Research* **28**(2), 171–195.

Falck-Zepeda, J. *et al.* (2013). The current status of the debate on socio-economic regulatory assessments: positions and policies in Canada, the USA, the EU and developing countries. *World Review of Science, Technology and Sustainable Development* **10**(4), 203–227.

Kathage, J. *et al.* (2015). *Framework for the Socio-Economic Analysis of the Cultivation of Genetically Modified Crops*. JRC Science and Policy Reports. Luxembourg: European Union.

Kelly, C. R. (2006). Power, linkage and accommodation: The WTO as an international actor and its influence on other actors and regimes. *Berkeley Journal of International Law* **24**(79), 79–50.

Kimera, H. R. and Mboyah, D. (2007). Stakeholder awareness and participation in biotechnology policy-making in Eastern African countries. In *Biotechnology: Eastern African Perspectives on Sustainable Development and Trade Policy*, ed. H. Baumüller and M. Bolo. Geneva: International Centre for Trade and Sustainable Development; and Nairobi: African Technology Policy Studies Network, pp. 44–57.

Komen, J. (2012). The emerging international regulatory framework for biotechnology. *GM Crops and Food: Biotechnology in Agriculture and the Food Chain* **3**(1), 78–84.

Morris, E. J. (2011). A semi-quantitative approach to GMO risk-benefit analysis. *Transgenic Research* **20**(5), 1055–1071.

Perron-Welch, F. (2012). Socioeconomics, biosafety, and sustainable development. *Asian Biotechnology and Development Review* **14**(3), 49–69.

Pew-MacArthur (2014). Evidence-based policymaking. A guide for effective government. Pew-MacArthur Results First Initiative. [Online]. Available from www.pewtrusts.org/~/media/assets/2014/11/evidencebasedpolicymakingaguideforeffectivegovernment.pdf

Pinski, F. (2012). New regulatory framework for agricultural biotechnology in Argentina. Buenos Aires: Biotechnology Directorate, Minister of Agriculture, Livestock and Fisheries, Argentina.

Racovita, M. *et al.* (2013). Can problem formulation help address socioeconomic considerations in GMO decision-making? Poster. 6th International Conference on Coexistence between Genetically Modified (GM) and Non-GM Based Agricultural Supply Chains, Lisbon.

Remondet, M. (2011). The French «High Council for Biotechnologies»: an innovative institution for GMOs assessment. Workshop on Capacity-Building for Research and Information Exchange on Socio-economic Impacts of Living Modified Organisms under the Cartagena Protocol on Biosafety. New Delhi, India. [Online]. Available from https://bch.cbd.int/protocol/socioeconomics/presentations/france.pdf

Smale, M. *et al.* (2008). Economic impact of transgenic crops in developing countries: a note on the methods. *International Journal of Biotechnology* **10**(6), 519–555.

Spök, A. (2010). Assessing socio-economic impacts of GMOs. Issues to consider for policy development. Vienna: Lebensministerium/Bundensministerium für Gesundheit.

Sutcliffe, S. and Court, J. (2005). *Evidence-Based Policymaking: What Is It? How Does It Work? What Relevance for Developing Countries*? London: ODI.

Tung, O. J. (2014). Transboundary movements of genetically modified organisms and the Cartagena Protocol: key issues and concerns. *PER: Potchefstroomse Elektroniese Regsblad* **17**(5), 1740–1787.

UNEP (2010). Summary Report on the Survey on the Application of and Experience in the Use of Socio-economic Considerations in Decision-Making on Living Modified Organisms. UNEP/CBD/BS/COP-MOP/5/INF/10. Nagoya: UNEP.

11 The Sharing of Information in Risk Assessment: How National Authorities Cooperate

Peter Kearns, Bertrand Dagallier and Takahiko Nikaido

Introduction

This chapter considers the risk/safety assessment of genetically modified organisms (GMOs) or genetically modified (GM) crops from a regulatory perspective, which is an important component of risk analysis. More specifically, it describes how national authorities have worked together through the Organisation for Economic Co-operation and Development (OECD) to draft documents in order to agree and share some of the information used in risk assessment.

The OECD is an intergovernmental organisation that has been active on the topic of the safety of GMOs since the mid 1980s (OECD, 1986). It provides a forum where governments compare policy experiences, seek answers to common problems, identify good practices and coordinate domestic and international policies.

Today, the OECD has 35 member countries, but it has increasingly developed joint programmes and other forms of cooperation with countries that are not members of the OECD. This is certainly the case in relation to the risk assessment of GMOs, which is a topic of global concern, and a number of non-member countries participate in OECD meetings and in the programmes and projects discussed in this chapter. These non-members include Argentina, Bangladesh, Brazil, Cameroon, China, India, Indonesia, the Philippines, the Russian Federation, South Africa, Thailand and Vietnam. These non-members are integrated into the process of drafting OECD documents and bring their perspective to the work. At the same time, the work has evolved to address GM crop species, which are of particular interest to developing countries and countries in transition.

The OECD's activities are organised in two programme areas: (i) the environmental risk/safety assessment of transgenic crops; and (ii) the safety assessment of foods and feeds derived from such crops. It is important to note that these programmes are designed and funded by governments in order to best respond to their needs. Two bodies have been established at the OECD to undertake this work. The Working Group on Harmonisation of Regulatory Oversight in Biotechnology addresses environmental risk/safety assessment of GMOs (OECD, 2000a), while the Task Force for the Safety of Novel Foods and Feeds addresses the safety assessment of foods and feeds derived from GMOs (OECD, 2000b). These bodies comprise delegates nominated by their governments, including non-members, who are from those ministries or agencies which have regulatory responsibility for GMOs.

Other stakeholders participate in the work so that relevant expertise is provided by participants from the Convention on Biological Diversity (CBD), the Food and Agriculture Organization of the UN (FAO), the World Health Organization (United Nations) and the New Partnership for Africa's Development (NEPAD), which is a technical body of the African Union. In addition, international research institutes, including those from the consortium of the Consultative Group for International Agricultural Research (CGIAR), have provided input into a number of projects, as has the Business and Industry Advisory Committee to the OECD (BIAC).

The main outputs of the work are practical tools, often referred to as consensus documents, which are designed by these bodies for use in risk assessment. It is the case that most countries, especially those which have approved the use of commercial applications involving GM crops or food and feed products derived from them, have regulatory systems in place designed to ensure the safe use of GM crops and products. The use of risk assessment is a key part of the regulatory process and, whilst differences between jurisdictions are often commented upon (e.g., Anderson and Jackson, 2003), it should be emphasised that there are significant similarities in the process of risk assessment as well as much of the information that is used in such assessments.

The Environmental Risk/Safety Assessment of GM Crops

Certain principles related to environmental risk assessment of GM crops are held in common by most of those engaged in regulatory risk assessment. It is commonly agreed, for example, that risk assessment should be based on sound science and information derived from scientific practice. It is also commonly agreed that risk assessment of GM crops should be undertaken on a case-by-case basis (OECD, 2000a).

An environmental risk assessment of a GM crop takes into account specific types of scientific information, including the biological properties of the host organism, which is usually a well-known crop species. It also includes information on the source of the gene(s) introduced, how the gene(s) is (are) expressed in the GMO and the nature of the eventual gene product. The characteristics of the organism are taken into account, as well as its likely performance and impact in the environment where it is to be released. For example, exposure and toxicity data are used to examine potential ecological effects relating to resident wildlife and biodiversity. In addition, information on the eventual use of the product is necessary to ensure a complete assessment (OECD, 2000a).

As a result of a similar approach to risk assessment, risk assessors from different countries often use similar information to one another. They not only use similar sets of information, but also agree such information through the OECD so that it can easily be shared among all those involved in risk assessment. This chapter will describe a number of examples. All crop species (which have been the subject of genetic modification) have biological characteristics that are of interest to risk assessors. For example, is the crop species under consideration capable of outcrossing; in other words, does it share its pollen with other plants of the same species and does this lead to viable offspring? Can it also share its pollen and hybridise with wild relatives found in the surrounding environment? What are the likely consequences of this from a risk assessment perspective? Can seeds derived from the crop species in question remain viable and lead to volunteers in subsequent crop rotations or even form possible invasive weed populations? These types of questions and the answers derived from them are relevant to all new varieties derived from the same crop species and depend on the biology of the crop species under consideration.

Through the OECD, regulatory authorities work together to prepare *consensus documents*, which compile information on the biology of a range of crop plant species that have been subject to genetic modification. These documents can be used by risk assessors when they need to consider information on the biology of a specific crop species when undertaking an environmental risk assessment. In the early years, most of the focus was on crop plants that are major trading commodities, such as maize (OECD, 2003), soybean (OECD, 2000c), cotton (OECD, 2008a) and rapeseed (OECD, 2012a).

In recent years an increasing emphasis has been placed on crops of interest to developing countries and countries in transition, including those with tropical climates. Documents on such crops have been published for cassava (OECD, 2014a), cowpea (OECD, 2015), sugarcane (OECD, 2013), cucurbits (OECD, 2012b), banana and plantains (OECD, 2009a), chili peppers (OECD, 2006a) and papaya (OECD, 2005a). This work has been enhanced by the increasing participation of developing countries and their experts in the work of the OECD. In any event, the objective is always the same for OECD members and non-members alike: to develop practical risk assessment tools with which to compile information on the biology of major crops. This is information agreed upon by consensus and addresses those topics which national authorities believe to be relevant to risk assessment when comparing new GM products with those of a more conventional origin.

The best way to illustrate the use of the consensus documents related to environmental risk assessment is to consider two examples. One of the early OECD documents was a Consensus Document on the Biology of *Zea mays* subspecies *mays* maize (OECD, 2003). This is a crop species of interest for two reasons: (i) it is the world's third leading cereal crop; and (ii) it is an important source of food (in countries such as Mexico and various sub-Saharan Africa countries) as well as a source of animal feed (in the USA, Europe and a number of Asian countries). It was one of the first crops for which GM varieties were approved for commercial planting as well as for food and feed use.

Modern maize is highly domesticated and cannot survive in the environment without the intervention of farmers. This fact related to the biology of maize is already of great interest to risk assessors. Maize does not have the biological attributes which would allow it to become a weed or persist in the environment without human intervention. As a result of its domestication, many races and varieties of maize have been developed. Maize is an open-pollinated cross-fertilising species and the various races and varieties may cross-pollinate.

The maize document notes that more than 40 land races have been identified in Mexico and other Latin American countries. It also describes the situation with respect to teosinte, which is an ancestor of maize and closely related; in fact, annual teosinte can readily hybridise with maize. Teosinte occurs in Mexico and some countries of Central America. The implications for a risk assessment are clear. If there is a concern as to the outcrossing of a new GM maize variety with either land races or teosinte, this can occur only in the countries where they exist, that is, Mexico and some other countries in Latin America.

A second example of an OECD document relevant to environmental risk assessment is the Consensus Document on the Biology of the Brassica Crops (*Brassica spp.*) (OECD, 2012a), which incorporated information published in an earlier Consensus Document on *Brassica napus* (oilseed rape).

The Brassica document is considerably more complex than that covering maize because the Brassicas are more diverse from a taxonomic perspective. The Brassica genus includes a large number of species of agricultural interest. Species grown as oilseeds include *B. napus*, *B. juncea*, *B. rapa* and *B. carinata*. The vegetable Brassicas include *B. napus*, *B. rapa*, *B. oleracea*, *Raphanus sativus*, *Lepidium sativum* and *Nasturtium officinale*. There are also some Brassicas used as condiments, minor potherbs and salad vegetables.

At the same time, there are also numerous weedy species among the Brassicas. Those of greatest interest with regard to cross-pollination with *B. napus* are *Sinapis arvensis*, *Raphanus raphanistrum*, *B. rapa* and *Hirschfeldia incana*. As described in the document, the taxonomic relationships in this group have become clearer over the years as a result of traditional techniques including cytology and crossing experiments as well as more modern molecular techniques.

A risk assessment involving a *Brassica sp.* could involve the consideration of much more information related to the genus as a whole than in the case of maize. The document gives a full description of the various species and varieties important in agriculture, including their geographical distribution, ecosystems and habitats as well as cultivation and agricultural management practices. The document also gives a description of the centres of origin and the origin of the ancestors of modern varieties.

The document also includes extensive information on the reproductive biology of the Brassica species, including pollination, pollen dispersal and viability. This is complemented by information on seed viability, longevity, dormancy and germination and seedling establishment. It addresses whether members of the Brassica genus have properties that could allow them to form feral populations. This is an important issue in Europe as rapeseed, which arose only about 2,000 years ago, could have the potential to form feral populations and possibly hybridise with wild relatives, albeit at low frequency.

The information included in the document is used by risk assessors considering a GM variety of a Brassica species. It is used to assess, amongst other things, whether or not a new genetically modified variety can outcross in the field, depending on the environment in which it is cultivated and the presence of compatible relatives as well as the management practices.

Consensus Documents Addressing Traits

The Working Group has also published documents on a number of traits that are commonly used across a range of transgenic crops. These 'trait documents' include topics such as tolerance of glyphosate herbicide; tolerance of phosphinothricin herbicides; virus resistance through coat protein gene-mediated protection; and insect control through *Bacillus thuringiensis* (Bt)-derived proteins. Similarly to the biology consensus documents on specific crop plants, the 'trait documents' address those traits which are most common in commercial applications and focus on issues of concern to risk assessors.

The Consensus Document on Safety Information on Transgenic Plants Expressing *Bacillus thuringiensis* (Bt) Derived Insect Control Proteins (OECD, 2007) can be considered in more detail as a pertinent example. As described in the document, naturally occurring isolates of the bacterium *B. thuringiensis* have been used for insect control for decades. Various isolates of *B. thuringiensis* have been reported to have pesticidal activity against insect species, primarily lepidopteran, coleopteran and dipteran species, as well as some non-insect invertebrates.

The document describes the main insecticidal components of *B. thuringiensis*, which are proteins known as δ-endotoxins. The types of δ-endotoxins vary enormously from one naturally occurring isolate of *B. thuringiensis* to another. Different δ-endotoxins are active against different insect taxa. Over recent decades, the introduction and expression of *B. thuringiensis* δ-endotoxin genes has been achieved in various crop plant species, and many commercialised examples have been developed and are in use today. The use of *B. thuringiensis* δ-endotoxins in GM crops raises questions such as whether or not the use of crop plants expressing δ-endotoxins could have a negative impact on non-target species.

The Bt consensus document includes a large amount of detail on the nomenclature and classification of the toxins and their genes. It discusses the mechanism of action of δ-endotoxins and their expression in plants, and devotes a specific section to the risk assessment of δ-endotoxins in crop plants. Much of this information relates to the effects of the toxins on non-target organisms, including mammals, fish and other vertebrates as well as invertebrates such as earthworms and non-target insects. This information is complemented by reference to a number of company-sponsored studies submitted to regulatory authorities, mainly the US EPA. In short, this document is a detailed description of information necessary for the risk assessment of genetically modified plants expressing insect-control proteins.

Three additional publications are important in understanding the process of how the consensus documents are developed and published, as well as the information they contain. First, an Introduction to the Biosafety Consensus documents (OECD, 2005b) describes what is meant by regulatory harmonisation in the context of the activities of the Working Group. The document considers the purpose of consensus documents and their intended users, that is, those involved in regulatory risk assessment.

Second, a Guide for the Preparation of Consensus Documents (OECD, 2008b) describes the role played by a lead country or lead countries that already have experience with the crop or trait in question and use this knowledge to prepare the initial draft of a consensus document. It is then circulated to all countries and other stakeholders for their input. This guide also points to potential sources of information for use in consensus documents, as well as providing guidance on style and layout as well as nomenclature. All documents are fully referenced with relevant scientific information.

Third, a document entitled Points to Consider for Consensus Documents was published (OECD, 2006b). This publication is important to an understanding of the types of information included in the consensus documents. For example, the document contains five sections as follows:

- species and taxonomic group
- reproductive biology
- genetics
- hybridisation and introgression
- interactions with other organisms.

It also offers a 'rationale' as to why these information elements are relevant to environmental risk assessment.

The Safety Assessment of Novel Foods and Feeds Derived from GM Plants

A key principle in the assessment of novel foods and feeds is the comparative approach. In the past, this process was referred to as substantial equivalence (OECD, 1993), but the phrase 'the comparative approach' is used more frequently today (Kok and Kuiper, 2003). The use of the comparative approach has been elaborated notably in principles (Codex Alimentarius, 2003a) and guidance (Codex Alimentarius, 2003b) developed by the Codex Alimentarius Commission:

> This approach is based on the principle that the safety of foods derived from new plant varieties, including recombinant deoxyribonucleic acid (DNA) plants, is assessed relative to the conventional counterpart having a history of safe use, taking into account both intended and unintended effects. Rather than trying to identify every hazard associated with a particular food, the intention is to identify new or altered hazards relative to the conventional counterpart. (Codex Alimentarius, 2003b)

The comparative approach involves a comparative analysis in which a novel food or feed is compared with an existing traditional counterpart for which there is a history of safe use. The comparative analysis addresses components of the food such as key nutrients, toxicants, anti-nutrients and allergens. This analysis is an important step in the safety assessment of a novel modified food or feed because it can indicate differences between the novel food and its traditional counterpart. If any differences are identified, then they will be the subject of further analysis in the safety assessment. The comparator used to detect unintended effects for all critical components should ideally be the near-isogenic parental line grown under identical conditions (OECD, 2006b).

The OECD's Task Force has developed consensus documents on specific crops from which food or feed is derived. Such documents compile information on those major components of these crop plants, such as key nutrients, toxicants, anti-nutrients and allergens, which national authorities believe are important to a safety assessment for a specific food or feed.

As with the consensus documents on environmental risk assessment, the major focus in the early years was on major trading commodities such as maize (OECD, 2002), soybean (OECD, 2012c), cotton (OECD, 2004) and low-erucic-acid rapeseed oil (OECD, 2011). More recently, documents on crops such as papaya (OECD, 2010a), sweet potato (OECD, 2010b), grain sorghum (OECD, 2010c) and cassava (OECD, 2009b), which are of particular interest to developing countries, have been published. Once again, this has been influenced by developing countries as they have engaged increasingly in OECD's work.

In 2002, the OECD published a Consensus Document on Compositional Components for New Varieties of Maize (*Zea mays*): Key Food and Feed Nutrients, Anti-Nutrients and Secondary Plant Metabolites (OECD, 2002). The document indicates those parameters which can be used in a comparative analysis for maize, including its key nutrients, anti-nutrients and secondary metabolites. The key nutrients are those components of maize which may have a substantial input in the overall diet. They include major constituents (fats, proteins and carbohydrates) and minor components (vitamins and minerals). The document addresses a number of known anti-nutrients in maize (phytic acid, DIMBOA and raffinose) as well as secondary plant metabolites (furfural, ferulic acid and *p*-coumaric acid). In presenting values for these various components the text takes into account the processing of maize as either a food or feed as well as in specific products derived from maize.

It is important to stress that a comparative analysis is one step in a food/feed safety assessment, but other forms of analysis may be necessary as part of the full safety assessment. If a transgenic crop produces a novel protein, for example, then there is the possibility of that protein being present in food and/or feed. For example, the OECD Bt consensus document (OECD, 2007) discussed above includes a large amount of data relevant to the potential toxicity of δ-endotoxins to humans and mammals. This type of information is relevant to a food/feed safety assessment involving Bt proteins. On some occasions, it may simply be the protein encoded by the transgene that is of interest, but on others it may be downstream metabolic products that can arise due to the activity of the transgene that are of interest.

A final publication, the Introduction to OECD's Food and Feed Safety Consensus Documents (OECD, 2006c), provides more detailed explanation of their content and development. It describes their purpose and their use as a practical contribution to the safety assessment of foods and feeds derived from transgenic organisms. In short, these documents are a means of implementing the comparative approach to foods and feeds derived from specific crop species.

Molecular Characterisation

In 2010, the OECD published a document on the Molecular Characterisation of Plants Derived from Modern Biotechnology (OECD, 2010d), which is a complementary text to the work on traits. This document is relevant to environmental risk assessment and food/feed safety assessment. It addresses the issues linked to molecular characterisation in a risk assessment. Molecular characterisation encompasses consideration of the transformation method employed, the inserted DNA and expressed material, as well as the inheritance and genetic stability of the inserted DNA. It can be important in focusing other components of the risk/safety assessment that assess the phenotype of the plant, such as characterisation of the levels of nutrients, anti-nutrients, endogenous toxicants or allergens, or changes in plant fitness. To date, the most appropriate available scientific procedures and technologies have been used in the molecular characterisation of GM crop plants. Experience derived from the use of these procedures and technologies forms the basis of this document. Given the current pace of technological advancement, it is expected that new methodologies may be applied to the molecular characterisation of GM crop plants should such technologies prove to have value as a mechanism of hazard identification in food, feed and environmental risk/safety assessments.

Organisms Other Than Crop Plants

The most prominent GMOs of commercial importance are crop plants; however, modern biotechnology has also been applied to a number of tree species, animals and microorganisms. This work, too, has led to OECD publications. The subject of one document is poplar trees (OECD, 2000d); a second addresses Eucalyptus species (OECD, 2014b). Transgenic Eucalyptus varieties have been approved for commercial use in the preparation of pulp, including in Brazil (Nature Biotechnology News, 2015). The OECD is currently considering two animal species. A document on Atlantic salmon is under preparation. Transgenic salmon have been developed and approved for food use (Ledford, 2015). Work is also under way on a document on the mosquito, *Aedes aegyptii*. Both these documents will be published during 2017.

Emerging Topics

As part of its work, the Working Group, together with non-members, has always aimed to address emerging issues of relevance to environmental risk assessment. It has, for example, been considering new plant breeding techniques (NPBTs) with a view to identifying whether or not they pose new issues for risk/safety assessment. A recent workshop report (OECD, 2016a) discusses seven techniques considered as NPBTs: (i) zinc-finger nucleases, (ii) oligonucleotide-directed mutagenesis, (iii) cisgenesis and intragenesis, (iv) reverse breeding, (v) RNA-directed DNA methylation, (vi) grafting on GM rootstock and (vii) agro-infiltration. Its main conclusion notes that

> Experience to date indicates that current guidance and tools for Environmental Risk/Safety Assessment of transgenic plants are applicable to plants developed using New Plant Breeding Techniques, where Environmental Risk/Safety Assessment of such plants may be required.

The OECD's Working Group will continue to consider new techniques as the technology develops and will probably consider CRISPR technology as one of the next steps.

Conclusion

This chapter shows how the OECD and its Working Group and Task Force have published a large number of documents compiling information that is used by those involved in the risk assessment of GM crops. This includes participants from developing countries and countries in transition who provide information in the drafting of OECD documents and are fully integrated in the process.

The value of these documents lies in the fact that they are developed by delegates from national authorities (with input from other stakeholders) who have access to a wide range of relevant expertise relevant to risk assessment. They give a major insight into those issues and the information that national authorities believe is important for safety.

These documents are not intended to be the complete answer to a risk assessment, because a regulatory assessment is always part of national decision-making processes, but they contain information used in the process that is similar among national authorities. So the same information does not need to be regenerated or re-examined anew each time a new GM crop is subjected to a risk assessment. For much of the information, it is simply necessary to refer to the relevant consensus document. These documents, therefore, are a means of facilitating the transportability of scientific data and information amongst authorities as well as other users; they save time and effort for those involved.

As an integral part of their work, the Working Group and Task Force consider evidence of how the documents are used at their meetings, and this information is used to set priorities for new documents or the updating of existing documents. This evidence comes from risk assessors who provide information on how the documents are used in their specific national circumstances. At the same time, the developers of GM crops offer information as to how they are used when they are seeking approvals.

In order to ensure wide dissemination and use of these outputs, all finalised documents are made freely available on the OECD website (OECD, 2016b).

Disclaimer and Acknowledgements

The opinions expressed and arguments employed herein are those of the authors and do not necessarily reflect the official views of the OECD or of the governments of its member countries.

The authors are grateful to all delegations to the Working Group and Task Force, all lead countries and their experts for their contributions to the OECD documents cited in this chapter.

References

Anderson, K. and Jackson, L. A. (2003). Why are US and EU policies toward GMOs so different? *AgBioForum* **6**(3), 95–100.

Codex Alimentarius (2003a). Principles for the risk analysis of foods derived from modern biotechnology. CAC/GL 44–2003. Rome: Joint FAO/WHO Food Standards Programme.

Codex Alimentarius (2003b). Guideline for the conduct of food safety assessment of foods derived from recombinant DNA plants. CAC/GL 45–2003. Rome: Joint FAO/WHO Food Standards Programme.

Kok, E. J. and Kuiper, H. A. (2003). Comparative safety assessment for biotech crops. *Trends in Biotechnology* **21**(10), 439–444.

Ledford, H. (2015). Salmon approval heralds rethink of transgenic animals. *Nature* **527**, 417–418.

Nature Biotechnology News (2015). Brazil approves transgenic Eucalyptus. *Nature Biotechnology* **33**, 577.

OECD (1986). *Recombinant DNA Safety Considerations*. Paris: OECD.

OECD (1993). *Safety Evaluation of Foods Derived by Modern Biotechnology*. Paris: OECD.

OECD (2000a). *Report of the Working Group on Harmonisation of Regulatory Oversight in Biotechnology (to the G8 Heads of State and Government)*. [C(2000)86/ADD2]. Paris: OECD.

OECD (2000b). *Report of the Task Force for the Safety of Novel Foods and Feeds (to the G8 Heads of State and Government)*[C(2000)86/ADD1]. Paris: OECD.

OECD (2000c). *Consensus Document on the Biology of Glycine max (L.) Merr. (Soybean)*. Paris: OECD.

OECD (2000d). *Consensus Document on the Biology of Populus L. (Poplars)*. Paris: OECD.

OECD (2002). *Consensus Document on Compositional Components for New Varieties of Maize (*Zea mays*): Key Food and Feed Nutrients, Anti-Nutrients and Secondary Plant Metabolites*. Paris: OECD.

OECD (2003). *Consensus Document on the Biology of* Zea mays *Subspecies* mays *(Maize)*. Paris: OECD.

OECD (2004). *Consensus Document on Compositional Considerations for New Varieties of Cotton (*Gossypium hirsutum *and* Gossypium barbadense*): Key Food and Feed Nutrients and Anti-nutrients*. Paris: OECD.

OECD (2005a). *Consensus Document on the Biology of Papaya (*Carica papaya*)*. Paris: OECD.

OECD (2005b). *Introduction to the Biosafety Consensus Documents*. Paris: OECD.

OECD (2006a). *Consensus Document on the Biology of* Capsicum annuum *Complex (Chili, Hot and Sweet Peppers)*. Paris: OECD.

OECD (2006b). *Points to Consider for Consensus Documents*. Paris: OECD.

OECD (2006c). *Introduction to OECD's Food and Feed Safety Consensus Documents*. Paris: OECD.

OECD (2007). *Consensus Document on Safety Information on Transgenic Plants Expressing* Bacillus thuringiensis *– Derived Insect Control Proteins*. Paris: OECD.

OECD (2008a). *Consensus Document on the Biology of Cotton (Gossypium spp.)* Paris: OECD.

OECD (2008b). *Guide for the Preparation of Consensus Documents*. Paris: OECD.

OECD (2009a). *Consensus Document on the Biology of Bananas and Plantains (Musa spp.)* Paris: OECD.

OECD (2009b). *Consensus Document on Compositional Considerations for New Varieties of Cassava (*Manihot esculenta *Crantz): Key Food and Feed Nutrients, Anti-nutrients, Toxicants and Allergens*. Paris: OECD.

OECD (2010a). *Consensus Document on Compositional Considerations for New Varieties of Papaya (*Carica papaya L.*): Key Food and Feed Nutrients and Anti-nutrients, Toxicants and Allergens*. Paris: OECD.

OECD (2010b). *Consensus Document on Compositional Considerations for New Varieties of Sweet Potato [*Ipomoea batatas (L.) *Lam.]: Key Food and Feed Nutrients, Anti-nutrients, Toxicants and Allergens*. Paris: OECD.

OECD (2010c). *Consensus Document on Compositional Considerations for New Varieties of Grain Sorghum [*Sorghum bicolor (L.) *Moench]: Key Food and Feed Nutrients and Anti-nutrients*. Paris: OECD.

OECD (2010d). *Molecular Characterisation of Plants Derived from Modern Biotechnology*. Paris: OECD.

OECD (2011). *Revised Consensus Document on Compositional Considerations for New Varieties of Low Erucic Acid Rapeseed (Canola): Key Food and Feed Nutrients, Anti-nutrients and Toxicants*. Paris: OECD.

OECD (2012a). *Consensus Document on the Biology of the Brassica Crops (Brassica spp.)*. Paris: OECD.

OECD (2012b). *Consensus Document on the Biology of Cucurbita L. (Squashes, Pumpkins, Zucchinis and Gourds)*. Paris: OECD.

OECD (2012c). *Revised Consensus Document on Compositional Considerations for New Varieties of Soybean [*Glycine max (L.) *Merr]: Key Food and Feed Nutrients, Anti-nutrients, Toxicants and Allergens*. Paris: OECD.

OECD (2013). *Consensus Document on the Biology of Sugarcane (Saccharum spp.)*. Paris: OECD.

OECD (2014a). *Consensus Document on the Biology of Cassava (*Manihot esculenta *Crantz)*. Paris: OECD.

OECD (2014b). *Consensus Document on the Biology of Eucalyptus spp*. Paris: OECD.

OECD (2015). *Consensus Document on the Biology of Cowpea (*Vigna unguiculata (L.) *Walp.)* Paris: OECD.

OECD (2016a). *Report of the OECD Workshop on Environmental Risk Assessment of Products Derived from New Plant Breeding Techniques*. Paris: OECD.

OECD (2016b). Documents on Harmonisation of Regulatory Oversight in Biotechnology and the Safety of Novel Foods and Feeds. [Online]. Available from http://www.oecd.org/env/ehs/biotrack/documentsonharmonisationofregulatoryoversightinbiotechnologyandthesafetyofnovelfoodsandfeeds.htm

12 Should GM Rice with Nutrition Benefits Be Deployed? Findings from Biotech and Socio-Economic Research

Hans De Steur, Dieter Blancquaert, Christophe Stove, Willy Lambert, Dominique Van Der Straeten and Xavier Gellynck

Introduction

As the public and scientific debate on genetically modified (GM) crops is almost solely focusing on already commercialised GM crops, in which GM technology is used to improve agronomic traits and productivity (i.e., first-generation GM crops), the majority of consumer and marketing research on biotech crops is targeted towards these farmer-orientated applications. Nevertheless, a second generation of GM crops is in the pipeline. These crops have been modified for the improvement of quality (e.g., micronutrient content, sensory attributes or shelf-life) rather than quantity traits (e.g., disease resistance, herbicide tolerance). Out of these consumer-oriented GM crops, transgenic 'biofortified' crops – crops that are engineered to contain a (higher) level of vitamins or minerals – are among the most examined second-generation GM crops, both at biotech level and at socio-economic research level (De Steur *et al.*, 2015). Given the high micronutrient concentrations obtained through transgenic breeding approaches, this type of biofortified crop can be considered a valuable complementary strategy when conventional biofortification or other micronutrient strategies, such as supplementation or fortification strategies, fall short (Ruel and Alderman, 2013).

But despite the rapid and global adoption of commercialised GM crops, with about 2 billion hectares of biotech crops having been cultivated between 1996 and 2015 (James, 2015), transgenic biofortified crops (Moghissi *et al.*, 2016) as well as other novel GM crops face difficulties with regard to obtaining approval (Jaffe, 2006). In this context, market research could play an important role. By providing insights into consumers' demand for GM crops with nutrition benefits, it aims to better inform decision-makers and health intervention planners evaluating and allocating resources to different micronutrient strategies.

Building upon the successful efforts in the field of transgenic biofortification of rice, i.e., one of the world's most important staple crops, this chapter summarises the evidence on consumer research and societal impact studies. The focus will be on two advanced cases: vitamin A-enriched 'Golden Rice' and Folate Biofortified Rice. In addition, we also reflect upon the progress on iron, vitamin E and multi-biofortified rice.

Golden Rice

Research, Development and the Road to Commercialisation

Golden Rice, which is rich in pro-vitamin A (β-carotene), is one of the most advanced examples of transgenic biofortification. Before their partnering with Syngenta, Golden Rice research was initiated by the teams of Ingo Potrykus (ETH Zürich) and Peter Beyer (University of Freiburg). The first-generation Golden Rice (GR1) was developed by introducing transgenes from daffodil and the *Pantoea* bacterium, formerly known as *Erwinia* (Ye *et al.*, 2000). As the former gene was limited for β-carotene accumulation, it was replaced by the phytoene synthesis '*psy*' gene from maize in the second-generation Golden Rice (GR2) (Paine *et al.*, 2005). As a consequence, the total increase in carotenoid shifted from 1.6 μg/g in GR1 to 37 μg/g in GR2, a 23-fold increase (The Golden Rice Project, 2015). Furthermore, feeding trials have demonstrated an effective conversion of β-carotene into vitamin A in adults (Tang *et al.*, 2009).

Since vitamin A is an essential nutrient for skin, bone and body growth, vision and the immune system, vitamin A deficiency is associated with various adverse health outcomes, such as measles, (night) blindness and corneal scarring. Worldwide, about 250 million pre-school children are estimated to be vitamin A-deficient, leading to an annual number of blind children between 250,000 and 500,000, especially from sub-Saharan Africa and Southern Asia (WHO, 2015).

In order to further tackle this major public health problem, the Golden Rice pro-vitamin A trait was and will be transferred – by using conventional and marker-assisted backcrossing – into high-yielding local commercial 'Indica' varieties popular in India, Bangladesh, the Philippines and Vietnam (Bertebos Foundation, 2008). Bangladesh has recently approved field trials, but the Philippines may still be the first in line to adopt Golden Rice. Once it becomes available in the marketplace, a humanitarian licence for adoption in developing countries, known as a Humanitarian Use Technology Transfer (HUTT), will be introduced. This will enable a free transfer of rights to public research institutions – currently established for the Philippines, India, Vietnam, Indonesia and China – and a royalty-free use of Golden Rice for resource-poor farmers in priority markets (Dawe and Unnevehr, 2007).

Health Impacts

Several studies have analysed the potential health impacts of Golden Rice adoption on a 30-year time-scale in three Asian priority countries: the Philippines (Zimmermann and Qaim, 2004), India (Stein *et al.*, 2006, 2008) and China (De Steur *et al.*, 2012b) (Table 12.1). The paper by Zimmermann and Qaim (2004) was the first health impact study on transgenic biofortified rice. By using the Philippines as a case study, they provide evidence of the potential of Golden Rice to alleviate the burden of adverse health outcomes associated with vitamin A deficiency. Even though the burden is lower than in the other studies, both the health benefits (6%–32% reduction of the burden of

Table 12.1 Overview of health impact and cost-effectiveness/benefits studies on Golden Rice

Health impacts and cost-effectiveness		Philippines (Zimmermann and Qaim, 2004)		India (Stein *et al.*, 2006)		China (De Steur *et al.* 2012b)	
Vitamin A-deficiency burden	DALYs lost, per year (millions)	0.27		2.33		1.99	
Scenario[a]		**Low**	**High**	**Low**	**High**	**Low**	**High**
Health impacts	DALYs saved, per year (millions)	0.02	0.08	0.20	1.38	0.33	1.20
Cost-effectiveness analysis[b]	US$/DALY saved, per year			19.4	3.1	18.1	5.0
Cost–benefit analysis[c]	IRR (%)	66	133	29; 35; 44	66; 77; 93		
	BCR			26; 52; 129	163; 327; 816		

Note: detailed figures are presented in De Steur *et al.* (2015). BCR, benefit–cost ratio; DALY, Disability-Adjusted Life-Year; IRR, internal rate of return.

a The low and high scenarios considered were based on pessimistic and optimistic efficacy and coverage estimates, respectively.

b Calculations based on a time horizon of 30 years.

c Figures based on a DALY valued at standard values for a DALY (US$500, US$1,000) or at the national per-capita income (US$2,500), in sequence from left to right.

vitamin A deficiency) and the cost-effectiveness/benefits figures lend support for Golden Rice as a valuable complementary health strategy.

Through a German–Indian collaboration, *Stein et al.* (2006, 2008) – who refined the Disability-Adjusted Life-Years (DALY) approach to biofortification as a health intervention in India (Stein *et al.*, 2005), reported a 9% reduction of the Indian Vitamin A deficiency burden (VAD) in a low-impact scenario, up to a reduction of no less than 59% in a high-impact scenario. Taking into account the implementation costs (US$21–28 million), introducing Golden Rice in India is considered highly cost-effective (US$3–19 per DALY saved). In an additional analysis, other cost–benefit indices, i.e., the internal rate of return (IRR) and benefit–cost ratio (BCR), underpin the conclusion that the large potential benefits justify the resources allocated.

In their single- and multi-biofortification analysis on China, the team of De Steur *et al.* (2012b) derived impact and cost-effectiveness figures for Golden Rice, which are in line with those from the aforementioned analyses.

The Consumer's View

Between 2001 and 2004, four studies examined consumer reactions towards Golden Rice in the developed world, of which three studies were carried out in the USA

Table 12.2 Overview of acceptance and willingness-to-pay studies on Golden Rice

Acceptance and willingness to pay		USA (Lusk, 2003; Rozan *et al.*, 2007; Hallman *et al.*, 2002)	China (Curtis and Moeltner, 2006; Li *et al.*, 2003)	France (Rozan *et al.*, 2007)	India (Deodhar *et al.*, 2008; Kajale and Becker, 2015)	Philippines (Corrigan *et al.*, 2009; Depositario *et al.*, 2009)
Acceptance	(% of the sample)	80.7		50.0	46.1[a]	
Willingness-to-pay	(% premium)	19.5; 38.0	38.0		4.0[a]; 19.5	33.3; 40.0; 66.7[a]
Intention to purchase	(% of the sample)		43.9	25.5		
Preference	(% of the sample)				65.9	

Note: detailed figures are presented in De Steur *et al.* (2015). This overview is an update.
[a] For a student sample.

(Lusk, 2003; Rozan *et al.*, 2007; Hallman *et al.*, 2002) and one in France (Rozan *et al.*, 2007) (Table 12.2). As expected, the acceptance study of Hallman *et al.* (2002) found that most US consumers are in favour of Golden Rice (81% of 1,203 consumers), while the premium levels derived from the valuation studies of Lusk (2003) and Lusk and Rozan (2005) vary between 20% ($n = 574$) and 38% ($n = 501$). According to a mail survey of 200 French consumers, about half the sample subjects accept Golden Rice, of whom 26% actually intend to purchase it (Rozan *et al.*, 2007).

In line with the aforementioned health impact studies, consumer studies in developing regions have been applied to the Asian context. In China, about 600 consumers were interviewed in 2002, of whom about 44% were willing to pay more for Golden Rice than for conventional rice, at an average premium of 38% (Curtis and Moeltner, 2006; Li *et al.*, 2003). A similar figure was observed in the choice experiments and experimental auctions in the Philippines, where premium levels in 2006 varied between 33% and 67% (Corrigan *et al.*, 2009; Depositario *et al.*, 2009). In India, a population-based survey ($n = 2937$) (Deodhar *et al.*, 2008) and a student survey ($n = 154$) (Kajale and Becker, 2015) were conducted, in 2006 and 2009, respectively. An average premium of 20% and a large preference over conventional rice (66%) were found in the former study, while the latter demonstrated that students were largely accepting of Golden Rice (46% of the sample), with a mean willingness-to-pay of 4%. These positive findings were mainly attributable to the (higher) micronutrient content, often reinforced by positive information on nutrition benefits and/or GM technology (unpublished work). These results differ from the discounts that are generally reported in consumer research on GM crops with improved agronomic traits.

While market research on GM biofortified crops is generally targeted towards the individual consumer, studies on farmers are clearly lacking. An exception is the exploratory study by Chong (2003), in which acceptance of Golden Rice by Philippine rice farmers was not an issue as long it could be demonstrated that Golden Rice brought agronomic benefits or cost savings in comparison with what was currently being grown, or was at least equal to the latter.

Folate Biofortified Rice

Research and Development

Folate (vitamin B_9)-enhanced rice was developed at Ghent University through the overexpression of Arabidopsis transgenes in rice endosperm (Storozhenko *et al.*, 2007). Although rice contains folate, unlike pro-vitamin A, folate levels are negligible and natural variation of folate in rice kernels is low, making conventional breeding not sufficient as a means to reach adequate nutrient intake levels (Blancquaert *et al.*, 2010). The first generation of Folate Biofortified Rice (FBR1) is characterised by folate levels that are 100-fold above that of wild-type rice (Storozhenko *et al.*, 2007). In a second generation (FBR2), the folate concentration in rice is close to 150 times the initial level, while long-term storage was ensured through enhanced folate stability (Blancquaert *et al.*, 2015).

Although no efficacy trials have been conducted, impact studies (see further on in this chapter) have indicated its potential to adequately reduce folate-related adverse health outcomes in poor priority regions. Maternal folate deficiency, for example, is known to increase the risk of various undesirable pregnancy outcomes and congenital malformations in particular, of which the link with neural-tube defects such as spina bifida is most established (Lassi *et al.*, 2013). When mapping the prevalence of neural-tube defects as the most common indicator of folate deficiency, Southeast Asia can be considered the most problematic region, with Cambodia, Bangladesh, Myanmar, Thailand and China being the most affected countries (Blancquaert *et al.*, 2014). Worldwide, Shanxi Province (northern China) and Uttar Pradesh (northwestern India) are highly relevant, given that they have some of the highest prevalent rates of folate deficiency and neural-tube defects (Cherian *et al.*, 2005; Gu *et al.*, 2007). Besides the impact of folate on birth defects, there is growing evidence of its relationship with cardiovascular disease, stroke, anaemia, depression, Alzheimer's disease, schizophrenia and cancer (Iyer and Tomar, 2009), which emphasises the need to further invest in folate strategies.

Health Impacts

The assessment of the potential benefits of introducing Folate Biofortified Rice as a health intervention was conducted using China as a case study (Table 12.3). De Steur and his colleagues conducted three different analyses: a health impact study on China and its 31 provinces (De Steur *et al.*, 2010a) and a cost-effectiveness analysis on China, on the one hand (De Steur *et al.*, 2012b); and a study in its high-risk region of folate deficiency, Shanxi Province, on the other (De Steur *et al.*, 2012a). The estimated potential

Table 12.3 Overview of health impact and cost-effectiveness studies on Folate Biofortified Rice

Health impacts and cost-effectiveness		China (De Steur *et al.*, 2012b; De Steur *et al.*, 2010a)		China (Shanxi Province)[a] (De Steur *et al.*, 2012a)	
Folate-deficiency burden	DALYs lost, per year (million)	0.31		0.11	
Scenario[b]		Low	High	Low	High
Health impacts	DALYs saved, per year (million)	0.12; 0.06	0.26; 0.19	0.02	0.07
	% burden reduction, per year	20.0; 36.9	60.0; 81.9	20.0	60.0
Cost-effectiveness analysis[c]	US$/DALY saved, per year	64.2	21.4	120.3	40.1

Note: detailed figures are presented in De Steur *et al.* (2015). DALY, Disability-Adjusted Life-Year.

[a] One of the world's most important risk regions for folate deficiency.

[b] The low and high scenarios considered were based on pessimistic and optimistic efficacy and coverage estimates, respectively.

[c] Calculations based on a time horizon of 30 years.

decrease in incidence of adverse health outcomes related to folate deficiency following its implementation varies between 20% and 60%, up to a level of 82% in one of the high-impact scenarios. Owing to the relatively small difference between the costs of regional and country-wide implementation (US$20 million and US$32 million, respectively), the cost-effectiveness of Folate Biofortified Rice at national level is about twice as high as at regional level. Nevertheless, according to updated World Bank (Stein *et al.*, 2005) and WHO standards (Marseille *et al.*, 2015), for both strategies micronutrient interventions can be considered highly cost-effective.

The Consumer's View

To measure consumers' acceptance of, and willingness to pay for, Folate Biofortified Rice, De Steur *et al.* (2013, 2014, 2012c, 2010b) conducted several studies in Shanxi Province, in northern China (Table 12.4). In most of these studies the focus was on women of childbearing age as one of the key beneficiaries of increased folate intake levels. The first consumer study on Folate Biofortified Rice, a population-based survey in 2008 ($n = 944$), found that a large proportion of Shanxi rice consumers (62%) had a positive view of this GM crop with health benefits (De Steur *et al.*, 2010b). In 2011, experimental auctions were conducted with the participation of 252 women of childbearing age to elicit their willingness to pay for Folate Biofortified Rice (De Steur *et al.*, 2012c, 2013, 2014). The premium levels obtained, which ranged from 26% (for students) to 34% (for adults), are similar to those for Golden Rice in Asia (see Table 12.2). Purchase intention levels were even higher than had been found in the 2008 study, which

Table 12.4 Overview of acceptance and willingness-to-pay studies on Folate Biofortified Rice

Acceptance and willingness to pay		China (Shanxi Province)[a] (De Steur *et al.*, 2010b, 2012c, 2013, 2014)
Acceptance	(% of the sample)	62.2
Willingness to pay	(% premium)	25.8[b,c]; 33.7[b], 34.0
Intention to purchase	(% of the sample)	77.5[b,c]; 82.5[b], 84.9[b]
Preference	(% of the sample)	66.7[b]

Note: detailed figures are presented in De Steur *et al.* (2015).
[a] One of the world's most important risk regions of folate deficiency.
[b] For a sample of women of childbearing age.
[c] For a sample of female students.

is partially related to the selection of a sample at high risk for the 2011 study. Finally, when consumers where offered the option of a non-GM alternative, i.e., folic acid supplements, GM rice enriched with folate was in general the preferred option (De Steur *et al.*, 2014).

Iron, Vitamin E and Multi-Biofortified Rice

Research and Development

While Golden Rice and Folate Biofortified Rice are single-biofortified crops developed through metabolic engineering, several researchers have targeted other micronutrients in rice. For an overview, see De Steur *et al.* (2015).

Zhang *et al.* (2013) overexpressed Arabidopsis γ-tocopherol methyltransferase in rice seeds to elevate the vitamin E content to 33 μg/g dry weight (a 3-fold increase). With respect to iron, in various studies the iron content of rice was successfully enhanced by a factor of 1.5 (Qu *et al.*, 2005), 2 (Nandi *et al.*, 2002; Lucca *et al.*, 2002; Vasconcelos *et al.*, 2003), 3 (Goto *et al.*, 1999; Johnson *et al.*, 2011; Lee *et al.*, 2009, 2012) or 6 (Wirth *et al.*, 2009; Masuda *et al.*, 2012). The content of iron in rice after biofortification varies from 7 to 64 μg/g dry weight. For zinc, initial levels were successfully enhanced by a factor of 1.6–2.2 (Vasconcelos *et al.*, 2003; Lee *et al.*, 2009; Johnson *et al.*, 2011; Masuda *et al.*, 2012). The zinc concentration in biofortified rice ranges from 20 to 91 μg/g dry weight. A similar increase (by a factor of 1.7) was obtained for copper, resulting in a 2.2 μg/g dry weight concentration (Lee *et al.*, 2009).

In addition to applications of single-biofortified rice, researchers were also successful in enhancing concentrations of several minerals and/or vitamins in the same plant, which is known as transgenic multi-biofortification (De Steur *et al.*, 2015). Regarding rice, only minerals have been stacked, namely through a combination of iron and zinc with (Lee *et al.*, 2009) or without (Vasconcelos *et al.*, 2003; Johnson *et al.*, 2011; Masuda *et al.*, 2012; Trijatmiko *et al.*, 2016) copper.

Health Impacts

An *ex-ante* evaluation study on multi-biofortified rice, which was conceptualised as rice enriched with pro-vitamin A, folate (transgenic traits), iron and zinc (conventional traits), estimated its cost-effectiveness in China (De Steur *et al.*, 2012b). When implemented on a national scale, stacking health traits would generate large health benefits, saving 1.2–4.9 million DALYs per year, which reduces the current burden of deficiencies of the four targeted micronutrients by 11%–46%. Given the relatively low costs of its introduction (i.e., US$85.4 million), it has been shown that multi-biofortified rice would be a highly cost-effective health intervention were it to be introduced in China: US$2.3–9.6 to save a DALY which is currently lost.

Discussion

This chapter describes the pipeline of transgenic biofortified rice research and provides an overview of *ex-ante* studies on consumer reactions and societal impacts. While the results are promising, both in terms of research and development and with regard to potential consumer demand and cost-effectiveness, the first transgenic biofortified (rice) crop has yet to be approved. Although Golden Rice is expected to be the very first application available for consumers, it has been stuck in the regulatory process for more than a decade (Potrykus, 2013), despite its relatively low cost (per DALY saved). In a recent study, Wesseler and Zilberman (2014) estimated the monetary consequences of its delay at about US$199 million per year.

In order to evaluate the rationale behind these ever-repeating delays, socio-economic research should shift its scope towards a stakeholder approach, i.e., analysing views from farm to fork to identify the bottlenecks and success factors on the road to sustainable implementation. Furthermore, even though biofortified crops should be marketed in the developing world, Western society should be involved because Western reluctance might affect policy decisions all over the world (Paarlberg, 2002). In addition, policymakers should also investigate whether and how socio-economic research could further support decision-making – whenever biosafety and other regulatory requirements are fulfilled. Especially if transgenic biofortified crops are at stake, it is deemed crucial that the discussions should deal with the specific characteristics of transgenic biofortification, focusing on a case-by-case evaluation, rather than a generalisation.

Aside from the positive findings at socio-economic research level, many post-approval challenges still have to be addressed (see Box 12.1 for a brief overview of potential strategies to anticipate post-approval challenges) (for detailed information on the Golden Rice case, see Bertebos Foundation, 2008; Beyer, 2010). Nevertheless, once Golden Rice or another GM rice crop with health benefits has been commercialised, it is expected to pave the way for future transgenic single- or multi-biofortified crops and, given their potential at consumer and societal level, bring about large economic effects by increasing wellbeing and welfare.

Box 12.1. Potential Strategies for a Successful Introduction of GM Biofortified Rice

- Careful selection of the *transgenic line* and insertion of the micronutrient trait into *local staple crop varieties* in order to reach key target populations.
- Development of a well-adapted and integrated *market strategy* to introduce transgenic biofortified crops as an agriculture-based nutrition intervention, by
  - ensuring accessibility and availability;
  - adequately allocating resources;
  - installing an efficient monitoring and supervising system;
  - defining efficient distribution/supply chain and effective social marketing strategies;
  - involving key stakeholders, e.g., relevant governmental agencies, farmers, millers, vendors, seed companies and other supply chain members, as well as scientists from multiple disciplines; and
  - interacting with NGOs, aid agencies and the media.
- Provision of an (economic) *incentive to farmers* for producing transgenic biofortified crops.
  - A price premium could convince farmers, but contradicts the rationale behind biofortification as a micronutrient intervention, particularly in the poorest regions.
  - An alternative solution is a humanitarian licence or the provision of governmentally subsidised credit to purchase seeds, as for 'future' Golden Rice in Asia (Dawe and Unnevehr, 2007) and Bt cotton in South Africa (Twardowski and Małyska, 2015).
  - The micronutrient trait should be crossed into high-yielding varieties, as is the intention with Golden Rice in the Philippines. Especially the improvement of both nutrient and agronomic traits will increase the likelihood of adoption by farmers (Chong, 2003), since such stacked crops deliver benefits both for consumers and, directly and indirectly, for farmers.
  - As farmers often find it difficult to discover attractive varieties, one could initiate contracts with health-orientated retailers or use public distribution systems for subsidising their production and consumption in poor regions.
  - When biofortified seeds are made available, they need to be incorporated into existing marketing and distribution systems.

References

Bertebos Foundation (2008). Golden Rice and other Biofortified Food Crops for Developing Countries. Challenges and Potential. Report from the Bertebos Conference in Falkenberg, Sweden, 7–9 September 2008. *Kungl. Skogs- och Lantbruksakademiens Tidskrift* **7**, 1–116.

Beyer, P. (2010). Golden Rice and 'Golden'crops for human nutrition. *New Biotechnology* **27**(5), 478–481.

Blancquaert, D. *et al.* (2010). Folates and folic acid: from fundamental research towards sustainable health. *Critical Reviews in Plant Sciences* **29**(1), 14–35.

Blancquaert, D. *et al.* (2014). Present and future of folate biofortification of crop plants. *Journal of Experimental Botany* **65**(4), 895–906.

Blancquaert, D. *et al.* (2015). Improving folate (vitamin B9) stability in biofortified rice through metabolic engineering. *Nature Biotechnology* **33**(10), 1076–1078.

Cherian, A. *et al.* (2005). Incidence of neural tube defects in the least-developed area of India: a population-based study. *Lancet* **366**(9489), 930–931.

Chong, M. (2003). Acceptance of golden rice in the Philippine 'rice bowl'. *Nature Biotechnology* **21**(9), 971–972.

Corrigan, J. R. *et al.* (2009). Comparing open-ended choice experiments and experimental auctions: an application to Golden Rice. *American Journal of Agricultural Economics* **91**(3), 837–853.

Curtis, K. R. and Moeltner, K. (2006). Genetically modified food market participation and consumer risk perceptions: a cross-country comparison. *Canadian Journal of Agricultural Economics* **54**(2), 289–310.

Dawe, D. and Unnevehr, L. (2007). Crop case study: GMO Golden Rice in Asia with enhanced vitamin A benefits for consumers. *AgBioForum* **10**(3), 154–160.

De Steur, H. *et al.* (2010a). Health impact in China of folate-biofortified rice. *Nature Biotechnology* **28**(6), 554–556.

De Steur, H. *et al.* (2010b). Willingness-to-accept and purchase genetically modified rice with high folate content in Shanxi Province, China. *Appetite* **54**(1), 118–125.

De Steur, H. *et al.* (2012a). *Ex-ante* evaluation of biotechnology innovations: the case of folate biofortified rice in China. *Current Pharmaceutical Biotechnology* **13**(15), 2751–2760.

De Steur, H. *et al.* (2012b). Potential impact and cost-effectiveness of multi-biofortified rice in China. *New Biotechnology* **29**(3), 432–442.

De Steur, H. *et al.* (2012c). Determinants of willingness-to-pay for GM rice with health benefits in a high-risk region: evidence from experimental auctions for folate biofortified rice in China. *Food Quality and Preference* **25**(2), 87–94.

De Steur, H. *et al.* (2013). Role of information on consumers' willingness-to-pay for GM rice with health benefits. An application to China. *Asian Economic Journal* **27**(4), 391–408.

De Steur, H. *et al.* (2014). Consumer preferences for micronutrient strategies in China. A comparison between folic acid supplementation and folate biofortification. *Public Health Nutrition* **17**(6), 1410–1420.

De Steur, H. *et al.* (2015). Status and market potential of transgenic biofortified crops. *Nature Biotechnology* **33**(1), 25–29.

Deodhar, S. Y. *et al.* (2008). Emerging markets for GM foods: an Indian perspective on consumer understanding and the willingness to pay. *International Journal of Biotechnology* **10**(6), 570–587.

Depositario, D. P. T. *et al.* (2009). Effects of information on consumers' willingness to pay for Golden Rice. *Asian Economic Journal* **23**(4), 457–476.

Goto, F. *et al.* (1999). Iron fortification of rice seed by the soybean ferritin gene. *Nature Biotechnology* **17**(3), 282–286.

Gu, X. *et al.* (2007). High prevalence of NTDs in Shanxi Province: a combined epidemiological approach. *Birth Defects Research Part A: Clinical and Molecular Teratology* **79**(10), 702–707.

Hallman, W. K. *et al.* (2002). Public Perceptions of Genetically Modified Foods: Americans Know Not What They Eat. [Online]. Available at https://works.bepress.com/johnlang/8/

Iyer, R. and Tomar, S. K. (2009). Folate: a functional food constituent. *Journal of Food Science* **74**(9), R114–R122.

Jaffe, G. (2006). Regulatory slowdown on GM crop decisions. *Nature Biotechnology* **24**, 748–749.

James, C. (2015). Global Status of Commercialized Biotech/GM Crops: 2014. ISAAA Brief No. 51. Ithaca, NY: ISAAA.

Johnson, A. A. *et al.* (2011). Constitutive overexpression of the OsNAS gene family reveals single-gene strategies for effective iron- and zinc-biofortification of rice endosperm. *PLoS ONE* **6**(9), e24476.

Kajale, D. B. and Becker, T. C. (2015). Willingness to pay for Golden Rice in India: a contingent valuation method analysis. *Journal of Food Products Marketing* **21**(4), 1–18.

Lassi, Z. S. *et al.* (2013). Folic acid supplementation during pregnancy for maternal health and pregnancy outcomes. Cochrane Database of Systematic Reviews. [Online]. Available from http://onlinelibrary.wiley.com/doi/10.1002/14651858.CD006896.pub2/pdf

Lee, S. *et al.* (2009). Iron fortification of rice seeds through activation of the nicotianamine synthase gene. *Proceedings of the National Academy of Sciences of the USA* **106**(51), 22014–22019.

Lee, S. *et al.* (2012). Activation of rice nicotianamine synthase 2 (OsNAS2) enhances iron availability for biofortification. *Molecules and Cells* **33**(3), 269–275.

Li, Q. *et al.* (2003). Consumer attitudes toward genetically modified foods in Beijing, China. *AgBioForum* **5**(4), 145–152.

Lucca, P. *et al.* (2002). Fighting iron deficiency anemia with iron-rich rice. *Journal of the American College of Nutrition* **21**(suppl. 3), 184S–190S.

Lusk, J. L. (2003). Effects of cheap talk on consumer willingness-to-pay for Golden Rice. *American Journal of Agricultural Economics* **85**(4), 840–856.

Lusk, J. L. and Rozan, A. (2005). Consumer acceptance of biotechnology and the role of second generation technologies in the USA and Europe. *Trends in Biotechnology* **23**(8), 386–387.

Marseille, E. *et al.* (2015). Thresholds for the cost-effectiveness of interventions: alternative approaches. *Bulletin of the World Health Organization* **93**(2), 118–124.

Masuda, H. *et al.* (2012). Iron biofortification in rice by the introduction of multiple genes involved in iron nutrition. *Nature Scientific Reports* **2**, 543. [Online]. Available from www.nature.com/articles/srep00543

Moghissi, A. A. *et al.* (2016). Golden Rice: scientific, regulatory and public information processes of a genetically modified organism. *Critical Reviews in Biotechnology* **36**(3), 535–541.

Nandi, S. *et al.* (2002). Expression of human lactoferrin in transgenic rice grains for the application in infant formula. *Plant Science* **163**(4), 713–722.

Paarlberg, R. (2002). The real threat to GM crops in poor countries: consumer and policy resistance to GM foods in rich countries. *Food Policy* **27**, 247–250.

Paine, J. A. *et al.* (2005). Improving the nutritional value of Golden Rice through increased pro-vitamin A content. *Nature Biotechnology* **23**(4), 482–487.

Potrykus, I. (2013). Unjustified regulation prevents use of GMO technology for public good. *Trends in Biotechnology* **31**(3), 131–133.

Qu, L. Q. *et al.* (2005). Iron accumulation does not parallel the high expression level of ferritin in transgenic rice seeds. *Planta* **222**(2), 225–233.

Rozan, A. *et al.* (2007). Consumer acceptance of a genetically modified organism of the second generation: the Golden Rice. *Revue d'Économie Politique* **117**(5), 843–852.

Ruel, M. T. and Alderman, H. (2013). Nutrition-sensitive interventions and programmes: how can they help to accelerate progress in improving maternal and child nutrition? *Lancet* **382**(9891), 536–551.

Stein, A. J. *et al.* (2005). *Analyzing the Health Benefits of Biofortified Staple Crops by Means of the Disability-Adjusted Life Years Approach: A Handbook Focusing on Iron, Zinc and Vitamin A*. HarvestPlus Technical Monograph 4. Washington, D.C.: International Food Policy Research Institute, p. 32.

Stein, A. J. *et al.* (2006). Potential impact and cost-effectiveness of Golden Rice. *Nature Biotechnology* **24**(10), 1200–1201.

Stein, A. J. *et al.* (2008). Genetic engineering for the poor: Golden Rice and public health in India. *World Development* **36**(1), 144–158.

Storozhenko, S. *et al.* (2007). Folate fortification of rice by metabolic engineering. *Nature Biotechnology* **25**(11), 1277–1279.

Tang, G. *et al.* (2009). Golden Rice is an effective source of vitamin A. *American Journal of Clinical Nutrition* **89**(6), 1776–1783.

The Golden Rice Project (2015). [Online]. Available from www.goldenrice.org

Trijatmiko, K. R. *et al.* (2016). Biofortified indica rice attains iron and zinc nutrition dietary targets in the field. *Nature Scientific Reports*, **6**, 19792. [Online]. Available from www.nature.com/articles/srep19792

Twardowski, T. and Małyska, A. (2015). Uninformed and disinformed society and the GMO market. *Trends in Biotechnology* **33**(1), 1–3.

Vasconcelos, M. *et al.* (2003). Enhanced iron and zinc accumulation in transgenic rice with the ferritin gene. *Plant Science* **164**(3), 371–378.

Wesseler, J. and Zilberman, D. (2014). The economic power of the Golden Rice opposition. *Environment and Development Economics* **19**(6), 724–742.

WHO (2015). *WHO Global Database on Vitamin A Deficiency*. Geneva: World Health Organization.

Wirth, J. *et al.* (2009). Rice endosperm iron biofortification by targeted and synergistic action of nicotianamine synthase and ferritin. *Plant Biotechnology Journal* **7**(7), 631–644.

Ye, X. *et al.* (2000). Engineering the provitamin A (β-carotene) biosynthetic pathway into (carotenoid-free) rice endosperm. *Science* **287**(5451), 303–305.

Zhang, G.-Y. *et al.* (2013). Increased α-tocotrienol content in seeds of transgenic rice overexpressing Arabidopsis γ-tocopherol methyltransferase. *Transgenic Research* **22**(1), 89–99.

Zimmermann, R. and M. Qaim (2004). Potential health benefits of Golden Rice: a Philippine case study. *Food Policy* **29**(2), 147–168.

Part III

Risk-Analysis-Based Regulatory Systems

13 The Risk Analysis Framework and Biosafety Policy

Peter W. B. Phillips and Stuart J. Smyth

Introduction

Sociologist Ulrich Beck (1992) suggests the hazards and insecurities introduced by modernisation and technology have created a form of 'risk society'. Risk analysis is now pervasive, with all manner of products and services analysed through some form of risk analysis. Almost all new technologies we use to produce the goods and services we consume are evaluated for their effect on the health of the environment or people. This chapter explores the challenging evolution of risk analysis following its formal establishment in 1983 and examines its role in biosafety policy.

The History of Risk Analysis and Biosafety

The effort to manage environment risk both generally and in agriculture has evolved in fits and starts. The Brundtland Commission (World Commission on Environment and Development, 1987) first introduced sustainable development as a core global policy goal, but we can trace the conceptual and procedural approaches to our system of nested national and international regulatory efforts to manage biosafety back to earlier periods.

Bernstein (1996) asserts that risk has only recently replaced our belief in fate. It was only in the past generation that a formal, structured risk analysis framework emerged. Some date the impetus for the modern application of risk analysis to Rachel Carson's book *Silent Spring* in 1962 that documented the detrimental effects of artificial pesticides on the environment, particularly on birds. As evidence accumulated, government agencies in the USA and elsewhere responded. The US Environmental Protection Agency (EPA), for example, was created in 1970 with an explicit mandate to regulate agricultural and industrial chemicals and by-product pollutants to 'protect human health and to safeguard the natural environment – air, water, and land – upon which life depends' (EPA, 2009). Although risk analysis was not yet a formally recognised process, the EPA completed its first risk assessment document in December 1975, issued its first directive on procedures and guidelines in 1976, and produced water quality criteria documents for 64 contaminants in 1980, the first EPA documents describing quantitative procedures used in risk assessment.

In 1983, the US National Academy of Sciences (NAS) published *Risk Assessment in the Federal Government: Managing the Process* (NRC, 1983), also known as the

'Red Book'. This laid out, for the first time, the context for a systemic, comprehensive process for risk analysis. The Red Book defined risk assessment as largely a science-driven process that quantitatively evaluates the probability of risk. Risk assessment thus is largely, but not entirely, removed from emotional factors that influence risk perception. Risk assessment flows in a logical, stepwise fashion from problem formulation through characterisation of the hazard and exposure analysis, culminating in a risk conclusion. The characterisation process is iterative and recursive; if any step yields unexplored concerns, the process loops back to develop or incorporate new science to refine the risk assessment. This process is carried out in stages, with different levels of potential hazards assessed proportionately.

In practical terms, the risk analysis framework is commonly divided into three distinct phases – risk assessment, risk management and risk communication – with different actors tasked with different aspects of the analysis. Obviously, these specific domains are embedded in, informed by and have influence on other parts of the development, commercialisation, production, marketing and consumption of products and services, but they operate somewhat independently of the broader social concerns. At least partly due to the diverse nature of the three phases, they emerged in different ways.

Risk assessment, as defined in the Red Book and standardised through the *Codex Alimentarius*, is 'a scientifically-based process consisting of the following steps: (i) hazard identification, (ii) hazard characterisation, (iii) exposure assessment, and (iv) risk characterisation'. Hazards are defined as 'biological, chemical, and physical agents capable of causing adverse . . . effects and which may be present'. Hazard identification involves examining whether the agent could cause an adverse effect, and is informed by laboratory and field observations of adverse effects due to similar agents. Hazard characterisation involves asking about the relationship between the dose of an agent and an adverse response in the environment, animals or humans. While no formal evaluation or decision flows from the assessment stage, the choices made and evidence gathered provide the context for the decision made in the risk management phase.

The NRC subsequently produced *Science and Judgment in Risk Assessment* (NRC, 1994), also known as the 'Blue Book', and the Orange Book (NRC, 1996), effectively shifting risk characterisation from a science-driven enterprise to an 'analytic–deliberative' process. In essence, the process is a 'decision-driven activity, driven toward informing choices and solving problems'. This requires a broad understanding of the harms or consequences of a technology or product on a defined population, which requires participation of risk analysts, decision-makers and the affected population; thus, public officials, natural and social scientists and affected parties were now viewed as legitimate and important actors in the analysis, deliberations, learning and feedback stages.

The role of the larger society was introduced in the Blue and Orange Books but left to Codex to explore further. Codex asserts that the 'process of weighing policy alternatives in the light of the results of risk assessment and, if required, selecting and implementing appropriate control options' inevitably involves the social and political realm. Science cannot judge when a technology or product is 'safe enough' – that is the role for the political system. Different governments in different societies will have different views, depending on a wide range of philosophical, historical, social and institutional factors. Within the Organisation for Economic Co-operation and Development (OECD), most member states have developed explicit rules and institutions to analyse and control market access to protect the environment, economy and society. However, most developing countries are not members of the OECD and many are still constructing their rules and institutions.

Risk communication, defined by Codex as 'the interactive exchange of information and opinions throughout the risk analysis process concerning risk, risk-related factors and risk perceptions, among risk assessors, risk managers, consumers, industry, the academic community and other interested parties, including the explanation of risk assessment findings and the basis of risk management decisions', is seldom formally included in guidelines. While the Orange Book recommended enhanced two-way communication about technologies and products under risk assessment, it offered little direction on how to communicate risk management decisions or how to undertake dialogue with users and consumers. At times of crisis, such as when a harm is found in the food or drug system, regulators have standard operating procedures for informing consumers and the public, but their success has varied widely. Phillips *et al.* (2006) examined the risk communications system and showed that, in spite of the efforts of regulators, many risks have been amplified by what Sandman (1994) called 'outrage', resulting in wide divergences in the perceptions of risk between regulators and citizens. It is sometimes hard to measure outrage – sometimes it is evident in the marketplace, but often states simply assume it is there and intervene between the consumer and the market. Regulatory delays, proactive marketing bans and different production, marketing and labelling rules for food stuffs using specific technologies (e.g., GMOs) or from specific countries often preclude any opportunity for consumers to make their own choice.

While the core of the risk analysis system has been fairly well articulated, the community of scholars and practitioners of risk analysis sees significant room for further development. In the first instance, most developing countries lack domestic regulatory competence, capacity and experience in assessing biosafety risks, which creates uneven and disruptive market conditions. Even for the OECD countries, which largely have functioning systems, there are many new technologies and product categories that raise both theoretical and operational issues.

Theoretical Challenges from Emerging Risks

Emerging risks – defined by the European Food Safety Authority (EFSA, 2007) as 'a newly identified hazard to which a significant exposure may occur or from an unexpected new or increased significant exposure and/or susceptibility to a known hazard' – present significant challenges to our understanding of knowledge, our articulation and balancing of errors in the risk analysis system and our capacity to deal with complexity. The single largest challenge to regulators everywhere is the suite of new plant-breeding technologies, both those that move genes between species and those that simply amplify and express genes that are already present in a plant species. While genetically modified (GM) crops have been safely produced for 20 years now, many environmental non-governmental organisations and critics of GM continue to argue that the technology is inherently risky, requiring increased regulation. This causes particular challenges for developing countries that are just now constructing domestic regulatory capacity.

Articulating and Balancing Errors in the Risk Analysis System

In a perfect world the risk system should approve safe technologies and products and reject unsafe ones; in practice we face uncertainty and are likely to make errors. Regulators fear type I errors, whereby they accept unsafe products, but often end up making type II errors, whereby either false positives or undue delays cause safe and efficacious products to fail to enter the market. It is impossible to minimise both risks simultaneously: one must choose what balance of type I and type II errors is acceptable (Cranor, 1997). Part of the problem is that it is impossible to set up any scientific experiment to prove something is safe, because the possible hazards of any product are infinite. Hence, the risk assessment process is set up to analyse whether a specific technology or product will induce a specific set of hazards in specific uses. Ultimately, most regulatory systems identify a range of likely or less likely hazards that could be related to a product and test for those.

Conceptually, risk analysis is about balancing incommensurate outcomes. States inevitably have to choose between two possible errors from any risk analysis system they create. They may err on the side of caution and do as many tests and procedures as they could imagine, thereby raising the probability that they will reject, or more often simply discourage the introduction of, safe and efficacious products. This is called a type II error, driven by false-positive results. This is the *de facto* effect of the European Union's treatment of GM crops. This policy approach has also been the basis for the Cartagena Protocol on Biosafety (CPB) and, hence, delays and rejections of GM crops in many developing countries.

The emergence of the precautionary approach (some call it the precautionary principle) has created room for regulators to delay or suspend judgement on a product that is suspected to pose unacceptable, irreversible risks even where the necessary science is not available. Because an unchecked precautionary orientation could exacerbate type II errors, most articulations of the concept involve limits (e.g., Commission of the European Communities (CEC), 2000, pp. 19–21). Nevertheless, the EU, the CPB and a number of nation states have framed precaution to justify delays in the introduction and diffusion of GMOs, even in cases where many assert the science shows products are probabilistically safe enough. Canada, the USA and most large-scale adopters of GMOs challenge those actions, asserting that this use of precaution is simply a trade barrier.

Normalising Knowledge

Risk analysis is about making practical decisions regarding technologies and products. Jasanoff (1987) posits that 'regulatory science' differs in several key ways from academic science. While the evidence submitted by an applicant as part of a risk assessment of a new product is usually articulated in a positivist, deductive, hypothesis-driven set of empirically validated studies, regulators then 'deconstruct' them in the regulatory review process to expose areas of weakness or uncertainty. At the end, regulators then 'reconstruct' the evidence to justify their decisions. In the process, they follow evidence and decision-making rules not supported solely by the underlying science. While most national systems articulate clear goals, targets or timelines for this process, few have adopted clear and transparent standard operating procedures that show the step-by-step activities involved in assembling the evidence (e.g., what is required to be submitted when), reviewing the evidence (e.g., who assesses the technical merits and how), making decisions (e.g., which decision framework is used) and resolving disputes (e.g., how are dissenting opinions accommodated) (Canadian Biotechnology Advisory Committee (CBAC), 2002).

As a result, some have reasoned that the system is not truly scientific or objective. While there have been some high-profile examples where science has simply been ignored (e.g., Kenya's ban on GM food imports, which was based on the discredited studies by Gilles-Eric Seralini), the criticism goes deeper. Brunk *et al.* (1991) undertook a seminal study of regulatory decisions related to the cancellation of the registration of alachlor in Canada in 1985, concluding that evaluation processes inherently involve value frameworks (is safety or economics more important?) that affect participants' views of how to deal with uncertainty and where to assign the burden of proof. Brunk *et al.* (1991) examined the evidentiary base for the decisions and concluded that each of the key actors – the regulator, Monsanto as the proponent and a special Review Board – was forced to make normative choices when faced with incomplete information that could not be remedied by further scientific investigation. Cranor (1997) argues that 'the very idea of a risk assessment is normatively laden. A risk is a probability of an undesired outcome. Thus a risk is the probability of an outcome that will be unfortunate or undesirable from some evaluative point of view'. As such, choices about experimental design are normatively based. Ultimately, the orientation of various actors depends on whether they are trying to minimise type I errors (i.e., accepting a dangerous product) or type II errors (i.e., rejecting a safe product). The strong normative component in the process is a major concern to many people. More recently we have found that simple distrust of companies, especially multinational companies, can sway opinions and decisions (Lassoued *et al.*, 2016).

Starting with the fundamental divergence between science and regulatory science, one can see how a lack of clear and unambiguous rules for normalising and acknowledging knowledge claims can lead to difficulty.

Emerging Complexity

The wide global distribution of the sources and uses of knowledge, technology and products creates real challenges for dealing with both well-known and emerging risks. In practice, risk analysis is a heavily networked activity, involving a mix of public regulators, private firms and various values-based, interests-based and epistemic authorities (Phillips, 2007).

Knowing who is in charge of biosafety is complicated because of the significant diffusion of responsibility and authority. In the first instance, regulatory authority vested in national agencies is often linked to other countries via mutual recognition agreements and to various international governmental organisations (such as the OECD, World Trade Organization, International Plant Protection Convention, Codex Alimentarius and Biosafety Clearing House). By extension, many professional and epistemic communities, functioning under the auspices of one or more of these national or international institutions, are heavily involved in defining the knowledge, procedures and protocols that make up the risk analysis system. Moreover, proponents are heavily involved in any assessment and through them a variety of other actors are engaged. All regulatory regimes require owners of a product to be their own proponents and to define the way a product is intended to be used. While regulators (alone or in collaboration with others) may define the variables of interest and the nature of the data needed to assess a product, the owner must undertake or contract with some third party to assemble the materials and data which the regulator will review and evaluate. In this sense, the owner is a key participant in the definition of the product and how it will be evaluated.

The internationalisation of both science and trade has led both to more coordination (i.e., a gradual narrowing of differences between systems that are based on voluntary international codes of practice) and harmonisation (standardisation of regulation in identical form) (Davies, 2002). Recent efforts among OECD nations to advance coordination and harmonisation of the regulation of GMOs in a range of preferential trade negotiations (i.e., the Canada–EU Comprehensive Economic and Trade Agreement (CETA), the Trans-Pacific Partnership (TPP) and the Transatlantic Trade and Investment Partnership (TTIP)) have engaged but made little progress. Interestingly, there is some evidence that groups of the developing world are having more success, perhaps because they are focused more on capacity building than on reconciling divergent intents. For example, the COMESA Council of Ministers in 2014 adopted a regional biosafety policy to assist member states to develop capacity (Waithaka *et al.*, 2015).

Institutional Architecture for Coordination and Harmonisation

At the operational level, coordination, communications and management challenges are amplified by emerging risks. The processes of coordination and harmonisation have varied widely in pace and scope over the past century. The creation of the World Health Organization (WHO) and the Food and Agriculture Organization (FAO) precipitated a significant effort to codify existing knowledge and develop international processes to further coordination and harmonisation. These institutions, for instance, developed the Codex Alimentarius Commission in 1962 to coordinate efforts to standardise food labelling and safety standards. Over the years membership in these organisations has grown to involve most of the countries in the world. At times the FAO and WHO have also commissioned expert panels of independent scientists to consider areas of uncertainty in the international food or health systems, seeking to contribute to greater understanding of how a risk or event might assert itself.

Other regional and functional configurations of countries and experts have also engaged in the debate about regulating risks (e.g., regional trade agreements such as COMESA). The OECD, in particular, has an important role to play. Apart from codifying practices and providing a forum for coordinating policy development, the OECD has created a new quasi-regulatory instrument in the biotechnology area called a *Consensus Document* (OECD, 2005) (see also Chapter 11 in this book). These mutually accepted scientific reports (50 as of May 2015) document the biology of a host crop plant or the characterisation of an introduced trait or transgenic method, thereby providing a common base to be used in regulatory assessment of agricultural or food products derived through biotechnology. While such documents underpin the science-based regulatory systems of GM-adopting nations such as Canada and the USA, they have a reduced role or degree of significance in many developing nations.

Harmonisation efforts got a major boost with the negotiation of the WTO in 1995. In the context of risk assessment, the WTO permits national standards or regulations for the classification, grading or marketing of commodities in international trade (Art. XI.2.b) and for the adoption or enforcement of measures necessary to protect human, animal or plant life or health (Art. XXb). That authority is then delimited in the sub-agreements related to sanitary and phytosanitary measures (the SPS Agreement) and technical barriers to trade (the TBT Agreement). The operation of these two sub-agreements has created a strong incentive for countries to conform to international standards, such as those provided by the IPPC or Codex Alimentarius. The potential for diverging scientific and procedural results is thereby sharply reduced, as those organisations generally negotiate and establish harmonised international standards by consensus.

There has been extensive effort to generate a similar level of interlocking international authorities and institutions to address biosafety. Negotiations that began in 1982 at the Third World Congress on National Parks and Protected Areas in Bali, Indonesia, culminated at the 'Earth Summit' in Rio de Janeiro in 1992 with the Convention on Biological Diversity (CBD), included as Agenda 21 of the United Nations Conference on Environment and Development. This later resulted in the CPB regulating the transboundary movement of living modified organisms. While each of these agreements involves processes that encourage and support greater coordination of policies and regulations related to the environment, none of them has the effective power of the WTO, with its dispute-settlement system (DSM) that has the ability to inflict economic pain on countries that do not conform (the WTO DSM is a double-edged sword for many countries, as it can resolve disputes fairly, but access to the system is prohibitively costly for potential litigants from many poorer countries in Africa and Asia). Asymmetrical membership creates further challenges. The USA, for example, is a key member of the OECD, WTO and most other international regulatory programmes, but is not a formal party to any of the CBD follow-on agreements because it has not ratified the 1992 Convention.

Conclusions and Recommendations

The biosafety regulatory system is a work in progress. Given the uneven and in places inconsistent regulatory goals and practices, and the absence of effective authority and capacity in many developing countries, gaps, overlaps and conflicts are inevitable. Our focus on emerging risks highlights how far the modern risk analysis system has come and how far it has to go to effectively deal with public concerns.

The regulation of new agricultural technologies, such as GM crops, is frequently a costly, time-consuming and highly constrained process, whether in a wealthy mega-adopting country or in a developing nation. While larger countries have the resources and networks to muddle through, developing countries face a range of political and capacity challenges. We posit that efforts to overcome or at least substantially reduce four main obstacles are required in order to ensure the efficient biosafety regulation of GM crop technologies in developing countries, thereby facilitating effective and efficient access to new safe and efficacious technologies. These four obstacles can be described as follows. First, if greater coordination and harmonisation can be realised, it should be possible for countries large and small to more efficiently and effectively assess the unique risks of new technologies in their economy, while relying on the foundational risk judgements of others. COMESA and others are showing this has promise. Second, increasing the scientific technical capacity of existing regulatory agencies, large and small, would greatly improve global risk assessment. The global risk system is only as sound as its weakest link. Governments, industries, learned societies and universities are engaged in extensive capacity building in many parts of the world. Third, a move to more regional assessments of biosafety hazards and risks through new partnership programmes involving developed and developing countries, in partnership with proponents, could help to more effectively assess and manage any large-scale or systemic effects of new technologies on the environment, while at the same time absolving each individual country from replicating all of the steps in the risk analysis of a new technology. Getting there will not be easy, and it might not work in all regions. There is scepticism about the role of industry in these large-scale projects. For example, while there is great interest among many about the Water Efficient Maize for Africa (WEMA) project, the African Centre for Biosafety has asserted that 'WEMA is a Trojan horse to pressurise participating governments to pass weak biosafety regulations and open the door to the proliferation of GMOs that will undermine food sovereignty'. Fourth, philanthropic organisations could and probably should be encouraged to redirect some of their resources to assisting with the technical development of regulatory systems and risk assessments in key target markets and regions. Investing scarce philanthropic resources in developing new technologies and foods is only half the challenge of addressing global food security – better governance is also needed. There are some efforts here to build on. The USAID Program for Biosafety Systems and Global Environmental Facility already provide funding to support capacity development. However, progress has been very slow and inconsistent, in part because recipient countries often have little or no existing capacity or ability to sustain the capacity without sustained international funding.

References

Beck, U. (1992). *Risk Society: Towards a New Modernity*. New Delhi: Sage (translated from the German *Risikogesellschaft. Auf dem Weg in eine andere Moderne*. Frankfurt: Suhrkamp, 1986).

Bernstein, P. (1996). *Against the Gods: The Remarkable Story of Risk*. New York, NY: Wiley.

Brunk, C. *et al.* (1991). *Value Assumptions in Risk Assessment: A Case Study of the Alachlor Controversy*. Waterloo: Wilfred Laurier University Press.

CBAC (2002). *Improving the Regulation of Genetically Modified Foods and Other Novel Foods in Canada: A Report to the Government of Canada*. Ottawa: Government of Canada. [Online]. Available from http://publications.gc.ca/collections/Collection/C2-589-2001E.pdf

CEC (2000). Communication from the Commission on the Precautionary Principle. COM 2000/0001. [Online]. Available from http://eur-lex.europa.eu/legal-content/EN/TXT/?uri=CELEX:52000DC0001

Cranor, C. (1997). The normative nature of risk assessment: features and possibilities. *RISK: Health, Safety & Environment Journal* **8**, 123.

Davies, L. (2002). Technical cooperation and the international coordination of biotechnology inventions. *Journal of Law and Society* **29**(1), 137–162.

EFSA (2007). Definition and Description of 'Emerging Risks' Within the EFSA's Mandate. EFSA/SC/415 Final. [Online]. Available from www.efsa.europa.eu/sites/default/files/scientific_output/files/main_documents/escoemriskdefinition.pdf

EPA (2009). Environmental Protection Agency Website. [Online]. Available from www.epa.gov/

Jasanoff, S. (1987). Contested boundaries in policy-relevant science. *Social Studies of Science* **17**(2), 195–230.

Lassoued, R. *et al.* (2016). Anchoring effects in regulatory decisions. Working paper. Saskatoon: University of Saskatchewan.

NRC (1983). *Risk Assessment in the Federal Government: Managing the Process*. Washington, D.C.: National Academy Press.

NRC (1994). *Science and Judgment in Risk Assessment*. Washington, D.C.: National Academy Press.

NRC (1996). *Understanding Risk: Informing Decisions in a Democratic Society*. Washington, D.C.: National Academy Press.

OECD (2005). *An Introduction to the Biosafety Consensus Documents of OECD's Working Group for Harmonization in Biotechnology*. Paris: OECD.

Phillips, P. W. B. (2007). *Governing Transformative Technological Innovation: Who's in Charge?* Oxford: Edward Elgar.

Phillips, P. W. B. *et al.*, eds. (2006). *Governing Risk in the 21st Century: Lessons from the World of Biotechnology*. New York, NY: Nova Science Publishers.

Sandman, P. (1994). Mass media and environmental risk: seven principles. *Risk: Health, Safety, and Environment* **5**, 251.

Waithaka, M. *et al.* (2015). Progress and challenges for implementation of the Common Market for Eastern and Southern Africa policy on biotechnology and biosafety. *Frontiers in Bioengineering and Biotechnology* **3**, 109. [Online]. Available from www.ncbi.nlm.nih.gov/pmc/articles/PMC4519696/

World Commission on Environment and Development 1987. *Report of the World Commission on Environment and Development: Our Common Future*. [Online]. Available at http://www.un-documents.net/our-common-future.pdf

14 The Precautionary Principle as a Barrier to GMO Risk Analysis: Elicitation of Experts' Viewpoints

Ademola A. Adenle

Introduction

Risk analysis procedures remain an integral part of the governance process before genetically modified (GM) organisms (GMOs) can be released because the procedures are key to assessing the benefits and safety issues relating to animal and human health and the environment. Such procedures are made up of steps and objectives that are based on the risk analysis framework to guide how risk assessment, risk management or risk communication is carried out. In many cases internationally agreed protocols on how risk analysis should be done are consulted before a decision is made. Gene technology, together with a few other emerging technologies, is overly regulated and highly controversial, partly due to the fragmented framework of authoritaries which has been set up for the risk analysis of GMOs, therefore it is worthwhile to understand some of the challenges and seek possible solutions.

This chapter focuses exclusively on barriers to GMO risk analysis by interviewing 16 experts (Table 14.1) based in various countries around the world. These experts are currently, or have been, involved in regulation and research activities pertaining to GMO risk analysis at the national, regional or international level. The experts were asked a series of questions, including (1) 'Why are GMOs regulated differently from other types of modern biotechnology?', (2) 'What are the main obstacles to risk analysis of GMOs at the national level?' and (3) 'How should international protocol guide the risk analysis of GMOs?' As the goal of the interviews was to gain a better understanding of GMO issues related to national regulation and international protocols and their impact on implementation of GMO policy, the chapter provides some insight into aspects of politics and challenges relating to GMOs and suggests a possible way forward.

Why Is the Precautionary Approach a Major Obstacle to GMO Regulation?

The precautionary principle has been criticised partly due to lack of clarity in the ways it has been interpreted or applied in protecting and regulating environmental and health safety issues since the approach emerged for the sustainable development concept at international meetings (Cooney and Dickson, 2005; Peterson, 2006). According to Cooney and Dickson (2005), the agreement on GM technology stands out among multilateral agreements as one where the precautionary principle remains highly contentious,

Table 14.1 List of experts interviewed

Expert	Affiliation	Position/expertise
1	Food and Agriculture Organization of United Nations, Rome, Italy	Senior Food Safety Officer
2	Oregon State University, USA	Professor – Applied Entomology
3	Clark University, USA	Senior Research Scientist – Technological Innovation
4	Massachusetts Institute of Technology, USA	Senior Research Scientist – Biology and GMOs
5	University of Oxford, UK	Professor – Food Systems
6	University of Georgia, USA	Professor – Plant Breeding and Genomics
7	Syracuse University, USA	Professor – Agricultural Biotechnology
8	The Sainsbury Laboratory, UK	Professor – Plant Pathogens and Transgenics
9	Julius Kühn-Institut, Germany	Professor – Biosafety and Plant Biotechnology
10	Golden Rice Project, Switzerland	Executive Secretary – Agricultural Biotechnology
11	Food Standards Australia, Australia	Principal Scientist – Safety Assessment of GM Foods
12	Free University of Brussels, Belgium	Professor – Biosafety and Law
13	Massey University, New Zealand	Professor – Agriculture and Environment
14	New Brunswick University, Canada	Professor – Biology
15	North Carolina State University, USA	Professor – GMOs
16	Queensland University of Technology, Australia	Professor – Agricultural Biotechnology

particularly with regard to regulating the technology. In fact, most of the experts interviewed felt that the precautionary principle still represents a significant challenge to the development and adoption of GM products around the world, particularly in developing countries. While this feeling has been widely reported in the existing literature, some of these experts shared their experiences as to why the precautionary principle has gained ground for a long time and why it may continue to remain controversial, hence making the overall framework of risk analysis of GMOs more complex at the international level, with adverse spill-over effects on developing countries. Experts who are conversant with the history of the precautionary principle attributed the GMO controversy not only to culture and differences in regulation between the USA and European Union (EU) but also to other factors that were not known to many interest groups, including the public. One expert argued that the disagreement between the US government and the French government has arisen because the latter will not be rational about GMO regulation because of the deal made with Greenpeace (a fiercely anti-GMO organisation) not to oppose nuclear programmes; this has been admitted by a French government official (PRRI, 2013). Another expert mentioned that scientific argument about whether to release GMOs is almost irrelevant without resolving what originally led to the impasse between the powerful stakeholders such as the USA and the EU.

This controversial opinion was supported by another five experts who argued that the precautionary principle was not originally designed to regulate GMOs but was signed into law to regulate environmental concerns under the Montreal Protocol and became a legally binding international treaty at the Rio Convention. However, a sudden, major change in the way the precautionary principle was interpreted and applied to environmental issues occurred between 2000 and 2003 through the Cartagena Protocol on Biosafety (CPB) of the Convention on Biological Diversity (CBD) under the United Nations Environment Programme (UNEP), which placed a restriction on the use of GMOs. One expert pointed out that the current definition of a GMO is not a biological or scientific concept but a legal concept that has set up a series of contradictions in the way the technology is being developed and adopted. Most of the experts suggested that current GMO regulation (via the precautionary principle) was premeditated, orchestrated and financed by the EU countries as disguised under the auspices of the CBD in an effort to undermine large-scale application of GMOs around the world. In fact, one expert stated that the introduction of the CPB was largely controlled and spearheaded by the majority of the representatives from the EU countries, which makes the whole process of GMO regulation suspicious at the international level. The same expert called the CPB an obfuscation built on a false premise to deceive the world about the true application of GMOs. The argument presented in this chapter is consistent with a study that argues that environmental organisations played a key role alongside the UNEP-CBD in influencing the biosafety policy in developing countries (Arts and Mack, 2003).

Following the CPB argument, some of the experts explained that many of the UNEP-Global Environment Facility (GEF) capacity-building programmes in developing countries had been set up with ulterior motives and encouraged further restriction on the application of GMOs. They explained also that many governments in developing countries have been deceived into believing that UNEP-GEF programmes aimed to help develop expertise and skills required for the safe application of GMOs as originally claimed. Their argument that the imposition of the precautionary principle on developing countries was a means to limit the application of GMOs was based on the following three points: (1) most developing countries (especially countries in Africa) that participated in the programme do not have a clear national biotechnology policy that can support and be aligned with the objectives of the precautionary principle, thereby creating room for manipulation; (2) limited expertise in risk analysis does not allow individual countries to come up with a framework for research and development, testing and release of GMO products, leading to a focus on the risks alone, without considering the benefits of the technology; and (3) the cost of regulation is too high and obtaining a permit for cultivation becomes almost impossible as a result of imposition of the precautionary principle. One biosafety expert argued that these three points are among the inherent factors which have slowed down the application of GMOs since many developing countries became signatories of the CPB. This argument is reinforced by the fact that the signing of the CPB by these countries has (in their view) shifted their responsibilities to the international organisation, thereby limiting the need for them to develop a functional biosafety policy to facilitate the application of GMOs. One key question asked is 'Why have so few African countries developed functional biosafety systems and adopted GM crops despite UNEP-GEF programmes?' The responses had a lot to do with the way people are trained to recognise the risks and danger GMOs may pose, which are clearly spelled out in many biosafety documents. It was mentioned that the importance of capacity and expertise that should have been emphasised in training programmes has been twisted and undermined by the precautionary approach. As one expert described it, 'the coordinators of the UNEP-GEF programmes know many African countries do not have certain systems in place, no experience, no technical know-how and won't be able to use procedures designed for that purpose, so they took advantage of the situation'. Another expert put it this way: 'I think it's had a huge impact, because a lot of countries have signed up for it and developed their national regulations as a result. It is a complete impediment to innovation and disallows plant breeders from using the full extent of techniques available to them.' However, three experts had different viewpoints on capacity development and training, since they mentioned that imposition of the precautionary principle might not have been responsible for weak biosafety training programmes. They felt that many research institutes and universities have limited facilities and resources and lack programmes that can support the training of scientists and facilitate expertise to conduct risk assessment and implement credible biosafety regulatory systems to accommodate the use of GMOs. For example, one expert mentioned that some countries did not see the need to enact biosafety regulation partly due to limited biotechnology activities and the fear that GMOs could be imposed on their countries by the multinationals. Other countries went ahead and started developing biosafety regulatory frameworks but stopped the processes as soon as UNEP-GEF programmes came to an end (see Chapter 9).

The controversy around the precautionary approach among the experts further led to an argument on the divergence between the USA and the EU on the intrinsic value of biodiversity conservation, organic agriculture and food production and supply. Two of the experts noted that EU countries have stricter regulations for organic farming than the USA, particularly in applying the precautionary approach. This is mainly because of a strong cultural heritage that is linked to organic farming in many EU countries and greater public resistance by EU consumers. For example, one of them further explained that the way 2,4-D and glyphosate are used in farming practices in the USA and in EU countries is very different, and the US regulations are seen to be more lenient than those in the EU. The EU tends to support the idea that people have a moral duty to protect and respect the intrinsic value of biodiversity by establishing more sustainable agro-ecosystems through its stricter regulatory mechanism. This is strongly reflected in sermons being preached by the most prominent non-governmental organisations (NGOs), such as Greenpeace and Friends of the Earth, in which the precautionary approach forms part of their strategy to halt the application of GMOs by promoting organic farming, which often generates an emotional and public reaction that nature must not be tampered with. However, some of the experts argued that the original excuse and a motivating factor for the action of environmental activists is fear of multinationals such as Monsanto dominating agribusiness and the food supply in Europe and the rest of the world. Specifically, it was mentioned that transformation of the food supply, intellectual property rights (IPR) and ownership control of genetic resources by multinational corporations represents a huge concern among anti-GMO groups, but they have not found a coherent way of selling this story to the public, leading them to choose more controversial approaches. An issue pointed out by some of these experts was that anti-GMO groups chose to use the precautionary approach to promote their agenda because they thought it would be easier to achieve their aim by convincing consumers and other stakeholders that GMOs pose more risks than benefits, rather than pursuing their line of argument about the domination of agribusiness by GMO industries. As a result consumers may hold negative attitudes towards GMOs, since they tend to have more confidence and trust in anti-GMO activists than they do in industrial concerns. In fact, these experts mentioned that anti-GMO activists are using health and environmental concerns as a club with which to beat GMO industries, although the activists are actually less concerned about risk than they are about the structure and economic interests of these industries. Three of the experts disputed this argument, because they felt that previous food safety problems such as bovine spongiform encephalopathy (BSE) and other food scandals may have reinforced concerns about GMO risk in the EU, with consumers showing very little trust in government. However, one of them acknowledged that environmental organisations may have taken advantage of the problem. This point is reinforced by the fact that anti-GMO activists will continue to increase their level of trust and confidence among consumers as long as the precautionary approach is being deployed for GMO regulation because it is driven by perception of risks and sentiments and absence of scientific data. Actually these activists are merely using arguments about the prevention of environmental problems or respecting the intrinsic value of ecosystem services and risk to health as an excuse to launch frequent attacks on GMOs, and this should be seen as their championing of a purely political cause driven by their selfish interest in maintaining the status quo.

Majority vs. Minority – the Problematic Nature of Risk Analysis of GMO Products

Building on the previous section, this section examines some political challenges behind safety assessment and criteria for guiding the regulatory and approval process relating to various GMO issues among stakeholder groups (Table 14.2).

The experts agreed that the precautionary approach triggered lingering debate about GMO risk assessment as a key component of risk analysis, particularly with regard to long-term effects and uncertainties of environmental and health risk. They felt strict regulation of GMOs is a deliberate attempt to undermine their application through the precautionary approach. Three experts argued that GM is the only technology on Earth for which there is an international treaty that compels regulatory compliance risk analysis for environmental release. The reason given was the role of the EU countries in making the CPB a legally binding international treaty for risk analysis and creating bureaucratic procedures for trading of GMO products.

In this case, the experts described majority group members, including multinational corporations, the large scientific community and independent and intergovernmental

Table 14.2 Trajectory of GMO regulation and risk analysis

Challenges of GMO risk analysis	Proponent of GMOs	Opponent of GMOs
Stakeholders	Majority – multinational corporations, a large scientific community, national government agencies, independent organisations and regulatory authorities	Minority – NGOs (e.g., Greenpeace), media, individuals and a few scientists
Evaluation criteria	Evidence-based rationality, scientific value	Strictly judged based on precautionary principle, irrationality and legal concept approach, uncertainty and social value
Codex guideline	Assessment based on substantial equivalence	No substantial equivalence, agitation for additional data
Ideology	Scientific and technology progress	Enemy of technological progress
Regulatory focus	Product-based, socio-economic consideration complex, approval subject to availability of appropriate indicators	Process-based, agitation for socio-economic consideration
Risk assessment strategy	Short-term testing	Long-term testing – a period of 3 months or longer
Benefit and risk emphasis	Assessment both of benefits and of potential risks (minimal)	Assessment of risk only (zero tolerance)
Labelling	Labelling discouraged to prevent product discrimination	Labelling encouraged to allow consumers to make choices
Organisational structure	Vertical	Horizontal

organisations, as those who argue that decisions on whether GMOs should be allowed must be based on scientific evidence rather than on the precautionary principle. This group often argues that more than two decades of scientific evidence has been accumulated to prove that GMOs are safe for human consumption and pose no bigger risk to the environment than do their traditional counterparts. The experts mentioned complex processes for the risk analysis of GMOs as one of the main obstacles to the development of harmonised regulatory frameworks. They pointed out that risk assessment processes, including risk identification, evaluation, approvals and regulation, vary from country to country, and also vary among various bodies and organisations that are involved in developing standards and guidelines. But competing interests within this group of people and organisations often make risk analysis procedures more complex, as reflected in regulatory mechanisms and procedures adopted in different countries. Here, the argument from the experts was based on lack of transparency, lack of clear strategy, bureaucracy and limited participation of key actors in risk analysis of GMOs. Four experts mentioned that the US government has allowed far too much autonomy and concentration of GM technology in the hands of multinationals such as Monsanto, which is reflected in the way protocols and procedures for risk analysis of GMOs are being dictated and manipulated in public research institutions and national regulatory agencies, including the US Department of Agriculture (USDA), Food and Drug Administration (FDA) and Environmental Protection Agency (EPA). For example, one expert argued that Monsanto came up with the tests that are going to prove that glyphosate-tolerant GMO products are safe, but their methodology could be flawed, which makes one suspicious as to whether appropriate risk assessment protocols have been developed to test the safety of the products. Similarly, one US expert, who is an entomologist, argued that IPR impedes risk assessment of GMOs due to the lack of exemption for research despite claims that multinational corporations allow scientists access to all relevant information. This argument was supported by another scientist, who mentioned that data sometimes become available only after GMO products have been released to the market.

Moreover, political interest at the EU level is also a huge factor in risk analysis of GMOs even among the majority group as regulators will disqualify scientific findings on political grounds. In contrast to the US experience, some of the experts explained that the European Food Safety Authority (EFSA), despite being a competent independent organisation, has struggled to get approval on purely scientific grounds for the risk assessment recommendations provided. One expert attributed the blockage to differences between the council ministers, member states and European Commission. According to the same expert, 'results of risk assessment cannot be forced onto member states because they have differences and similarities and can opt out even though the GMO has passed all these assessments'.

On the other hand, the experts described minority group members, including environmental activists, a small group of scientists and individuals, as being passionately against the use of GMOs despite a high level of awareness of the scientific evidence on the safety and benefits of the technology. The minority group members were described as a bad influence on the public and consumers as they intensify their campaign against the application of GMOs whilst citing supposed environmental and health risks that are based on irrational reasons to dissuade developing countries from adopting GMO products. They alluded to the fact that the minority group members have funded a number of notable animal studies for food safety assessment with huge media attention to stop the use of GMOs over the past two decades but have failed in these attempts to strengthen their propaganda. For example, Pusztai claimed that in a rat study GM potato had negative effects on gut morphology and damaged the immune system – he was criticised for announcing the results before completing the experiment and later suspended by his institution for misconduct. Another famous controversial example was the French scientist Seralini, who also claimed rats fed with GM maize developed cancerous tumours – he was severely criticised by top experts in this field for being dishonest about his methodology and on account of the strains of rats he used for his experiment, which later resulted in the paper being retracted from where it had originally been published. Following on the two controversial studies, some of the interviewed experts criticised the use of animal studies for the risk assessment of GMOs because they felt animal studies are more aligned with the EU's process-based approach for regulation of GMOs as outlined in the precautionary principle. These experts argued that assessments that are based on feeding complex food substances to animals using this experimental approach can be problematic due to lack of sensitivity, which may undermine the interpretation of the results. According to one expert, 'if you get a negative result in an animal study it does not exactly tell you that the food is safe, so we prefer to rely on this other approach where we focus on the composition of the food using sensitive chemical analyses and where we look at the changes that we have actually introduced'. Another expert concluded that only certain GMO products should be regulated, not the process, as the risks come with the product. For example, according to the same expert, risk assessment should focus on whether there is a change in phenotype of a genetically modified product compared with the non-modified product. However, some of the experts did not rule out the possibility that animal studies can be useful for food safety assessment, particularly when there is insufficient information to guide decision-making in the first place, as a carefully designed experimental approach with consistent and reproducible methodology can play an important role in reaching a successful conclusion.

Are Gene-Editing Techniques Ways to Resolve GMO Regulation?

A promising modern biotechnology called gene editing also received attention among the experts interviewed on this topic, particularly with regard to the possibility of emerging technology resolving the regulatory hurdles for GMOs that have been around for a long time. While these experts appreciate the benefits and advantages of gene editing, including better precision, better prediction and limited off-target effects for introducing DNA into the genome, they argued that some of the gene-editing techniques are novel and similar to GMOs and traditional biotechnology, which raises questions about regulation. For example, one expert explained that, if gene-editing techniques were used to introduce or insert a new gene into an organism, it could be called a GMO and therefore may be subject to regulation due to potential unintended effects. This opinion was shared by other experts, who felt that more insertions could occur despite the intention of carrying out a single gene insertion. However, many of these experts also argued that, if gene editing is used to copy what nature does, the organisms cannot be called GMOs, since the process is more like conventional breeding, with better precision and less likelihood of unintended effects. Moreover, three experts argued that it might be difficult to justify regulating gene editing, given that there are different types of gene-editing techniques and the majority of these techniques may not generate risks or concerns that will involve greater scrutiny like that applied to GMOs.

Regulation aside, these experts felt that IPR constraints, international harmonisation, NGOs' opposition, lack of public acceptance and limited scientific capacity in developing countries could be a major challenge in introducing the technology in the future in view of current experience with the use of GMOs. Despite overwhelming agreement among the experts that with gene editing there will be lesser risk and greater insertion control, they felt gene editing will still be subject to open debate and dialogue among relevant stakeholders as to whether it should be regulated or not. This is reinforced by the belief that the concerns for gene editing may not differ dramatically from those for

GMOs, but the discussion will provide an opportunity to work with scientists, government, the private sector and others on how to better govern the new technology without subjecting it to strict risk regulation like GMOs.

Discussion

Do concerns about GMOs and risk analysis have to do more with the linkage of the precautionary principle to political agendas among member states, including the USA, than with anything else? While it may be difficult to establish an affirmative answer to this question as fact, findings show considerable variations in terms of implementation of GMO regulation between the EU countries and the USA, resulting in a trade conflict shaped by two different international policies, with the World Trade Organization (WTO) favouring the USA and the precautionary principle favouring the EU (Lieberman and Gray, 2008). In fact, an evidence- or risk-based approach that is aligned with the WTO seems to support the GMO policy stance of other UN organisations such as the World Health Organization (WHO), Food and Agriculture Organization (FAO) and United Nations Development Programme (UNDP), leaving the CBD precautionary principle isolated (FAO/WHO, 2001; Potrykus and Ammann, 2010; UNDP, 2001). A further argument against the instigators of the precautionary principle was a failed attempt to manipulate Codex as a stricter international standard and create further trade barriers without scientific risk assessment. This difference created by some member states and lobbying groups (e.g., NGOs) has raised a lot of questions about divides and inconsistency among UN agencies in terms of how they prioritise technological innovation, including GMO technology, in tackling malnutrition, food insecurity, low agricultural productivity and climate change in developing countries. Some of the interviewed experts argued that, if political activism by the NGOs were motivated by genuine concern, their attention should have been focused more on rainforest deforestation, which results in unquantifiable destruction of biodiversity and ecosystem services. One expert called it a high level of moral hypocrisy where attention is focused on issues with small or no risk but issues with big risk are ignored.

Moreover, evidence suggests that the current GMO regulatory process is underpinned by the precautionary principle, which inevitably slows down research and development of various GMO products, including GM pro-vitamin A biofortified rice, popularly known as Golden Rice, whose introduction has been held back for more than 15 years (Dubock, 2014; Potrykus and Ammann, 2010). The experts interviewed argued that failure to address anomalies caused by the precautionary principle may lead to further delays in releasing other biofortified products, such as cassava, cowpea and banana, and therefore limit the opportunity to stimulate innovation that has great potential in solving the problem of vitamin A deficiency, which kills an average of 3 million children per year (WHO, 2016). In view of this challenge, it was mentioned that international regulatory mechanisms for GMOs should be relaxed because they have overshadowed the good work of scientists who have worked hard for many years to provide evidence regarding the safety of GMOs and the voices of many organisations that are interested in promoting the technology to address some of the challenges of sustainable development in developing countries.

The risk analysis of GMOs under the current international framework of CBD-CPB has been politicised as decisions made for the regulation and management of GMO products tend to affect different levels of society in many different ways, particularly in developing countries. Bringing GMO risk issues into the political agenda in the light of the precautionary principle has become a common means of gaining support from the public as well as the media, thereby undermining public policies aimed at developing functional biosafety systems to regulate GMOs (Adenle, 2014). Scholars have cited different interpretations of the precautionary principle as a big influence in the governance process of risk analysis of GMOs from the beginning. This was echoed by one of the interviewed experts, who stated that GMO regulation is proportional to the risk and benefit assessment, but the implementation process is disproportional and varies with the spirit in which it was drawn up. As a consequence, the majority vs. minority problem has led to national and regional governments reacting by establishing and conducting independent risk assessments in order to prevent undue influence and imposition. This has never helped, but rather has worsened the situation as it is reflected in different ways of conducting risk assessments at different levels of government without observing the right protocols and deliberately ignoring certain steps in the governance process, thereby leading to insufficient guidance in making decisions, particularly in developing countries. This raises questions about 'who' should be involved in risk analysis procedures and processes, at 'what stage' and 'how'. Each step is important in stakeholder engagement for a transparent risk analysis framework. The need for resolution is considered urgent, following more evidence from the latest report by the National Academies of Science, Engineering, and Medicine (NASEM, 2016) that risks from GMOs are no different from those of traditional plant biotechnology. Some of the experts interviewed called for more understanding and a new evaluation approach on the values that shape perceptions, interests and concerns of various stakeholder groups regarding the application of GMOs. Specifically, a rigorous methodology that captures both minority and majority concerns is likely to influence the debate on the benefits and potential risks of GMOs, hence paving the way for the best practices of risk analysis of GMOs.

Without this kind of approach, experts felt gene editing may suffer a similar fate as there is ongoing debate as to how/whether gene editing should be regulated. In fact, environmental activists have started demanding that gene editing be treated the same as GMOs, namely by applying the precautionary approach, because they feel that gene-edited products could still be harmful to health and the environment (Abbott, 2015). While many countries around the world have yet to take the steps required for regulating gene editing, in the USA, gene-edited products have been approved without passing through regulatory oversight (Wolt *et al.*, 2015). The US government has been criticised for the lack of mandatory oversight and failing to engage key actors, given the controversy surrounding the use of GMOs in the country. Nevertheless, the country is currently reviewing its regulation, which suggests that the USA might end up regulating gene editing. In the EU countries, the process of coming to a decision on gene-edited products has been slow, particularly at the European Commission level, although some

Box 14.1. Recommendations for Resolution of Conflict on GMOs

(1) Review CBD-CPB. Most of the experts argued that the CPB should be reviewed because disagreements on risk analysis of GMOs mainly come from the environmental organisations and those countries which are signatories to the CPB. The need to review the CPB is justified by the inability of many developing countries to make decisions that are based on scientific evidence. They felt that encouraging scientific evidence in a new CPB that engages all relevant stakeholders in a transparent manner will pave the way for the assessment of benefits of GMOs in such a way that developing countries will be able to make informed decisions.

(2) Abolish international regulation of GMOs. Six of the experts argued that there is no need for a uniform international protocol in view of existing regional differences regarding GMO regulation and advocated the abolition of the CPB. The following five reasons why the international protocol should be abolished were cited: (1) there is enough scientific evidence to show that GMOs are safe for animals, humans and the environment; (2) continued implementation of the CPB may encourage the creation of new international protocols, and more protocols will mean more barriers, not solutions; (3) it will give developing countries the opportunity to develop standard risk assessment models that are based on their agricultural and environmental practices (which vary from country to country) as many of these countries at present follow or adapt a Western model for the risk assessment; (4) it will limit the influence of environmental organisations and member states with the political agenda of halting the application of GMOs; and (5) it will encourage use of emerging technologies, such as gene editing, without subjecting the technology to strict regulation.

(3) Encourage accountability and greater transparency. The role of individual actors in the governance process must be made clear. All actors, including industries, regulators, governments and researchers, that are responsible for risk analysis of GMOs must engage with one another in a transparent and open manner. While the policy influence of industries in the field of biotechnology rests on their control of patents, which often impedes risk assessment, governments must ensure that researchers from public institutions are protected from external pressures so that all actors feel safe enough to share information that will create a better atmosphere in which to conduct risk assessment.

(4) Establish regional centres for risk analysis procedures. The experts agreed that developing countries should establish regional centres, rather than national ones, that encourage independent risk analysis of products that may raise health and environmental concerns in the future. One reason here is that it will help such countries make independent decisions that are based on a good framework and available scientific evidence, led by their own experts and avoiding external recommendation. In the case of controversial GMOs, national and regional experts can come together to quantify both positive and negative effects, and bring the results to the knowledge of policymakers and the public without relying on outside information. The experts acknowledged institutional and financial problems as one of the major barriers to this approach. But, with support from donors, especially those advocating the use of GMOs in African countries, and the leadership of the African Union, experts felt that the regional centres for risk analysis could become a reality.

EU countries are conducting field trials for certain gene-edited crops (Abbott, 2015), suggesting that individual EU countries are making decisions on a case-by-case basis.

Recommendations and Conclusion

The results from this elicitation of experts' views reveal some insights into how challenges facing GMO regulation and risk analysis can be overcome in order to govern the technology effectively. One challenge that featured prominently during the interviews was the difference between the EU and US GMO policies, which remains an obstacle at the international policymaking level. The difference is that the USA demands a scientific basis for its policies, while the EU formulates policy more on the basis of public opinion, rather than scientific merit. One of the experts explained that the reason for this is that policy in the USA is set by professional regulators. In the EU, it is set by politicians, who must get re-elected. Most of the experts argued that those who are opposing the deployment of GM technology should be engaged in constructive and intellectual debate where all stakeholders are allowed to participate in an open discussion that looks critically into evidence on the risks and benefits of the technology. Juggling their viewpoints together to achieve a common goal could lead to a better outcome. Furthermore, the outcome should be communicated in understandable language with an emphasis on facts, openness and sharing of the information that addresses social concerns. It is believed that an analysis of risk and benefits that is underpinned by more systematic evaluation of the science with better communication could offer a solution. In addition, four recommendations were suggested as possible solutions (Box 14.1).

References

Abbott, A. (2015). Europe's genetically edited plants stuck in legal limbo. *Nature* **528**, 319–320.

Adenle, A. A. (2014). Stakeholders' perceptions of GM technology in West Africa: assessing the responses of policymakers and scientists in Ghana and Nigeria. *Journal of Agricultural and Environmental Ethics* **27**, 241–263.

Arts, B. and Mack, S. (2003). Environmental NGOs and the Biosafety Protocol: a case study on political influence. *European Environment* **13**, 19–33.

Cooney, R. and Dickson, B. (2005). Precautionary principle, precautionary practice: lessons and insights. In *Biodiversity and the Precautionary Principle: Risk, Uncertainty and Practice in Conservation and Sustainable Use*, ed. R. Cooney and B. Dickson. London: Earthscan, Chapter 18.

Dubock, A. (2014). The politics of Golden Rice. *GM Crops & Food* **5**, 210–222.

FAO/WHO (2001). Evaluation of allergenicity of genetically modified foods. Report of a joint FAO (Food and Agriculture Organization of the United Nations/WHO (World Health Organization) expert consultation on allergenicity of foods derived from biotechnology. [Online]. Available from ftp://ftp.fao.org/es/esn/food/allergygm.pdf

Lieberman, S. and Gray, T. (2008). The World Trade Organization's report on the EU's moratorium on biotech products: the wisdom of the US challenge to the EU in the WTO. *Global Environmental Politics* **8**(1), 33–52.

NASEM (2016). *Genetically Engineered Crops: Experiences and Prospects*. Washington, D.C.: The National Academies Press.

Peterson, D. (2006). Precaution: principles and practice in Australian environmental and natural resource management. Productivity Commission. Presidential Address, 50th Annual Australian Agricultural and Resource Economic Society Conference, Manly, New South Wales.

Potrykus, I. and Ammann, K. (2010). Proceedings of a Study Week of the Pontifical Academy of Sciences. *New Biotechnology* **27**(5), 445–717.

PRRI (2013). The French deal: GM technology traded off for nuclear energy. [Online]. Available from www.prri.net/french-deal-gmos/

UNDP (2001). *The Human Development Report 2001*. New York: Oxford University Press. [Online]. Available from http://hdr.undp.org/en/content/human-development-report-2001

WHO (2016). Micronutrient deficiencies. [Online]. Available from www.who.int/nutrition/topics/vad/en/

Wolt, J. D. *et al*. (2015). The regulatory status of genome edited crops. *Plant Biotechnology Journal* **14**(2), 510–518.

15 How Do Socio-Political Disputes Concerning GMOs Influence Decision-Making in Developing Countries?

Hossein Azadi, Ademola A. Adenle and Klaus Ammann

Introduction

While the advent of genetic modification (GM) technology has resulted in different policy responses in various countries, the disputes that surround the possible risks of GM foods have also expanded into the public domain and concern. In some cases, the national responses have led to trade disputes between members of the World Trade Organization (WTO). It is recognised that developing countries, as either importers or exporters, will be affected by the GM policies utilised in countries with which they are trading. In a few countries, especially in Europe, the strong consumer scepticism regarding genetic engineering will also affect the trading environment in which developing countries have to compete. Wolson (2007) found that some urban consumers and environmental interest communities in South Africa have lately become sceptical of GM crops, reflecting popular views in the UK. It is therefore important to note and discuss how the socio-political disputes in the northern countries can significantly influence the regulation and decision-making policies on issues relating to genetically modified organisms (GMOs) in the south.

For most developing countries, the USA and the EU are the major trading parties, aid donors and providers of direct foreign investment. The size of the European and North American markets means that they strongly influence global food and feed production and marketing. For these reasons, among others, their policies and decisions on biotechnology and agricultural trade can considerably affect the policies of many countries in the south (Glover, 2003). Although the stance in the EU on GM crops probably affects the trade and adoption of GM crops in developing countries, the policies and socio-economic backgrounds differ among nations, which also has a definite influence on the utilisation of GMOs. Indeed, such differences indicate that numerous other factors influence developing countries' perceptions of GM crops besides the perception of GM crops in the EU.

Owing to these many influencing factors, developing countries are currently debating the acceptance or rejection of GMOs. Producers of GM foods have their obvious reasons for exportation, and importing countries have their own reasons to accept them. Yet there is no unified reason for their acceptance or rejection. Some countries may accept or reject GMOs for economic, social and/or political reasons, while other countries may accept or reject them because of environmental and health concerns. However, an inevitable factor of influence in the decision-making process of developing countries is the stances of some countries such as the USA and supranational powers such as the EU. When it comes to GM foods, the USA and the EU are on opposite sides of the spectrum and, due to their dominance in global trade, the ongoing conflict between the EU and the USA has huge implications for weaker nations (Gonzalez, 2009). Yet there is another factor of relevance in the ongoing dispute, which is that GMOs can play a significant role in developing countries' food safety and agricultural development requirements, and their ability to cope with associated problems of hunger, disease, malnutrition, immoderate increase in population and poverty levels. For instance, Golden Rice, which is genetically enriched rice with improved amounts of vitamin A, could reduce vitamin A deficiency in Africa and Southeast Asia, where vitamin A and malnutrition issues are common amongst children (Adenle, 2011a; Potrykus, 2010). There is also a growing body of literature that shows the positive impacts of GM crops in relation to food security and poverty reduction (Azadi *et al.*, 2015). A good example of this can be found in India, where Qaim and Kouser (2013) analysed the food security impacts of GM crops on a micro scale. They concluded that the adoption of GM cotton raised family incomes and has consequently substantially enhanced calorie consumption and the quality of daily diets. This technology has decreased food insecurity by 15%–20% among households that were producing cotton. Similarly, through conducting a case study in Pakistan, Ali and Abdulai (2010) found that adopting Bt cotton had a positive and significant impact on yield, household income and poverty reduction.

GMO Risks and Benefits for Developing Nations

For the first time, in 2012 developing countries planted more GM crops than developed nations, with 52% of global biotech crops. By 2015, of the 28 countries which were growing biotech crops, 20 were developing whereas only 8 were developed countries (James, 2015). Latin American, African and Asian farmers together grew 97.1 million hectares (54%) of the global 179.7 million biotech hectares (compared with 53% in 2014), whereas industrial countries planted 83 million hectares or 46% (versus 47% in 2014). In fact, developing nations planted 14.1 million hectares more crops than the industrial countries, a trend that is predicted to continue. This contradicts the pessimistic assumption (before the commercialisation of biotechnology in 1996) that GM crops could only be effective and utilised in industrial countries and would not be accepted and adopted by developing countries (James, 2014). In 2014, farmers in developing countries earned 46% of the farm income benefits. Over 19 years (1996–2014), the cumulative farm income gain derived by the farmers was 50.6% (US$76.06 billion). Moreover, the annual updated analysis shows that there continue to be very significant net economic benefits at farm level, amounting to US$17.7 billion in 2014 and US$150.3 billion for the 19-year period 1996–2014. These economic gains have

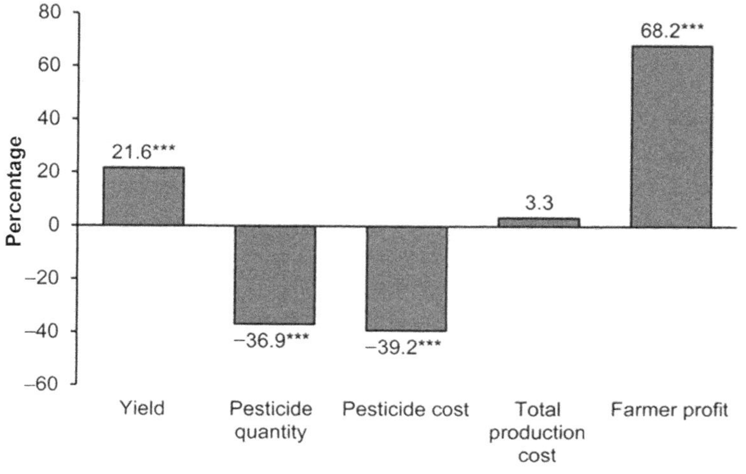

Figure 15.1. Impacts of GM crop adoption. Average percentage differences between GM and non-GM crops are illustrated. Results refer to all GM crops, encompassing herbicide-tolerant and insect-resistant traits. Adapted from Klümper and Qaim (2014). *** Significance at the 1% level.

been divided roughly 50% each between farmers in developed and developing countries (Brookes and Barfoot, 2016).

In general, GM crops have been shown to have many economic and environmental benefits, as well as risks and concerns related to consumers' health and welfare. From the economic point of view, a meta-analysis conducted by Klümper and Qaim (2014) on the economic effects of GM crops demonstrates that, on average, there are major agronomic and economic profits from GM products. Though effects differ according to the GM crop varieties in general, and the geographical location of the adoption in particular, they found that GM crops have raised yields by an average of 22% (Figure 15.1). This improvement in income has occurred not as a result of the larger capability of GM technology in generating benefits, but instead is largely due to pest management programmes, leading to less crop loss (Qaim and Zilberman, 2003). Simultaneously, the shift to GM crops has decreased pesticide use by 37% and its cost by 39%. GM seeds cost more than non-GM seeds; however, the extra seed costs are balanced by the decreased use of mechanical and chemical means of pest control, and the average profit earned by farmers who adopt GM crops has been shown to increase by 69%. According to the findings of Klümper and Qaim (2014), the income and profit from utilising GM crops are greater in developing than in developed countries.

In contrast, GM technologies may also pose significant economic risks, such as the failure of the technology to deliver the expected benefits, and the loss of export markets due to ongoing consumer resistance to GM foods and regulatory restrictions.

Environmentally speaking, GM technology may reduce agricultural clearing and thereby decrease habitat loss and the subsequent damage to biodiversity (Adler, 2000;

Carpenter, 2011). Use of pest-resistant GM crops can also lessen pesticide use, providing health benefits for agricultural workers and environmental benefits for non-target species. Although the potential for such methods in developing countries remains largely unevaluated, GM technology may make the use of low-till agricultural techniques easier, which, in turn, may reduce pollution run-off, erosion and the release of greenhouse gases captured in the soil (Paarlberg, 2000). In South Africa smallholder farmers have benefited from minimum tillage with herbicide-tolerant GM maize (Gouse *et al.*, 2006). A study by Gouse (2012) showed that the adoption of herbicide-tolerant maize by commercial farmers in South Africa has been impressive, increasing from 5% to almost 80% within five years, because they benefit through saving on weed-control activities and expenditures (herbicide, labour, machinery and fuel). Another study on GM technology between 1996 and 2014 documented that the use of insecticides has decreased by 51.6% (79.7 million kg) and the environmental impact from insecticides applied to the maize crop has fallen by 55.7% (Brookes and Barfoot, 2016). The authors argue that this is the equivalent of removing 11.88 million cars from the road in terms of the significant reduction of greenhouse-gas emissions. However, there are also potentially significant environmental hazards from the use of GM crops. A GM crop could transfer modified genes to wild relatives and potentially generate 'superweeds', or could itself become a weed that could become a major threat to biodiversity (but still considerably less of a threat than conventional herbicides (International Survey of Herbicide Resistant Weeds (ISHRW), 2014)). In addition, use of pest-resistant GMOs may hasten the development of pest resistance (Meijer and Stewart, 2004), but see Tabashnik *et al.*, (2013, 2015) with some more optimistic perspectives.

Furthermore, an important obstacle that must be considered is the limited capability of developing countries to assess the environmental and economic risks of GMOs. One substantial reason for this limited capacity can be associated with the powerful impact external elements have on decision-making in developing countries (Nuffield Council on Bioethics, 2003), meaning that there is a need for careful case-by-case evaluation and management of such risks (Bazuin *et al.*, 2011). The modern agricultural biotechnologies, including GMOs, are perceived as a motivating and valuable development by some people, whereas others strongly oppose their use, raising concerns about both the environment and food security (Azadi and Ho, 2010). Such concerns are forcing consumers and policymakers in numerous countries around the world to respond and, in some situations, to over-react. In debates on GMOs, because of institutional and agenda differences since the mid 1990s, scientific uncertainty and ethical concern have been considered differently in the EU and USA due to the 'Genomic Misconception' that there is a basic difference between conventional and transgenic crops ruling the EU's risk assessment (Peterson and White, 2010).

The Case of the US and European GMO Regulations and the WTO

Both the EU and the USA are willing to establish a safe food supply and valid regulatory systems. Nevertheless, they have utilised extremely different regulatory approaches to cope with the growing numbers of GM products, crops and animal feeds that have been introduced onto the market. As a result, they have faced many conflicts with each other over issues relating to GM food (Patterson and Josling, 2002). The divergence in regulatory approaches between the USA and the EU is confusing, and there are many similarities between the economic structures of the two sides. In this section we examine the content of the regulatory systems and the nature of the social, political and institutional elements that have exerted influences contributing to the development of these approaches. We then explore the potential impact of these two regulatory systems on developing countries and international trade (Ammann, 2014).

Differences in Regulatory Approaches between the EU and the USA

The USA adheres to the concept of substantial equivalence in relation to GM products, affirming that, if a GM product is essentially the same as its conventional counterpart, it should be regulated in just the same way as the conventional product. The EU, on the other hand, takes a 'precautionary approach'. This emphasises that, if the scientific evidence regarding a practice or a product's possible harms to human or environmental health is insufficient or unconvincing, it should be more strongly regulated or even forbidden if there are plausible reasonable concerns, thus providing protection against future unexpected problems. Under this approach, GM products are assumed to be, by their nature, different from their conventional counterparts and hence need a separate regulatory system (Hanrahan, 2010). The precautionary principle, which originated from the German environmental legislation of the 1960s, 'emphasizes a cautious approach to adopting a new technology when existing scientific understanding is incomplete or when there is not a consensus about the nature of the threat' (Young, 2003, p. 464). Accordingly, the precautionary principle shifts the burden of proof to demonstrate the safety of any new technology to the technology's developer. In this context, there are another three fundamental differences that separate the EU and US approaches to GMOs (European Union Center of North Carolina, 2007). First, the EU approval process is not based on scientific evidence alone, but also considers other, non-scientific, factors, such as 'societal, economic, traditional, ethical and environmental factors' in all risk management decisions (European Union Center of North Carolina, 2007). Second, compared with the USA, the EU grants consumers greater rights to access information. Accordingly, strict rules on labelling and traceability of GMOs have been enforced in the EU since the adoption of the 2001 legislation. Under these rules, labelling is mandatory for all authorised GM products, applying to all products that contain a trace of more than 1% of GMO content. Finally, the 'safeguard clause' in the EU legislation (Article 23 in Directive 2001/18/EC) governing plant biotechnology allows member states to ban the cultivation of biotech crops in their territories, if new scientific evidence suggests that such cultivation could be harmful to the environment, or to human or animal health (Salmon, 2015). Currently, 19 EU countries have 'opted out' of growing GMO crops within all or part of their territories (Walia, 2015). Besides, the labelling of GM products in the EU has caused some consternation in the USA and Canada. Bitter divisions between states led to a conflict over the implications for international trade in GM products and a dispute at the WTO. The USA complained that the EU approval system for biotech products was so slow that it amounted to a moratorium (Josling, 2015). The WTO rules become more directly relevant when there is trade in GM products or ingredients between countries. In the current EU legislation, the responsibility for allowing a GM product to enter the market lies with the EU institutions and not the member states. However, the EU is now reportedly reviewing its biotechnology approval process, and considering giving member states the option to ban the import of GM food within their territory (Josling, 2015).

Consequences for Developing Countries

There are five major components of international regulation that relate to GM crop usage, trade and research (Nuffield Council on Bioethics, 2003):

- The WTO agreements aiming to manage international trade constraints. Through this framework, as described above, the USA has challenged the EU over the GM crop regulations and authority (Jonquieres *et al.*, 2003).
- The *Codex Alimentarius*, a range of international codes of actions, directives and guidelines related to food safety. The WTO relies on the Codex for developing its regulations and decisions.
- The *Cartagena Protocol on Biosafety* under the *Convention on Biological Diversity*, a multilateral agreement that governs GMO practices applied over national borders that may have a negative impact on biodiversity.
- A multilateral agreement by the UN FAO called the *International Treaty on Plant Genetic Resources for Food and Agriculture*, which applies to any GM plant that is of major importance for agriculture and food.
- Regulations and guidelines by the EU and its related policies on environment, agriculture and GMOs at regional levels.

Such institutions at international level have made major contributions towards the approval of GM crops and biosafety processes. The most direct effect international regulations have on developing countries comes from the application of relevant international agreements. As mentioned above, the Codex Commission provided recommendations for GM food approval, and the Cartagena Protocol, which has been ratified by 170 countries, developed guidelines for regulating environmental aspects of GM crops. Yet, there are no distinct directions in any international agreement regarding the regulation of marketing issues (e.g., labelling, traceability and segregation) (Gruère, 2006). Developing countries that are not signatories to any of these agreements can be affected by international regulations due to the links within international trade systems, while those developing countries that are members of the UN Codex Alimentarius, the Biosafety Protocol or the WTO have the right to participate in designing new rules within these agreements. Ideally, most developing countries should actively be contributing to the design and implementation of international agreements. Reports have also revealed that many countries in the developing world pay special attention to their business connections with the EU prior to the adoption of GM crops, since they fear losing access to those markets (Glover, 2003; Meijer and Stewart, 2004; Gruère and Sengupta, 2009). Furthermore, southern African countries affected by famine are concerned about receiving GM food aid and about planting GM crops, fearing that it could be considered a negative point for exporting GM products to the EU. A possible reason for such concerns could be the idea amongst traders in developing countries that 'current markets in Europe (and Japan in some cases) are the only markets for exports' (Gruère and Sengupta, 2009, p. 18). As a result, developing countries are likely to display a 'go slow' approach with regard to GM technology, resulting in the delayed confirmation of biosafety regulations and the lack of a biosafety approach in many countries (Adenle, 2012). Other developing countries, such as Brazil, may have the impression that they can benefit from the difficulties of producers in the USA by responding to the high demand from the EU for GM and non-GM crops such as soybeans. According to Bennett (2013), 85% of Brazil's soybean crop is already GM, and the EU is the main importer of Brazilian soybeans as the EU's soybean imports from Brazil (used to make animal feed) nearly tripled during this decade (Brown-Lima *et al.*, 2012). Nonetheless, decision-makers on agricultural biotechnology in many developing countries are not knowledgeable about market-related issues. Furthermore, they tend to be more dependent on (but also more influenced by) export considerations, especially as related to their market access in Europe (Gruère and Sengupta, 2009). Some developing countries, to some extent, may be insulated from the impacts of the EU–USA controversy. However, national and international organisations such as the WTO do not clearly mediate the disputes between the USA and the EU (Reddy *et al.*, 2013; Adenle, 2011b). This is mainly because the various international organisations engaged in development assistance (such as alleviating poverty and improving economic opportunities through trade policy) pursue their own agendas, and there seems to be no cooperation on GM-related issues between the USA and the EU. This lack of cooperation affects many developing nations, hindering the making of rational decisions about adopting GM agricultural technologies (Adenle, 2011b; Meijer and Stewart, 2004).

The Role of National and International Organisations

Decisions over the growing, regulation and development of GM crops can be made at various levels and are affected by international agendas and domestic policies. They are also formulated by sub-national administrations, local authorities and, eventually, by farmers. Developing countries usually must address the concerns that GM crop policies raise within the context of their specific development requirements, together with the complicated system of international governance that has been established for GM crops (Nuffield Council on Bioethics, 2003). Thereby, governments of developing countries should have primary authority over decisions on GMO policies for their own countries, according to their local requirements and conditions, and international bodies should confirm and simplify this role. The international order, however, has been significantly unsuccessful in fulfilling this duty (Adenle, 2011a; Meijer and Stewart, 2004). This is a possible result of the lack of a common interest among international bodies to enhance and encourage GM technology due to the fact that each organisation is following its own goals and agenda. Furthermore, the fragmentation of international development assistance activities results in a situation that forces the ministries and governmental bodies of developing countries to become clients of various donor nations or organisations, making it harder for developing countries' governments to effectively decide on GMO issues (Meijer and Stewart, 2004).

At the international level, diverse international organisations, including the Organisation for Economic Co-operation and Development (OECD), United Nations Environment Programme (UNEP), WTO and Codex, are engaged in GMO issues through different programmes in order to assist developing countries in agricultural development, comprising growing and using GMOs, and in creating regulatory systems for biosafety. However, a lack of international harmonisation with respect to GM policy still undermines the application of GMOs in many of these countries. While international harmonisation may be a good idea, the processes over the past decade have been cumbersome and lack transparency, particularly in view of differences among developed and developing countries, in setting out clear international protocols for regulating GMOs. In this case, some developed countries (e.g., EU countries) have seized the opportunity to hijack the process at the international and national levels. This view is reflected in Chapter 14, where the author argues that the establishment of regional centres to manage GMO regulation and risk analysis in developing countries should be encouraged.

Potential Solutions Addressing the Developing Countries' GMO Dilemma

It is important to note that developing countries should be able to hold their own positions reflecting the various general economic, environmental, health and social risks and benefits of GMOs and the different GM crops and traits in a country-specific context. The types of regulatory decisions that governments make on agricultural biotechnology in all countries should take into consideration the following factors: food security and economic growth, stakeholder opinions, legal issues, liability and compensation, environmental and health risks and international agreements on GM. Accordingly, Gruère (2006) developed a diagram showing that policymakers should be able to consider a wide set of objectives when deciding whether to adopt selected GM crops or to introduce any type of biosafety or marketing regulation. He represented a possible set of four policy objectives for a typical developing country as shown in Figure 15.2. He assumes that most countries try to '(1) maintain or increase their market access to all importing countries, including those with stringent regulations on GM food, (2) find solutions to increase agricultural productivity, including the use of beneficial GM crops, (3) maintain low consumer price and sufficient food availability for consumers, and (4) manage biosafety risks (food safety and environmental) related to GM crop technology, while satisfying their international obligations'. The four numbered zones in Figure 15.2 represent areas where countries have achieved two of the four objectives and the four zones defined by letters in Figure 15.2 represent areas where countries may have achieved three objectives out of four.

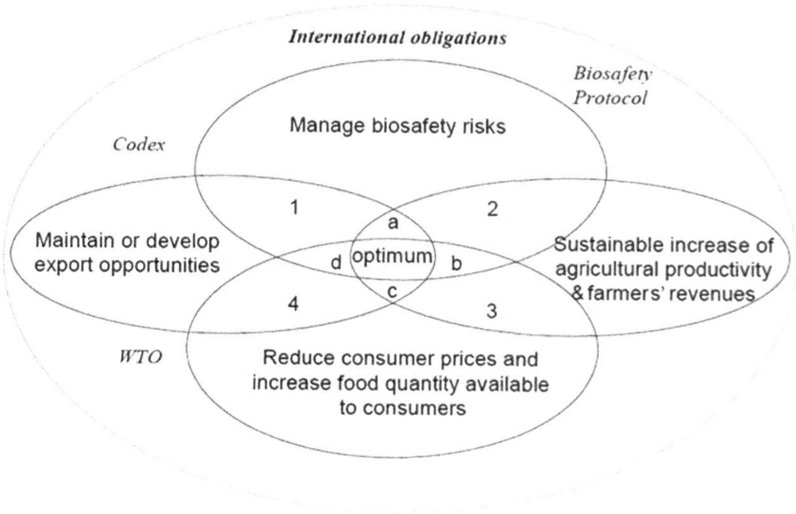

Figure 15.2. Combining policy objectives of developing countries for achieving optimum decisions about the adoption and developing of regulations on GMOs. Adapted from Gruère (2006).

Another necessary condition for achieving the desired result of adopting GM crops is to open the decision-making processes up to the public, which may help ensure that decision-makers have the best possible information at their disposal so that they can evaluate the benefits and risks of GMOs more accurately. Public participation could also help ensure better transparency and accountability in the decision-making process (Glowka, 2003). Many international legal instruments address the public's access to information in relation to GMOs, though few specifically address public participation in decision-making about GMOs (FAO, 2005). This has raised the question of the role and positions of UN organisations in facilitating the application of GMOs in developing countries, especially those in Africa (Adenle *et al.*, 2013). Governments and international bodies also need to take public concerns into account when elaborating risk-related regulations and when dealing with trade disputes (Marris, 2004). It is recommended that the international and suitable non-governmental organisations which observe the agricultural policy development of developing countries should evaluate the outcomes and effects of regulatory policies for the adoption of GM crops in developing countries. This will probably require an international organisation to take a leadership role as well as a holistic and even-handed approach to GMO issues so as to establish a procedure by means of which to report the impact of the regulations accordingly (Nuffield Council on Bioethics, 2003; Meijer and Stewart, 2004; Adenle *et al.*, 2013).

References

Adenle, A. A. (2011a). Global capture of crop biotechnology in developing world over a decade. *Journal of Genetic Engineering and Biotechnology* **9**(2), 83–95.

Adenle, A. A. (2011b). Response to issues on GM agriculture in Africa: are transgenic crops safe? *BMC Research Notes* **4**, 388.

Adenle, A. A. (2012). Are transgenic crops safe? GM agriculture in Africa. United Nations University, Japan. [Online]. Available from http://unu.edu/publications/articles/are-transgenic-crops-safe-gm-agriculture-in-africa.html

Adenle, A. A. *et al.* (2013). Status of development, regulation and adoption of GM agriculture in Africa: views and positions of stakeholder groups. *Food Policy* **43**, 159–166.

Adler, J. H. (2000). Cartagena Protocol: biosafe or bio-sorry? *Georgetown International Environmental Law Review* **761**, 772.

Ali, A. and Abdulai, A. (2010). The adoption of genetically modified cotton and poverty reduction in Pakistan. *Journal of Agricultural Economics* **61**(1), 175–192.

Ammann, K. (2014). Genomic misconception: a fresh look at the biosafety of transgenic and conventional crops. A plea for a process agnostic regulation. *New Biotechnology* **31**(1), 1–17. [Online]. Available from www.ncbi.nlm.nih.gov/pubmed/23684919

Azadi, H. and Ho, P. (2010). Genetically modified and organic crops in developing countries: a review of options for food security. *Biotechnology Advances* **28**(1), 160–168.

Azadi, H. *et al.* (2015). Genetically modified crops: towards agricultural growth, agricultural development or agricultural sustainability? *Food Reviews International* **31**(3), 195–221.

Bazuin, S. *et al.* (2011). Application of GM crops in Sub-Saharan Africa: lessons learned from Green Revolution. *Biotechnology Advances* **29**, 908–912.

Bennett, D. (2013). Brazil Says 'Yes' to Genetically Modified Foods. Mexico Says 'No'. [Online]. Available from www.bloomberg.com/news/articles/2013–10–30/brazil-says-yes-to-genetically-modified-foods-dot-mexico-says-no

Brookes, G. and Barfoot, P. (2016). GM crops: global socio-economic and environmental impacts 1996–2014. PG Economics Ltd, UK. [Online]. Available from www.pgeconomics.co.uk/pdf/2016globalimpactstudymay2016.pdf

Brown-Lima, C. *et al.* (2012). An overview of the Brazil–China soybean trade and its strategic implications for conservation. [Online]. Available from www.nature.org/ourinitiatives/regions/southamerica/brazil/explore/brazil-china-soybean-trade.pdf

Carpenter, J. E. (2011). Impact of GM crops on biodiversity. *GM Crops* **1**, 7–23.

European Union Center of North Carolina (2007). Policy Area: GMOs. [Online]. Available from http://europe.unc.edu/wp-content/uploads/2013/08/Brief0705-GMOs.pdf

FAO (2005). Public participation in decision-making regarding GMOs in developing countries: how to effectively involve rural people. Summary Document to Conference 12 of the FAO Biotechnology Forum. 17 January to 13 February 2005, Rome, Italy. [Online]. Available from www.fao.org/biotech/logs/C12/summary.htm

Glover, D. (2003). *GMOs and the Politics of International Trade. Democratising Biotechnology.* Genetically Modified Crops in Developing Countries Briefing Series. Briefing 5. Brighton: Institute of Development Studies.

Glowka, L. (2003). Law and modern biotechnology: selected issues of relevance to food and agriculture. FAO Legislative Study 78. [Online]. Available from www.fao.org/docrep/006/Y4839E/Y4839E00.HTM

Gonzalez, X. (2009). The politics of genetically modified organisms: global rules, local needs. The University of Texas at Austin. SIT Spring class 2009. International Studies Organizations and Social Justice, 39 pp.

Gouse, M. (2012). GM maize as subsistence crop: the South African smallholder experience. *AgBioForum* **15**(2), 163–174. [Online]. Available from www.agbioforum.org/v15n2/v15n2a05-gouse.htm

Gouse, M. *et al.* (2006). Output and labour effects of GM maize and minimum tillage in a communal area of KwaZulu Natal. *Development Perspectives* **2**(2), 192–207.

Gruère, G. P. (2006). An analysis of trade related international regulations of genetically modified food and their effects on developing countries. International Food Policy Research Institute (IFPRI), Environment and Production Technology Division (EPT) Discussion Paper 147. Washington, D.C., 78 pp.

Gruère, G. and Sengupta, D. (2009). Biosafety decisions and perceived commercial risks. The role of GM-free private standards. Environment and Production Technology Division, The International Food Policy Research Institute (IFPRI) Discussion Paper 00847. [Online]. Available from www.ifpri.org/sites/default/files/publications/ifpridp00847.pdf

Hanrahan, C. E. (2010). Agricultural Biotechnology: The U.S.–EU Dispute. Congressional Research Service Reports. Paper 69. [Online]. Available from http://digitalcommons.unl.edu/crsdocs/69

ISHRW (2014). International Weed Science. [Online]. Available from www.weedscience.org

James, C. (2014). Global Status of Commercialized Biotech/GM Crops: 2014. ISAAA Brief No. 49. Ithaca, NY: ISAAA.

James, C. (2015). Global Status of Commercialized Biotech/GM Crops: 2015. ISAAA Brief 50–2015: Executive Summary. Ithaca, NY: ISAA.

Jonquieres, G. *et al.* (2003). Sowing discord: after Iraq, the US and Europe head for a showdown over genetically modified crops. *Financial Times* 14 May 2003, p. 21.

Josling, T. (2015). A review of WTO rules and GMO trade. *Biotechnology* **9**(3). [Online]. Available from www.ictsd.org/bridges-news/biores/news/a-review-of-wto-rules-and-gmo-trade

Klümper, W. and Qaim, M. (2014). A meta-analysis of the impacts of genetically modified crops. *PLoS ONE* **9**(11), e111629, doi:10.1371/journal.pone.0111629.

Marris, C. (2004). Issues concerning public awareness and attitudes towards genetically modified bananas and tropical fruits. Third Session of the Intergovernmental Group on Bananas and Tropical Fruits, Puerto de la Cruz, Spain, 22–26 March 2004. [Online]. Available from ftp://193.43.36.93/unfao/bodies/ccp/ba-tf/04/j0803e.pdf

Meijer, E. and Stewart, R. (2004). The GM cold war: how developing countries can go from being dominos to being players. *RECIEL* **13**(3), 247–262.

Nuffield Council on Bioethics (2003). Chapter 5 – Governance in the use of genetically modified crops in developing countries. A Follow-up Discussion Paper, pp. 63–82. [Online]. Available from www.conacyt.gob.mx/cibiogem/images/cibiogem/comunicacion/publicaciones/Nuffield_Council-GMOs-for-dev-countries.pdf

Paarlberg, R. (2000). Agrobiotechnology choices in developing countries. *International Journal of Biotechnology* **2**(1–3), 167–168.

Patterson, L. A. and Josling, T. (2002). Regulating biotechnology: comparing EU and US approaches. European Policy Papers #8. [Online]. Available from http://aei.pitt.edu/28/1/TransatlanticBiotech.pdf

Peterson, M. J. and White, P. A. (2010). The EU–US Dispute over Regulation of Genetically Modified Organisms, Plants, Feeds, and Foods. International Dimensions of Ethics Education in Science and Engineering. [Online]. Available from http://scholarworks.umass.edu/cgi/viewcontent.cgi?article=1007&context=edethicsinscience

Potrykus, I. (2010). Lessons from the 'Humanitarian Golden Rice' project: regulation prevents development of public good genetically engineered crop products. *New Biotechnology* **27**, 466–472.

Qaim, M. and Kouser, S. (2013). Genetically modified crops and food security. *PLoS ONE* **8**(6), e64879, doi:10.1371/journal.pone.0064879.

Qaim, M., and Zilberman D. (2003). Yield effects of genetically modified crops in developing countries. *Science* **299**, 900–902.

Reddy, P. B. *et al.* (2013). GM technology can solve the world food crisis. *Science Secure Journal of Biotechnology* **2**(2), 36–40.

Salmon, D. G. (2015). Agricultural biotechnology annual. Global Agricultural Information Network. [Online]. Available from http://gain.fas.usda.gov/Recent%20GAIN%20Publications/Agricultural%20Biotechnology%20Annual_Paris_EU-28_7-23-2015.pdf

Tabashnik, B. E. *et al.* (2013). Insect resistance to Bt crops: lessons from the first billion acres. *Nature Biotechnology* **31**(6), 510–521. [Online]. Available from www.ask-force.org/web/Bt1/Tabashnik-Insect-Resistance-First-Billion-Acres-2013.pdf

Tabashnik, B. E. *et al.* (2015). Dual mode of action of Bt proteins: protoxin efficacy against resistant insects. *Nature Scientific Reports* **5**, 15107. [Online]. Available from www.ask-force.org/web/Bt1/Tabashnik-Dual-mode-action-Bt-proteins-2015.pdf

Walia, A. (2015). Here's why 19 countries in Europe just completely banned genetically modified crops. [Online]. Available from www.collective-evolution.com/2015/10/07/heres-why-19-countries-in-europe-just-completely-banned-genetically-modified-crops/

Wolson, R. A. (2007). Assessing the prospects for the adoption of biofortified crops in South Africa. *AgBioForum* **10**, 184–191. [Online]. Available from www.agbioforum.org/v10n3/v10n3a08-wolson.htm

Young, A. R. (2003). Political transfer and 'trading up'? Transatlantic trade in genetically modified food and US politics. *World Politics* **55**(4), 457–484.

16 Biosafety Communication: Beyond Risk Communication

Andrea Sonnino and Sandra Sharry

'O Tiger-lily', said Alice, addressing herself to one that was waving gracefully about in the wind, 'I wish you could talk!' 'We can talk', said the Tiger-lily: 'when there's anybody worth talking to'.

(Lewis Carroll, 1871)

Introduction

In spite of the dramatic rise of the global acreage cultivated with genetically modified organisms (GMOs) during the last two decades, they continue to be the subject of a harsh controversy (Ruane and Sonnino, 2011). The debate is equally polarised in industrialised countries (Gaskell *et al.*, 2011; Pew Research Center, 2015), emerging economies (Macnaghten and Carro-Ripalda, 2016) and the least-developed countries (Whitty *et al.*, 2013). Public opposition to GMOs does not seem to be diminishing in spite of increasing scientific evidence that they are safe to eat and not harmful to the environment (Blancke *et al.*, 2015).

Recent studies indicate that the lack of transparency and public consultation in the GMO approval process is among the most important triggers of public concern in Brazil, India and Mexico (Macnaghten and Carro-Ripalda, 2016). This result confirms the authors' experience in many developing countries, including Bolivia, the Dominican Republic and Grenada (Sonnino and Sharry, unpublished results), where communication challenges and issues are very specific. There is therefore a critical need to understand how the release of GMOs can be managed in a more inclusive manner.

This chapter aims to propose a more open and participatory approach to decision-making about GMO releases. The proposal is based on the lessons learnt from a very extended field experience (Sensi *et al.*, 2009), even though it has not yet been tested in its entirety. The first part of the chapter analyses the international framework, which establishes agreed principles related to communication and public participation in biosafety decision-making processes. Subsequently, we examine risk communication as a key component of biosafety risk analysis and describe the two distinct but complementary functions of access to information and public participation. The next section critically reviews different communication approaches and suggests adopting a participatory approach, the advantages and challenges of which are critically considered in the final part.

The International Legislative Framework

The Cartagena Protocol on Biosafety (Convention on Biological Diversity, 2000) (Article 23) requires that parties promote and facilitate public awareness, education and participation, and that they must 'consult the public in the decision-making process regarding Living Modified Organisms (LMOs) and shall make the results of such decisions available to the public'.

The document adopted by the Codex Alimentarius Commission (CAC, 2003) stipulates that 'Effective risk communication is essential at all phases of risk assessment and risk management.' It adds that 'Risk communication should include transparent safety assessment and risk management decision-making processes. These processes should be fully documented at all stages and open to public scrutiny' and concludes that 'Effective risk communication should include responsive consultation processes. Consultation processes should be interactive. The views of all interested parties should be sought and relevant food safety and nutritional issues that are raised during consultation should be addressed during the risk analysis process.'

The Convention on access to information, public participation in decision-making and access to justice in environmental matters of the United Nations Economic Commission for Europe (UNECE) addresses, *inter alia*, the people's right of access to public participation and information on environmental issues and the authorities' obligation to make such information available for public scrutiny. Article 6b (the Almaty Amendment) stipulates that 'the parties must provide effective information and public participation before making decisions about the release or marketing of GMO' (UNECE, 2005). The Almaty Amendment was adopted by the Meeting of the Parties to the CBD in 2005 and will enter into force when it has been ratified by at least three quarters of the Parties. The UNECE (2003) published the 'Guidelines on access to information, public participation and access to justice with respect to genetically modified organisms', known as the Lucca guidelines.

The Universal Declaration on Bioethics and Human Rights (UNESCO, 2006), adopted in Paris in 2005, recognises the need for biosafety and the importance of biosafety; emphasises the need for transparency in the decision-making processes; and affirms the need for biosafety communication and the principle of democratic participation in decision-making.

The FAO international conference on Agricultural Biotechnologies in Developing Countries, held in Guadalajara in 2010, advocated that effective communication and participation strategies are necessary to encourage and promote public involvement and empowerment in decision-making processes (FAO, 2011).

In summary, international declarations, conventions and treaties unanimously support two general principles: (i) the process of evaluation and management of the risks inherent to the release of GMOs should be transparent and should grant full access to any related information; and (ii) public participation in the decision-making processes related to the release and commercialisation of GMOs should be pursued.

Figure 16.1. Diagram of risk analysis components.

Risk Communication as an Integral Component of Risk Analysis

Risk analysis is defined as 'a process consisting of three components: risk assessment, risk management and risk communication performed to understand the nature of unwanted, negative consequences to human and animal health, or the environment' (FAO, 2001). Definitions proposed by other organisations, including the Australian Government (2013) and CAC (2008), state that each component is integral to the overall risk analysis process (Figure 16.1).

Risk assessment is a process designed to estimate the risks to human health and the environment associated with the use of a GMO or its products. Science-based assessments are then integrated with political and social considerations in order to design and implement risk management strategies able to prevent, reduce or eliminate such risks. Risk communication is a continual and iterative process that at the same time is fed by and provides feedback to both risk assessment and risk management and helps to define the issues and to link the different phases of risk analysis to each other (Brandenberg *et al.*, 2011).

Risk communication is used to explain both how and why decisions are taken, acknowledge any concern raised by the stakeholders and incorporate these concerns and any other information into the decision-making process (Figure 16.2) (FAO, 2008). As the CAC (2008) asserts, 'its major function should be to ensure that all information and opinion required for effective risk management is incorporated into the decision-making process'. Risk communication should therefore include mechanisms for inviting and considering public comments on the biosafety reports.

In addition, communication with the public can also contribute to a risk management plan by explaining the planned measures to people responsible for their implementation

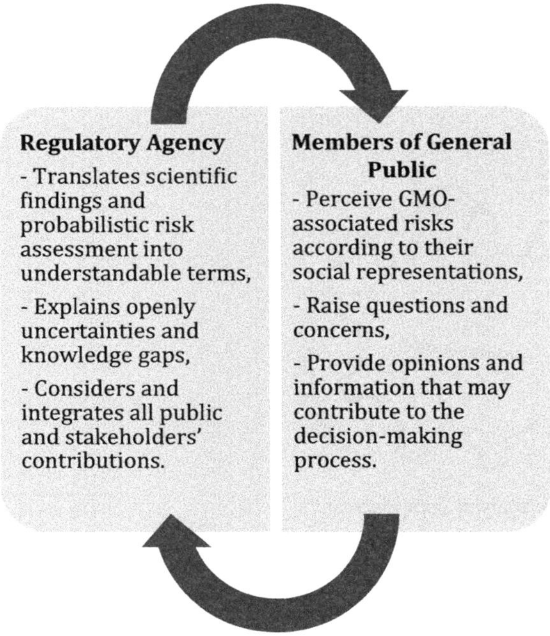

Figure 16.2. Bidirectional risk communication flow chart.

or providing information to people about risks they can control by themselves, or dissuading people from engaging in risky behaviours. Food labels that contain information about the presence of allergens are a good example of risk communication as an integral component of risk management.

Access to Information and Public Participation

According to the international instruments described above, risk analysis includes two distinct but complementary communication components: (i) regulatory risk communication, which consists mainly of means and tools to ensure open access to regulatory information, including judicial and administrative proceedings; and (ii) promotion of public participation in decision-making.

Access to information ensures transparency and openness to the entire risk analysis cycle and addresses the right of the public to examine the decision-making process about authorisation of GMO releases and risk management measures. It is used as a means of holding competent biosafety authorities accountable to the general public and building credibility and trust. In that respect, it must be accurate, honest, truthful, timely and complete in order to convince people that the decisions taken are based on the best available evidence and that the public interest is being considered (Brandenberg *et al.*, 2011).

Table 16.1 Stakeholders involved at various levels of decision-making

Level of decision-making	Stakeholders involved
Adoption of a biotechnology policy and related regulatory framework	General public, research centres, famers' organisations, political parties and movements, NGOs, biotech companies
Strategic planning of biotechnology research	General public, research centres, famers' organisations
Approval of contained GMO experimentation	Scientists (developers and biosafety evaluators), regulators
Approval of confined GMO experimentation and multi-site trials	Scientists (developers and biosafety evaluators), regulators, inhabitants of experiment sites
Approval of GMO seed commercialisation	Scientists (developers and biosafety evaluators), regulators, farmers' organisations, seed producers, dealers in other inputs
Approval of GMO commercialisation for human consumption	General public, including consumers' groups, retailers, supermarket chains, scientists, regulators, farmers' organisations

Many signatories to the Cartagena Protocol set up specific websites or national biosafety clearing-houses to enable access to information on biosafety (UNEP-GEF Biosafety Unit, 2006). In many instances, web-based communication reaches only a limited audience, for many reasons, including poor connectivity, connection cost, computer illiteracy and logging-in difficulty. Other media used include newspapers, leaflets, radio and TV, with variable levels of reach. The authors' experience is that many national biosafety authorities are still reluctant to disclose all the information related to the GMO approval process and that in some instances this attitude has hampered public acceptance of GMO-related decisions. Access to information is obviously a necessary but not sufficient condition for public participation (Sharry, 2013).

Public participation is 'the equitable and active involvement of all stakeholders in the formulation of development policies and strategies and in the analysis, planning, implementation, monitoring and evaluation of development activities' (FAO, 2004). The stakeholders who should be engaged vary according to the decision-making level (Table 16.1) (Ruane and Sonnino, 2006). The concept of public participation recognises that everybody has the right to participate in making decisions that can affect their life (Oughton, 2005) and responds to the notion that decisions are perceived as wrong or right more according to how they are made than according to what was decided (Rawls, 1971). Provisions for public participation are enshrined in the national biosafety regulatory frameworks of most countries. For instance, members of the public are normally offered the opportunity to provide comments, opinions and additional information in response to any biosafety report within a specified timeframe, while some countries consult interested parties through public hearings (UNEP-GEF Biosafety Unit, 2006). Other countries met requirements for public participation by including stakeholders in

Table 16.2 Synopsis of communication approaches for GMO risk communication

Approach	Main purpose	Type of communication	Timing of communication	Characteristics of target groups considered
Elitist	To inform about decisions taken and defend them	One-way, top-down	After decision-making	None
Educational	To provide people with better knowledge about GMOs	One-way, top-down	After decision-making	Level of knowledge
Marketing	To convince people of benefits of GMOs	One-way, top-down	After decision-making	Concerns
Open access	To inform about decision-making process and gather information	Two-way, asymmetric	During and after decision-making	Opinions, factual information
Participatory	Engage people in decision-making about GMO release	Two-way, symmetric	Before, during and after decision-making	Knowledge, skills, attitudes, practices

national or institutional biosafety committees, where their voices are heard to variable degrees. It is, anyway, very important that the sectors of society which are not organised in non-governmental organizations (NGOs) or civil society groups are not excluded (Chee and Lim, 2005). Full enforcement of public participation often does not go beyond the formal application of biosafety regulations and remains therefore one of the biggest challenges of biosafety implementation in most developing countries (UNEP-GEF Biosafety Unit, 2006).

Approaches to Communication of GMO Risks: A Retrospective Glance

The persistent controversy that surrounds the release of GMOs confirms that the hitherto-adopted communication approaches failed to achieve the expected goals. This section analyses the evolution of communication approaches – and their underlying principles (Table 16.2).

The '**elitist or technocratic approach**' is based on the notion that the general public lacks a firm understanding of basic scientific facts and concepts (National Science Board, 2004). As a consequence of this knowledge deficit, decisions on GMO release should be delegated to scientists and other specialists, who are endowed with all the necessary technical knowledge and skills (see for instance messages by Djoulde, Infante and Kumbizmbi to the FAO e-conference on Public Participation in Decision-Making Regarding GMOs in Developing Countries (Ruane and Sonnino, 2006)). According to this approach, GMO-related decisions must merely be communicated and need be defended only if they elicit opposition (the 'Decide–Announce–Defend' (DAD) model

(Oughton, 2005)). According to Gibbons (1999), this approach derives from the prevailing contract between science and society, with science expected to produce 'reliable knowledge', provided merely that it communicates its discoveries to society.

The elitist model did not work. To overcome the opposition to GMO release, deemed to be generated by ignorance, it was proposed to educate the general public (Hilgartner, 1990). Also this '**educational (or pedagogic) approach**' (Sinemus and Egelhofer, 2007) is based on the science literacy/knowledge-deficit model, but here the assumption is that more information related to science and its application would allow people to develop well-informed perceptions about the impact of GMOs (Traynor *et al.*, 2007). This approach, although it is clearly asymmetric, recognises the necessity of communication, albeit only one-way and top-down, between those who know and those who do not. Communication is therefore directed to fill the knowledge gap, assuming that, if people learn to understand science, they will automatically share the scientists' opinions (Bubela *et al.*, 2009). Certainly information is a key element, but some studies found that factual knowledge and education level have little influence on acceptance of GMOs (Sinemus and Egelhofer, 2007; Gaskell *et al.*, 2011; Sorgo *et al.*, 2011); other authors found a positive correlation between better knowledge of genetic engineering and more positive attitudes towards it (Casanoves *et al.*, 2015). The educational approach ignores the fact that opinions are based only partially on knowledge, while ideologies, social identity, emotions, trust and other cultural factors (Lucht, 2015; Macnaghten and Carro-Ripalda, 2016) play a more important role. The technical information on how a GMO is obtained is not essential for the public to understand and form an opinion about the social and ethical implications of its adoption (Arantes *et al.*, 2011).

According to the '**marketing approach**', if better public education does not necessarily trigger a better acceptance of GMO release, communication should be proactively directed towards 'convincing' people of the benefits of the adoption of GMOs (Scholderer and Frewer, 2003). This approach considers only one-way communication, as does the educational approach, but recognises that the messages have to be tailored to the specific culture, values and social representations of the groups to which they are directed. The results obtained applying the marketing approach are not significantly better than the results obtained through other approaches: Scholderer and Frewer (2003) found that this technology-driven communication approach had a uniform attitude-activation effect that decreased consumers' acceptance of GMOs. The nutritional enhancement of GM food, which should be perceived as a benefit, does not seem to shift the attitude of reluctant consumers (Canavari and Nayga, 2009).

The already-cited paper by Gibbons (1999) argues that the prevailing contract between science and society should be substituted by a new contract ensuring that scientific knowledge is 'socially robust'. According to the new contract, knowledge is jointly produced by science and society, so that scientific production is perceived by society as both transparent and participative. It will not become so until the group of peer scientists 'is broadened to take into account the perspectives and concerns of a much wider section of the community'. This '**participatory approach**' moves away from the false premise of a knowledge deficit and recognises the symmetry of knowledge – and the mirroring symmetry of ignorance – among scientists, decision-makers and the general public; and proposes a multidirectional model of communication that leads to involvement in the decision-making process of all the interested parties, each bringing its multiple forms of knowledge, including beliefs, perceptions and concerns (Bubela *et al.*, 2009). The participatory approach shifts from the DAD model to the Meet–Understand–Modify (MUM) scheme (Oughton, 2005), moving to a 'deliberative democracy' model, the legitimacy of which is created by giving all the actors an 'equal chance to have their voices heard' (Nielsen *et al.*, 2004).

The participatory approach is not immune to criticism. If too much attention is paid to public opinion, and too little to expert views, there is a risk that the process will lead to decisions that are not scientifically and technically sound and not in the public interest. As stated by Galilei (1623), 'If reasoning were like hauling I should agree that several reasoners would be worth more than one, just as several horses can haul more sacks of grain than one can. But reasoning is like racing, and not like hauling, and a single Arabian steed can outrun a hundred plowhorses.'

On the other hand, if the participative decision-making process is not – or is perceived as not being – genuine, if, in other words, stakeholders' opinions do not influence the final decisions, biosafety communication can be considered a subtle form of propaganda aimed at manipulating the public into acceptance of already-taken decisions (Oughton, 2005).

Public engagement is the core of the concept of Responsible Research and Innovation recently adopted by the European Commission's Science in Society Programme (Owen *et al.*, 2012) to better align both research and its outcomes with the values, needs and expectations of society (EC, 2012). Consideration of societal needs in research and innovation projects proved to be successful for the development of bio-based products and for medical imaging (EC, 2013). Stakeholders' engagement and public dialogue were sought in relation to a major climate-engineering project (Stilgoe *et al.*, 2013) and the introduction of GM salmon in Northern Europe (Bremer *et al.*, 2015).

Biosafety Communication

The promotion of public participation in decision-making processes exceeds the function of the communication of risks inherent to GMO release, assuming a broader meaning, better captured by the term 'biosafety communication'. It has a two-pronged specific objective: (i) to facilitate the adoption of technically sound decisions about the assessment and management of GMO-related risks; and (ii) to build trust in the institutions and make socially robust decisions. Public participation is, anyway, a transformative process that creates relationships, strengthens capacities, generates collective learning, nourishes empowerment, leads to social change and creates opportunities for development. The benefits of biosafety communication therefore go beyond its specific objective.

The following are its most important features.

Duration. Biosafety communication should be viewed as a continuous process – the dialogue should be promoted as early as possible (Oughton, 2005) and continued throughout the entire decision-making process for approval of GMO release and beyond, during the post-release monitoring phase.

Object. Risk communication is based on identified risks and their management. The questions asked are 'is it safe?' and 'how can its safety be optimised?' Biosafety communication addresses also concerns, including non-evidence-based beliefs, pre-conceptions and perceptions, and includes the potentially generated benefits too. The questions asked are therefore 'what is in it for me?' and 'can I use it without generating harm?' Risk communication is based on science, while biosafety communication is centred on people.

Breadth of the target. Public dialogue should involve all stakeholders in order to solicit a broad range of opinions and incorporate laymen's concerns into decision-making processes. The breadth of the target is therefore the result of the trade-off between suitability and costs. This aspect brings in the problem of the representativeness of the audience involved, because the most vocal groups do not necessarily represent the majority of the interested parties (Oughton, 2005).

Nature of the target. Biosafety communication is directed to an audience that is diverse in terms of knowledge, education level, attitudes, beliefs and values (Table 16.1). Communication strategies must therefore be tailored to the specific characteristics of each target group, which can be understood through specific formative research.

Prior formative research. Strategic biosafety communication should be based on a thorough appraisal of the context and on a systematic empirical understanding of the stakeholders' values, knowledge and attitudes (Nisbet and Scheufele, 2009). Acunzo *et al.* (2014) proposed a participatory communication appraisal that comprises a situation analysis, to identify key issues, analyse policies and programmes and map interested organisations and institutions, and what they called a Field Participatory Rural Communication Appraisal to prioritise communication issues, identify the stakeholders and analyse their knowledge, skills, attitudes and practices, consider communication resources and map the social networks. Biosafety capacity-development projects in Bolivia, the Dominican Republic and Grenada used the Knowledge, Attitude and Practices (KAP) survey as a first step towards the formulation of a communication and public awareness strategy (Sensi *et al.*, 2009). The KAP survey was used also in a study of public attitudes towards the use of GM forest trees in 16 European and non-European countries (Kazana *et al.*, 2015). A KAP survey is a representative study of a specific population used to collect information on what is known, believed and done in relation to a particular topic, in this case biosafety. The interviewer verbally collects data using a structured and standardised questionnaire. Data can be analysed either quantitatively or qualitatively, depending upon the objectives and design of the study. Unlike simple surveys, the KAP survey considers broader cultural issues through questions about practices and beliefs in general (Sharry, 2013).

Participatory methods. Nielsen *et al.* (2004) discuss the landscape of participatory arrangements, namely their set-up, procedure and mode of participation, including consensus conferences, citizens' fora, future workshops, scenario workshops, Delphi technology studies, public hearings, focus groups and a referendum.

Participatory methods were successfully applied to develop biotechnology policies and plans in Bangladesh, Nicaragua, Paraguay and Sri Lanka (Sensi *et al.*, 2009).

Diversity of media. Biosafety communication should use a mix of different media, including the internet, brochures, flyers, educational games, radio and TV, workshops and face-to-face communication, as well as fairs and exhibitions. Dramas, games and songs have proved to be very effective, for instance in Mexico (de la Peña Wing, personal communication; Cordo, 2008) and Grenada (Protz, personal communication). Official languages are not always understood by all, so translations into indigenous languages can be necessary (Sensi *et al.*, 2009) while cartoons, storytelling and oral communication tools can be used where the illiteracy level is high. Positive experiences in that sense have been gained in Latin America by the project BIOEDUCAR.

Attributes. Biosafety communication should be clear, understandable, complete, honest, reliable and punctual, should build trust and mutual respect among all interested parties, and should influence the final decisions. Biosafety information should be communicated by authoritative and neutral sources, with no vested interests.

Conclusions

In many developing countries open access to biosafety-related reports is ineffectively ensured and public participation is hardly promoted. The initiative is therefore left in the hands of GMO developers and anti-GMO activists. The prevailing debate on GMOs is dominated by risk-associated issues: whether they are safe, as proponents maintain, or hazardous if not harmful, as opponents argue. Yet, the factors fuelling the controversy are much wider than that and involve cultural, political and social processes (Macnaghten and Carro-Ripalda, 2016).

The proposal presented here tries to develop a more reflexive approach, which includes scientists, stakeholders and the general public in a collective process of socially robust as well as scientifically sound technology development and uptake. The necessary mind-set change is not trivial, and its accomplishment requires time and commitment, making biosafety communication a difficult process, especially where the local cultures do not have a tradition of participatory decision-making.

Biosafety communication activities involve a cost. The return will be technically sounder and socially more robust decisions, which will be more widely accepted and more smoothly implemented. In contrast, taking decisions without involving all the stakeholders can prove cheaper, but the decisions might not be enforceable, or could be expensive in terms of social consent.

Biosafety communication, in conclusion, is not a panacea, but, if correctly utilised, can ensure a smoother implementation of biosafety regulation and help to achieve the stakeholders' participation and the wider public's acceptance of biosafety-related decisions.

References

Acunzo, M. *et al.* (2014). *Communication for Rural Development Sourcebook.* Rome: Food and Agriculture Organization.

Arantes, O. *et al.* (2011). *Desenvolvimento de comunicação estratégica sobre biossegurança de plantas geneticamente modificadas – o caso do projeto LAC – Biosafety no Brasil.* Aguariúna, SP: Embrapa Meio Ambiente, p. 33.

Australian Government (2013). *The Risk Analysis Framework 2013.* Canberra: Department of Health and Ageing, Office of the Gene Technology Regulator.

Blancke, S. *et al.* (2015). Fatal attraction: the intuitive appeal of GMO opposition. *Trends in Plant Science* **20**(7), 414–418.

Brandenberg, O. *et al.* (2011). *Biosafety Resource Book* (5 volumes). Rome: Food and Agriculture Organization.

Bremer, S. *et al.* (2015). Responsible techno-innovation in aquaculture: employing ethical engagement to explore attitudes to GM salmon in Northern Europe. *Aquaculture* **437**, 370–381.

Bubela, T. *et al.* (2009). Science communication reconsidered. *Nature Biotechnology* **27**(6), 514–518.

CAC (2003). Report of the Twenty-Sixth Session. Rome. 30 June–7 July 2003. Appendix IV. Working Principles for Risk Analysis for Application in the Framework of the Codex Alimentarius. Joint FAO/WHO Food Standards Programme. Rome: Food and Agriculture Organization.

CAC (2008). *Codex Alimentarius Commission Procedural Manual.* Eighteenth edition. Rome. Food and Agriculture Organization.

CAC (2011). Working principles for risk analysis for application in the framework of the Codex Alimentarius. In *Codex Alimentarius Commission Procedural Manual.* Twentieth edition. Rome: Joint FAO/WHO Food Standards Programme.

Canavari, M. and Nayga, R. (2009). On consumers' willingness to purchase nutritionally enhanced genetically modified food. *Applied Economics* **41**(1), 125–137.

Carroll, L. (1871). *Through the Looking Glass and What Alice Found There.* London: Macmillan.

Casanoves, M. *et al.* (2015). Knowledge and attitudes towards biotechnology of elementary education preservice teachers: the first Spanish experience. *International Journal of Science Education* **37**(17), 2923–2941.

Chee, Y. L. and Lim, L. L. (2005). Public Participation in the Implementation of the Biosafety Protocol, Third World Network Biosafety Briefing. [Online]. Available from www.biosafety-info.net/file_dir/1410148856539c28e5.pdf

Convention on Biological Diversity (2000). Cartagena Protocol on Biosafety to the Convention on Biological Diversity. Text and Annexes. [Online]. Available from www.cbd.int/doc/legal/cartagena-protocol-en.pdf

Cordo, A. (2008). Nuevos métodos de comunicación: ciencia a la mexicana. [Online]. Available from www.surysur.net/nuevos-metodos-de-comunicacion-ciencia-a-la-mexicana/

EC (2012). Responsible Research and Innovation: Europe's Ability Responds to Societal Challenges. [Online]. Available from https://ec.europa.eu/research/swafs/pdf/pub_public_engagement/responsible-research-and-innovation-leaflet_en.pdf

EC (2013). *Options for Strengthening Responsible Research and Innovation.* Luxemburg: Publication Office of the European Union; Brussels: European Commission.

FAO (2001). *Glossary of Biotechnology for Food and Agriculture.* Rome: Food and Agriculture Organization.

FAO (2004). *Participation: Sharing Our Resources. Informal Working Group on Participatory Approaches and Methods to Support Sustainable Livelihoods and Food Security*. Rome: Food and Agriculture Organization.

FAO (2008). *GM Food Safety Assessment: Tools for Trainers*. Rome: Food and Agriculture Organization.

FAO (2011). *Agricultural Biotechnologies in Developing Countries: Options and Opportunities in Crops, Forestry, Livestock, Fisheries and Agro-industry to Face the Challenges of Food Insecurity and Climate Change (ABDC-10), 2010*. Rome: Food and Agriculture Organization.

Galilei, G. (1623). The Assayer, translated by S. Drake and C. D. O'Malley. In *The Controversy on the Comets of 1618* (1960). Philadelphia, PA: University of Pennsylvania Press.

Gaskell, G. *et al.* (2011). The 2010 Eurobarometer on the life sciences. *Nature Biotechnology* **29**(2), 113–114.

Gibbons, M. (1999). Science's new social contract with society. *Nature* **402** (Suppl.), C81–C84.

Hilgartner, S. (1990). The dominant view of popularisation: conceptual problems, political uses. *Social Studies of Science* **20**, 519–539.

Kazana, V. *et al.* (2015). Public attitudes towards the use of transgenic forest trees: a cross-country pilot survey. [Online]. www.sisef.it/iforest/contents?id=ifor1441-008

Lucht, J. M. (2015). Public acceptance of plant biotechnology and GM crops. *Viruses* **7**(8), 4254–4281.

Macnaghten, P. and Carro-Ripalda, S., eds. (2016). *Governing Agricultural Sustainability: Global Lessons from GM Crops*. London: Routledge.

National Science Board (2004). Science and technology: public attitudes and understanding. In *Science and Engineering Indicators 2004*. Arlington, VA: National Science Foundation, Chapter 7.

Nielsen, A. P. *et al.* (2004). Involving the public – participatory methods and democratic ideals. *Global Bioethics* **17**, 191–201.

Nisbet, M. C. and Scheufele, D. A. (2009). What's next for science communication? Promising directions and lingering distractions. *American Journal of Botany* **96**(10), 1767–1778.

Oughton, D. (2005). The promises and pitfalls of participation. In *Biotechnology-Ethics. An Introduction*, ed. L. Landerweerd *et al.* Florence: Angelo Pontecorboli Editore, pp. 305–313.

Owen, R. *et al.* (2012). Responsible research and innovation: from science in society to science for society, with society. *Science and Public Policy* **39**(6), 751–760.

Pew Research Center (2015). Public and Scientists' Views on Science and Society. [Online]. Available from www.pewinternet.org/files/2015/01/PI_ScienceandSociety_Report_012915.pdf

Rawls, J. (1971). *A Theory of Justice*. Cambridge, MA: Belknap Press.

Ruane, J. and Sonnino, A. (2006). *Results from the FAO Biotechnology Forum: Background and Dialogue on Selected Issues*. Rome: Food and Agriculture Organization.

Ruane, J. and Sonnino, A. (2011). Agricultural biotechnologies in developing countries and their possible contribution to food security. *Journal of Biotechnology* **156**, 356–363.

Scholderer, J. and Frewer, L. J. (2003). The biotechnology communication paradox: experimental evidence and the need for a new strategy. *Journal of Consumer Policy* **2**, 125–157.

Sensi, A. *et al.* (2009). *Building Biosafety Capacities: FAO's Experience and Outlook*. Rome: Food and Agriculture Organization.

Sharry, S. (2013). Communicating biosafety – a new approach for agrobiotechnology adoption. *Agrotechnology* **2**, e107.

Sinemus, K. and Egelhofer, M. (2007). Transparent communication strategy on GMOs: will it change public opinion? *Biotechnology Journal* **2**, 1041–1146.

Sorgo, A. *et al.* (2011). Knowledge about and acceptance of genetically modified organisms among pre-service teachers: a comparative study of Turkey and Slovenia. *Electronic Journal of Biotechnology* **14**(4). [Online]. Available from www.scielo.cl/scielo.php?script=sci_arttext&pid=S0717-34582011000400005

Stilgoe, J. *et al.* (2013). Developing a framework for responsible innovation. *Research Policy* **42**(9), 1568–1580.

Traynor, P. *et al.* (2007). Strategic approaches to informing the public about biotechnology in Latin America. *Electronic Journal of Biotechnology* **10**(2), 169–177.

UNECE (2003). Guidelines on access to information, public participation and access to justice with respect to genetically modified organisms. MP.PP/2003/3, KIEV.CONF/2003/INF/7.

UNECE (2005). Almaty Declaration. Report of the Second Meeting of the Parties. Almaty, Kazakhstan.

UNEP-GEF Biosafety Unit (2006). A comparative analysis of experiences and lessons from the UNEP-GEF biosafety projects. [Online]. Available from www.unep.ch/biosafety/old_site/development/devdocuments/UNEPGEFBiosafety_comp_analysisDec2006.pdf

UNESCO (2006). Universal Declaration on Bioethics and Human Rights. [Online]. Available from http://unesdoc.unesco.org/images/0014/001461/146180E.pdf

Whitty, C. J. M. *et al.* (2013) Africa and Asia need a rational debate on GM crops. *Nature* **497**, 31–33.

17 The Role of Mass Media and Lobbies in the Formulation of GMO Regulations

Mauro Vigani

Introduction

The adoption of new crop varieties obtained through genetic modification (GM) as a means of enhancing agricultural productivity and food security is determined by the domestic biotechnology and biosafety regulations on the cultivation and commercialisation of genetically modified organisms (GMOs). Most of the developed countries are already equipped with full and detailed GMO regulations, but many less developed countries (LDCs) are still in the process of formulating such regulations.

While deciding about GMO regulations, governments are driven by several socio-economic factors and by the preferences of different groups in society. The political equilibrium is mediated by the mass media and more specifically by the structure of the media market (Vigani and Olper, 2015). Not only do the mass media play a pivotal role in influencing regulation preferences of consumers, but also the structure of the media market can affect which social groups are targeted by the media outlets (Vigani and Olper, 2013).

This double aspect of the mediation role of mass media in shaping GMO regulation is studied in a literature that cuts across two main disciplines. The influencing role of mass media on consumer preferences is mainly covered by social science analyses (e.g., Frewer *et al.*, 2003; Vilella-Vila and Costa-Font, 2008). The link of mass media markets with the political equilibrium between the different groups of a society is captured by theoretical and empirical political economy analyses (e.g., McCluskey and Swinnen, 2004; Vigani and Olper, 2013).

This chapter discusses the role of mass media in the formulation of GMO regulation in developed and developing countries, distinguishing two major approaches: the social science and the political economy approach. The chapter also presents socio-economic indicators on quality of regulation, mass media and the restrictiveness of GMO regulations in developed countries and in LDCs.

A Critical Review of the Theoretical Approaches: Social Science vs. Political Economy

The Social Science Approach

The social science approach uses the type and amount of news consumed by the public to study how the communication strategies of mass media influence the public perception of risks and the formation of attitudes towards GMOs.

In the social science framework, mass media have an intermediary role. They operate as preferential two-way 'channels' of communication: from government to the public and from the public to government (Gaskell and Bauer, 2001). This intermediary role makes the media a major forum of the public sphere in modern societies.

The intensity and duration of the influence of mass media on public attitudes towards GMOs is a function of five main factors:

(i) public trust in the institutions and information sources
(ii) prior attitudes of the public on GMOs
(iii) the media bias in reporting risks
(iv) the heterogeneity of media reporting
(v) the importance of communicating the benefits of GMOs along with their safety.

Public Trust in Institutions and Information Sources

If those responsible for risk assessment are distrusted, members of the public feel insecure about the effectiveness of risk detection and management even though the procedures implemented may be scientifically rigorous. The issue of public trust does not concern exclusively the authorities, but also any source delivering information on potential risks associated with GMOs, such as industry, NGOs and consumers' groups.

Public trust is a function of two factors (Frewer *et al.*, 2003): the 'competence' of the information source (the expertise of the communicator and his/her ability to disseminate information) and its 'honesty' (the extent to which a communicator will be truthful).

Mass media can affect public trust if they fail to communicate the competence and honesty of the information source. For example, media can choose to deliver knowledge through experts and by demonstrating a genuine concern for the public welfare over special interests (Kim *et al.*, 2015; Gaskell *et al.*, 2004).

However, public distrust is associated with the perception that an information source is deliberately distorting information, promoting biased views of relevant issues or protecting certain interests. Information sources such as consumer organisations, environmental groups and scientists are often considered more trustworthy than the biotech industry and government (Costa-Font *et al.*, 2008).

Prior Attitudes on GMOs

Strong and established prior attitudes help to explain the stability and resistance to change in attitudes towards GMOs. If the public has pre-formed attitudes about a new technology, this can influence the interpretation of the technology's potential risks as communicated by the source of information (Frewer *et al.*, 2003), despite the competence and honesty of the source.

Regarding GM crops, prior scepticism can derive from people questioning the point of modifying the genetic makeup of plants. 'Is it necessary when there is plenty of food in the shops?' 'Why change the character of food when it is already good and wholesome?' The answers to these questions are difficult to convey to people who do not experience scarcity in quantity and quality of food (Gaskell *et al.*, 2004).

Instead of being convinced of a new point of view, people are more likely to assess information to see whether it aligns with the view they already have. If it does not, it is more likely that they will change their opinion about the information source rather than change their attitudes towards the technology (Frewer *et al.*, 2003). Petts *et al.* (2001) suggest that the media can amplify or attenuate risk perception only if the message captures or resonates with a pre-existing public mood.

Media Bias in Reporting Risks

The 'pluralist theory of the media' (Harrop, 1987) suggests that the media's reporting of news is biased by its own agenda, meaning that people working in the media determine 'what is/what is not news' and prioritise particular items of news. The media bias is linked to several aspects of a media outlet's agenda. Journalists and editors may adjust the story frame to their ideology and knowledge limitations, thereby characterising the potential socio-political and ethical implications of a story (McCluskey *et al.*, 2016).

A second potential source of media bias is incorporated in the number and type of stories published. Negative media coverage of biotechnology, for example, was related to an increase in negative public opinion in the USA (Nisbet and Lewenstein, 2001), Spain (Vilella-Vila and Costa-Font, 2008) and Kenya (Lore *et al.*, 2013). Therefore, the media can promote a few sets of ideas over others, effectively limiting the debate by making particular choices (Botelho and Kurtz, 2008).

Also the wording and the structure of the storytelling may contribute to the production of biases. Journalists are seldom entirely neutral in their choice of events and words (Holmgreen and Vestergaard, 2009), and those choices can contribute to the formation of public attitudes (Lore *et al.*, 2013). This is particularly relevant when considering information on biotechnology, which is a field characterised by great controversy and strong conceptualisations, with extensive use of metaphors (Ventura *et al.*, 2016). For instance, the discourse used by opponents of GMOs frequently invokes the contrast between natural (good) and artificial (bad), using metaphors of disease and pollution, and deploys terms such as 'infected' and 'polluted' against 'pure' and 'free' (Cook, 2004; Holmgreen and Vestergaard, 2009).

Media can be exploited by groups skilled in creating media-friendly events and campaigns. Activists can strategically portray environmental and food safety issues as 'emotional dramas', in order to attract public attention and harness the eagerness of politicians to please those activists who portray themselves as acting in the public interest. The objective is to shift consumers' preferences and to build consensus in order to be able to exert more pressure on governments, as well as obtaining external funding (Aerni *et al.*, 2015, 2016).

In the case of GMOs the media are often saturated with messages from lobby groups, leading to the publication of stories that can be contradictory to their normal stance. For example, in the UK some newspapers that are normally very pro-business, such as the *Daily Mail*, can adopt anti-GM rhetoric, while independent pro-science media organisations such as the Science Media Centre generally adopt a more pro-GM view.

Heterogeneity in Media Reporting

There are two main sources of heterogeneity: the heterogeneity within the public and that across geographical areas.

Consumers of mass media have heterogeneous perceptions of risks and benefits depending on individual characteristics, e.g., age, gender, education, income level, religion and ethnicity. These characteristics can influence an individual's level of scientific knowledge, their propensity to use particular technologies, and their moral and ethical considerations (Costa-Font *et al.*, 2008). Although differences among individuals are evident across countries and cultures, consumer heterogeneity can be even larger within the same country and among the different groups within a society (Frewer *et al.*, 2003).

Also the narrative structure, or information frame (Vilella-Vila and Costa-Font, 2008), which influences public perception, can be different across countries and regions, owing to geographical and historical circumstances. Differences in media framing have been reported between the USA and the EU (Gaskell *et al.*, 1999) and between member states within the EU (Vilella-Vila and Costa-Font, 2008), but also within single countries such as the UK (Botelho and Kurtz, 2008), South Korea (Kim *et al.*, 2015) and Kenya (Lore *et al.*, 2013; DeRosier *et al.*, 2015).

Communicating Benefits to Increase Acceptance

Public attitudes towards a given product or technology are linked to its perceived risks and benefits. If the risks are not intolerably high, the perception of a new technology will predominantly reflect the benefits of its adoption. Thus, benefits are more important than risks in determining willingness to consume new products or adopt technologies (Costa-Font *et al.*, 2008). People will tolerate risks, however, if they perceive direct benefit to themselves, but not to other groups, such as producers or industry (Frewer *et al.*, 2003).

In the case of agro-biotechnologies, it is frequently argued that consumer rejection comes from the type of GM products initially introduced on the market, such as herbicide-tolerant and insect-resistant crops. These crops have direct benefits to farmers, such as higher productivity and lower use of inputs, but consumer benefits are mainly indirect (such as reduced food costs) and might not be sufficient to counterbalance the perceived risks.

The benefits with the greatest potential to improve consumer acceptance of GM crops are sustainability (less use of energy and release of pollutants during production), food security (new crop varieties to reduce hunger in LDCs) and health claims (the presence of functional ingredients). However, these benefits are not always efficiently communicated, and consumers often see only the potential risks of GMOs. Without the perception of an improvement in terms of quality, price or other attributes, there is simply no incentive to accept GMOs (Gaskell *et al.*, 2004).

Framed in another way, at the base of the public rejection of GM crops and food is not so much the misperception of scientific risks, but rather the perceived absence of benefit for consumers.

The Political Economy Approach and the Importance of the Media Structure

When it comes to deciding on new regulations, individual attitudes are translated into preferences for more permissive or restrictive rules on the use and consumption of GMOs. Individuals are organised in different groups (e.g., consumers, producers, industry), each with different policy preferences (derived from the perceived benefits of GMOs) and lobbying strength. Moreover, policymakers can be driven by egoistic objectives, such as maximising the number of votes they can attract, instead of aiming for public benefits and safety.

During the policymaking process, the mass media transfer the preferences of the civil society to governments. Simultaneously, governments use the mass media in a top-down manner to seek public support and to avoid public dissatisfaction. Mass media are, therefore, bidirectional mediators between policymakers and voters.

Mass Media Competition

As with any other commercial enterprise, a private mass media company aims to maximise profits, attempting to increase the amount of news sold by providing consumers with information that is of interest to them. Mass media companies operate under increasing economies of scale as the average cost decreases with the amount of news sold. Once the fixed cost of producing a TV programme, a website or the first newspaper has been borne, the variable costs of an additional viewer or reader are relatively small. These economic conditions induce private media to transmit news that is of interest to large groups of media consumers, which is also valuable to advertisers (Strömberg, 2001). Therefore, the policy programmes targeting the largest group of voters are diffused in greater quantity and more detail, potentially altering the political competition and ultimately introducing a bias in public policy (Strömberg, 2004).

In developed countries the producers' group (farmers) is typically smaller than the consumers' group, which is mainly concerned with food safety issues. In (agriculture-based) developing economies the farmers' population is relatively large, and private media are more likely to promote agricultural innovations and policies that favour farmers (Vigani and Olper, 2014).

The 'Bad News' Hypothesis

In order to increase sales, mass media companies can adapt the narrative of news in such a way as to make their news products more attractive to consumers. McCluskey and Swinnen (2004) observed that the marginal value of an item of news with negative welfare effects is higher than the marginal value of an item of news with positive welfare effects, meaning that private media have incentives to deliver 'bad' news, which will increase the consumption (and hence profit) of media more than will 'good' news.

In many LDCs mass media are often owned by the state, and governments have incentives to reduce public concern and risk perception in order to reduce public discontent and opposition (Curtis *et al.*, 2008). Consequently, consumers in LDCs may be less exposed to negative media coverage concerning GMOs, while in rich countries consumers are more exposed to coverage of risks related to GMOs. This has been confirmed empirically: positive media coverage of GMOs is found in Kenya (Lore *et al.*, 2013; DeRosier *et al.*, 2015), but negative coverage in Spain, the UK and South Korea (Vilella-Vila and Costa-Font, 2008; Botelho and Kurtz, 2008; Kim *et al.*, 2015).

It is important to mention that there are significant exceptions to the 'bad news hypothesis'. Lobby groups can deliberately manipulate media in order to build public support in their campaigns. To this end, negative information can be overhyped on certain innovations (e.g., GM crops), while positive information can be overhyped on others (e.g., new medical biotechnologies).

The Internet as a Source of Information

For a large proportion of the population, especially the youngest group, the internet is the main source of information. So it must be noted that the internet has some peculiar characteristics.

In the internet borders are removed and news is globally accessible, provided that there is no language barrier. Internet users are not only 'consumers' of news, but also have the opportunity to engage with sources through commentary sections or social networks. Moreover, the extremely low initial costs of a website allow the creation of a vast variety of sources: established mass media companies (e.g., newspapers, TV stations), NGOs and private individuals.

Although anybody can put information on the internet, not everybody is a trustworthy source. The numerous possibilities for interacting can facilitate the diffusion of ideological and unverified information, and the fact that it is difficult to assess the source's competence can reduce trust in internet news, especially by the older generations. This suggests that, despite it being cheap to get started, it takes time to become an established and trusted internet source, unless the source has already acquired credibility through traditional media.

Moreover, the low initial costs and the diffusion of portable devices (e.g., tablets, smartphones) allow a mixed form of communication, which can be written text, images or videos. Information biases can easily be introduced by negative images that accompany unbiased text (Ventura *et al.*, 2016).

GMO Regulation, Media Freedom and Quality of Regulation

This section discusses the international differences in GMO regulation, considering the quality of regulatory systems and media freedom for a sample of 15 LDCs (Argentina, Bangladesh, Brazil, China, Colombia, Guatemala, India, Indonesia, Kenya, Peru, the

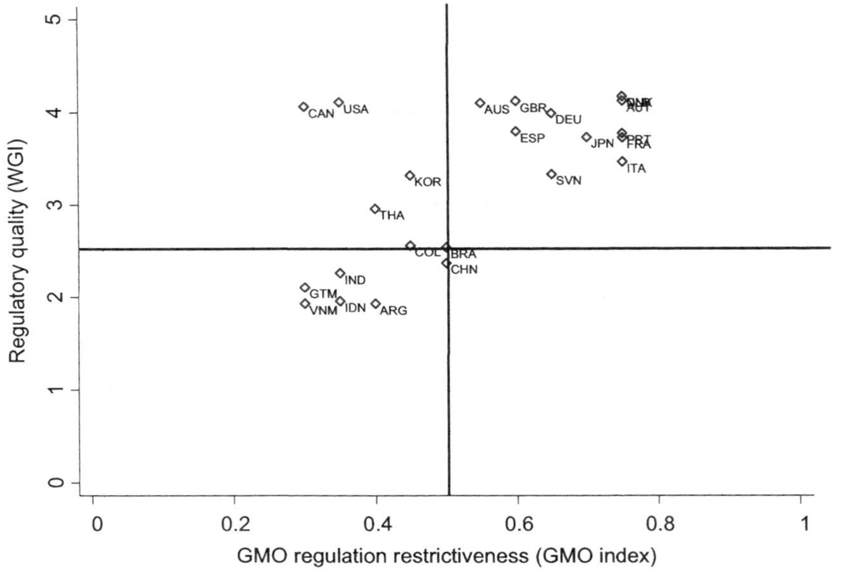

Figure 17.1. Distribution of countries by regulatory quality and restrictiveness of GMO regulation in 2009. (Regulatory quality index from Worldwide Governance Indicators (http://info.worldbank.org/governance/wgi/index.aspx#home); GMO index from Vigani and Olper (2013).)

Philippines, Thailand, Vietnam, Zimbabwe and Zambia) and 15 developed countries (Australia, Austria, Canada, Denmark, France, Germany, Italy, Japan, the Netherlands, Portugal, Slovenia, South Korea, Spain, the UK and the USA).

GMO regulation has several components: approval, traceability, coexistence and labelling. Approval can concern the cultivation of GM crops and/or commercialisation of GM food and feed, and it can include a comprehensive environmental and health risk assessment (e.g., China and Guatemala). Traceability can be limited to ensuring the preservation of the identity of products (e.g., Australia and the USA) or it can include records tracking GM products along the entire supply chain (e.g., Japan). In order to ensure the coexistence of traditional, GM and organic crops, some countries provide detailed rules on segregation of fields (e.g., the EU). Labelling of food and feed products can be voluntary (e.g., the Philippines) or mandatory. When it is mandatory, labels can be applied to products containing less (e.g., Russia) or more (e.g., South Africa) than 1% of GM ingredients.

Figure 17.1 shows the distribution of countries by regulatory quality and restrictiveness of the GMO regulation. The index of regulatory quality captures the public perception of the ability of the government to formulate and implement sound policies and regulations (Worldwide Governance Indicators). It can be used as a proxy for the quality of institutions and government responsiveness. The GMO index measures the restrictiveness of GMO regulations (Vigani and Olper, 2013).

The distribution of the countries in the two upper quadrants of Figure 17.1 shows a contraposition between North American and EU countries. Despite having institutions of similar (high) quality, they have opposite GMO regulations: voluntary labelling

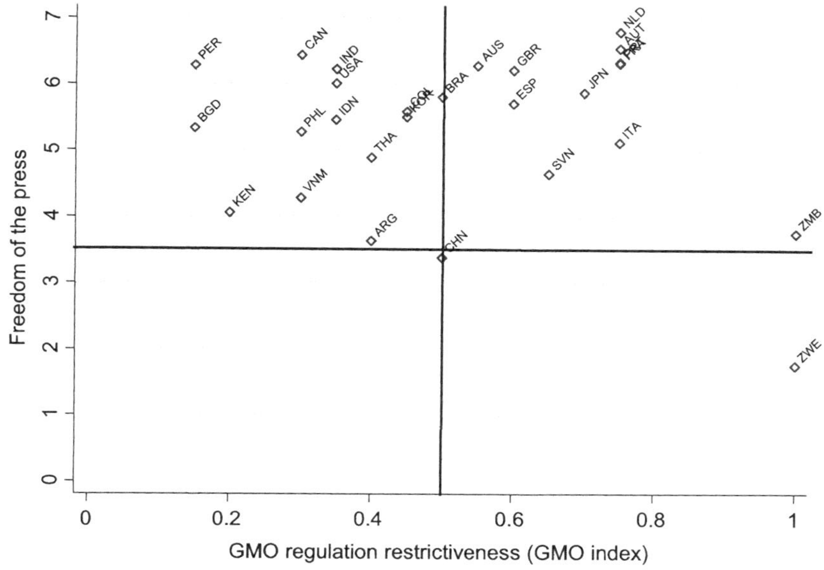

Figure 17.2. Distribution of countries by freedom of the press and restrictiveness of GMO regulation in 2009. (Freedom of the press index from Freedom House (https://freedomhouse.org); GMO index from Vigani and Olper (2013).)

and no coexistence rules in the USA and Canada; full traceability, coexistence rules and labelling at a threshold of 0.9% in the EU. This reflects the opposing approaches of these two blocks of countries towards GMOs (substantial equivalence to non-GMO products in North America; while there is strong opposition by European consumers), which indirectly can affect the GMO regulation applied by governments in trade partners, especially in those LDCs which are seeking access to the EU market.

The countries at the intersection of the two axes of Figure 17.1 (Colombia, Brazil and China) have a relatively low quality of regulation, but a relatively high restrictiveness of GMO regulation. In these countries the perception that the government has a relatively low level of ability to formulate and implement policies can affect the level of public trust in the competence and honesty of institutions, making the public ask for more restrictive rules governing approval and risk assessment.

In the bottom-left quadrant of Figure 17.1 there is a cluster of LDCs with GMO regulations that impose few restrictions. As stressed by the political economy literature, in LDCs farmers are the largest group, and media can have incentives to provide information on agricultural technologies with a positive coverage.

In Figure 17.2, countries are distributed by the GMO index and the freedom of the press index, which captures the degree of freedom of mass media in each country, and the efforts made by the authorities to ensure this freedom (Reporters Without Borders).

Almost all the countries are in the two upper quadrants, but at a similar level of press freedom the EU countries have more restrictive GMO regulations than the rest of the world. This contraposition can reflect the media market structure. In the top-right quadrant there are mainly developed countries in which private media have the majority of the market share. Private media outlets can pursue the objective of maximising their profit by providing 'bad' news that confirms and exacerbates the prior negative attitude towards GMOs. This can lead the majority of consumers to ask governments for more restrictive labelling and coexistence regulations. In the top-left quadrant there are mainly LDCs, in which the media market is typically dominated by public outlets controlled by governments. Governments can have incentives to reduce public concern and opposition to GMOs that can be beneficial for the agricultural sector.

Along with LDCs, the USA and Canada are in the top-left quadrant of Figure 17.2. Their less restrictive regulation of GMOs could be due to the fact that the influence of agricultural lobbying groups on the political outcome is stronger in these countries than it is in the EU, and such groups consequently have more influence on the media sector, which can thus be persuaded to inform civil society about the benefits of GMOs. The better-informed society can then ask the government for regulations benefiting farmers. In contrast, in the EU the lobbies that seem to be the most effective are consumers' groups, green organisations and associations of organic producers, which mainly act in favour of restrictive GMO standards (Aerni *et al.*, 2015; Vigani and Olper, 2013).

New Challenges: The Communication and Regulation of New Plant Breeding Techniques

New plant breeding techniques (NPBTs), especially genome editing, are emerging, and their regulation has yet to be decided in many countries. A key aspect for regulating genome-edited crops is the need to consider their differences from GMOs. While conventional genetic modification relies on the insertion of exogenous DNA to obtain new characteristics in crop varieties, genome editing exploits new 'molecular scissors' to cut and alter the existing genome of plants (Jones, 2015). The main question for regulators is whether genome editing differs from existing techniques and how the resulting products should be classified according to current definitions of a GMO (Lusser *et al.*, 2012).

The importance of regulating these innovative techniques involves actors at different stages in the food chain. Plant breeders can have an enlarged breeding toolbox, meaning a faster breeding process and earlier returns from R&D investments (Lusser *et al.*, 2012). Consumers can receive foods with improved nutritional value, fewer allergens and a longer shelf-life, reducing food waste. Farmers can obtain high-quality plant varieties tolerant of climatic stresses or diseases, reducing the need for crop protection and contributing to more sustainable production.

Mass media attention to genome editing was triggered in April 2016, when the US regulators allowed the cultivation and sale of a white button mushroom and a corn variety obtained with genome editing. According to the US authorities, these crops are not GMOs, hence they did not fall under the restrictive GMO regulation.

The debate in the EU is still largely unresolved, and it is expected that the EU's position will influence the policy outcome of many other countries that intend to align their legislation with that of the EU. The fact that the US decision could have set a precedent for other NPBT crops opened the media's gates for stakeholders who want their voices to be heard by the EU policymakers. Two groups of stakeholders are facing each other in the media arena.

The first group is mainly composed of green NGOs and organic associations. They posit an equivalence between NPBTs and genetic modification, using the expression 'new GMOs' to evoke the negative perception that the term GMO has on some consumers, and they would like the EU GMO law to be fully applied to NPBTs. The arguments used to oppose NPBTs are not different from the anti-GMO rhetoric: 'NPBTs are just another attempt at selling GMOs to consumers'; 'NPBTs present a real risk to the environment and human health'; 'an intense lobbying intrusion by US representatives pushed Brussels to disregard GMO rules' (EurActive, 2016a, 2016b).

The second group is composed of scientists, farmers' associations and agribusiness representatives, aiming for a case-by-case risk assessment that is based on scientific criteria and proportionate to potential risks. This group uses a more articulated communication to support NPBTs, addressing a variety of socio-economic arguments. First of all, classifying NPBTs as GMOs is inappropriate because there is no introduction or presence of exogenous DNA in the final plant (Jones, 2015). Moreover, the mutations generated by NPBTs are indistinguishable from those arising in nature (Jones, 2015), therefore the new mutation techniques should fall under the already-existing exemptions for classical mutagenesis. Second, the agricultural sector is in need of innovative traits to ensure farmers' competitiveness and global food security. Third, biotechnology regulation in the EU should evolve hand-in-hand with the scientific knowledge of plant genomes, otherwise plant research and modern crop breeding in the EU will continue to fall behind the rest of the world. If NPBTs do fall under the GMO regulations, it would create a further barrier to innovation in plant breeding, with serious negative impacts on the EU's small and medium enterprises engaged in plant breeding, leading to loss of jobs and growth (EurActive, 2016a, 2016b).

The current media interest in NPBTs is based on the assumption that commercialisation is imminent. However, data on actual adoption of NPBTs are still lacking (Lusser *et al.*, 2012). Apart from the two US crops mentioned above, to date there has only been a herbicide-tolerant canola obtained through genome editing approved in the USA, which is expected to enter the Canadian market in 2017 (Jones, 2015).

Conclusions and Recommendations

The news delivered by the media can affect the capacity of consumers to make informed choices about GMO regulations. In order to improve public attitudes, a well-targeted strategy of communication of risks and benefits is required. It is erroneous to assume that the benefits of GMOs are self-evident; on the contrary, there are benefits to consumers that should be better communicated, especially the lower pressure on natural resources and the increased content of functional ingredients and nutrients for food and nutrition security.

However, it is important to note that the effects of mass media tend to be temporary and limited in magnitude. Therefore, providing information on the benefits of GMOs might not be sufficient, but this can be coupled with best practice in science communication that takes into account the level of public trust in institutions and the lobbying activity of organised groups.

Past experiences of communication on GMOs can be useful for responding to future challenges to plant breeding and biotechnologies, especially given that nowadays key decisions need to be taken with regard to the regulation of NPBT in many countries.

To summarise, some suggestions for a good media communication strategy are as follows: (i) people should be informed not only about potential risks, but also about benefits, in an objective manner; (ii) consumers should obtain their information from honest and competent organisations not representing special interests; and (iii) information should be provided by authoritative public figures in a clear and persuasive manner. Appropriate communication strategies are even more sensitive today, given the enhanced accessibility of information provided by the internet.

References

Aerni, P. *et al.* (2015). Agricultural biotechnology and public attitudes: an attempt to explain the mismatch between experience and perception. In *Genetically Modified Organisms in Food*, ed. R. R. Watson and V. R. Preedy. New York, NY: Academic Press, pp. 149–156.

Aerni, P. *et al.* (2016). The role of biotechnology in combating climate change: a question of politics? *Science and Public Policy* **43**, 13–28.

Botelho, D. and Kurtz, H. (2008). The introduction of genetically modified food in the United States and the United Kingdom: a news analysis. *The Social Science Journal* **45**, 13–27.

Cook, G. (2004). *Genetically Modified Language: The Discourse of Arguments for GM Crops and Food.* London: Routledge.

Costa-Font, M. *et al.* (2008). Consumer acceptance, valuation of and attitudes towards genetically modified food: review and implications for food policy. *Food Policy* **33**, 99–111.

Curtis, K. *et al.* (2008). Differences in global risk perceptions of biotechnology and the political economy of the media. *International Journal of Global Environmental Issues* **8**(1–2), 77–89.

DeRosier, C. *et al.* (2015). A comparative analysis of media reporting of perceived risks and benefits of genetically modified crops and foods in Kenyan and international newspapers. *Public Understanding of Science* **24**(5), 563–581.

EurActive (2016a). Decision on new plant breeding techniques further delayed. 31 March 2016.

EurActive (2016b). Trade body: don't lump 'new plant breeding techniques' in with GMOs. 3 May 2016.

Frewer, L. J. *et al.* (2003). Communicating about the risks and benefits of genetically modified food: the mediating role of trust. *Risk Analysis* **23**(6), 1117–1133.

Gaskell, G. and Bauer, M. W. (2001). *Biotechnology 1996–2000: The Years of Controversy*. London: Science Museum.

Gaskell, G. *et al.* (1999). Worlds apart? The reception of genetically modified foods in Europe and the U.S. *Science* **285**, 384–388.

Gaskell, G. *et al.* (2004). GM foods and the misperception of risk perception. *Risk Analysis* **24**(1), 185–194.

Harrop, M. 1987. Voters and the media. In *The Media in British Politics*, ed. J. Seaton and B. Pimlott. Aldershot: Gower.

Holmgreen, L. L. and Vestergaard, T. (2009). Evaluation and audience acceptance in biotech news texts. *Journal of Pragmatics* **41**, 586–601.

Jones, H. D. (2015). Regulatory uncertainty over genome editing. *Nature Plants*, 8 January 2015. [Online]. Available from www.nature.com/articles/nplants201411

Kim, R. *et al.* (2015). Risk communication for GM foods in South Korea: the role of media and government. *International Food Research Journal* **22**(5), 1878–1882.

Lore, T. *et al.* (2013). A framing analysis of newspaper coverage of genetically modified crops in Kenya. *Journal of Agricultural & Food Information* **14**(2), 132–150.

Lusser, M. *et al.* (2012). Deployment of new biotechnologies in plant breeding. *Nature Biotechnology* **30**(3), 231–239.

McCluskey, J. J. and Swinnen, J. F. M. (2004). Political economy of the media and consumer perceptions of biotechnology. *American Journal of Agricultural Economics* **86**(5), 1230–1237.

McCluskey, J. *et al.* (2016). Media coverage, public perceptions, and consumer behavior: insights from new food technologies. *Annual Review of Resource Economics* **8**, 467–486.

Nisbet, M. and Lewenstein, B. V. (2001). A Comparison of U.S. Media Coverage of Biotechnology with Public Perceptions of Genetic Engineering 1995–1999. International Public Communication of Science and Technology Conference, Geneva, Switzerland, 1–3 February, 2001.

Petts, J. *et al.* (2001). Social Amplification of Risk: The Media and the Public. Contract Research Report 329/2001. Health and Safety Executive.

Strömberg, D. (2001). Mass media and public policy. *European Economic Review* **45**, 652–663.

Strömberg, D. (2004). Mass media competition, political competition, and public policy. *Review of Economic Studies* **71**, 265–284.

Ventura, V. *et al.* (2016). How scary! An analysis of visual communication concerning genetically modified organisms in Italy. *Public Understanding of Science*, 1 April 2016.

Vigani, M. and Olper, A. (2013). GMO standards, endogenous policy and the market for information. *Food Policy* **43**, 32–43.

Vigani, M. and Olper, A. (2014). GM-free private standards, public regulation of GM products and mass media. *Environment and Development Economics* **19**(6), 743–768.

Vigani, M. and Olper, A. (2015). Patterns and determinants of GMO regulations: an overview of recent evidence. *AgBioForum* **18**(1), article 6.

Vilella-Vila, M. and Costa-Font, J. (2008). Press media reporting effects on risk perceptions and attitudes towards genetically modified (GM) food. *The Journal of Socio-Economics* **37**, 2095–2106.

Part IV

Case Studies from Developing Countries

18 Risk Regulation of Agricultural GMOs in China: Challenges and Prospects

Wen Xiang

Introduction

Food security has ranked as one of the major concerns for China, due to its huge population as well as the limited amount of cultivated land. Brown made a prediction in 1995 that China would face a severe food shortage crisis in the future, because he estimated that China's grain production would fall by no less than 20%, from 340 million tonnes in 1990 to 272 million tonnes by 2030, but China's population would have reached approximately 1.6 billion by then (Brown, 1995). Since then, there has been a call for increased agricultural productivity. The emergence and fast development of biotechnology helped to solve the food shortage crisis, but the potential uncertainties regarding the effects of biotechnology on human beings or the environment have held back research on genetically modified organisms (GMOs) and delayed their application. This is especially true in China at present, where such problems as the relatively weak legal framework of regulation, insufficient transparency in decision-making and the absence of broad participation of the general public are prevalent (Zhao and Ho, 2005). It has been noted that politics, law, the sciences, industrial concerns and social values need to be better organised in order to solve the problems (Beck, 1992a, 1992b).

This chapter aims to provide an analysis of risk regulation of agricultural GMOs in China and to specify the challenges brought by the development of biotechnology in China. It highlights several aspects of the risk regulation system for agricultural GMOs in China, including risk assessment, the approval process and the labelling system as well as public participation. By virtue of the case of Bt rice, it concludes that risk regulation of agricultural GMOs poses unique challenges for China, and China is called upon to develop a legitimised regime that is accountable to its citizens, as well as consistent with its international obligations.

Risk Assessment

The main agricultural GMO regulations and implementing measures are listed in Table 18.1. According to RSAAGMOs and MSEAAGMOs, the Ministry of Agriculture (MoA) is the primary institution in charge of implementing the agricultural biosafety regulations. The governing body under the MoA is the Leading Group on Agricultural GMO Biosafety Management (LGAGMOBM), which oversees the Agricultural

Table 18.1 Main agricultural GMO regulations and implementing measures

	Regulation on the Safety Administration of Agricultural Genetically Modified Organisms (RSAAGMOs)	Measures on the Safety Evaluation Administration of Agricultural Genetically Modified Organisms (MSEAAGMOs)	Measures on Administration of the Import Safety of Agricultural Genetically Modified Organisms (MAISAGMOs)	Measures on Administration of Labelling of Agricultural Genetically Modified Organisms (MALAGMOs)
Issuing authority	State Council	Ministry of Agriculture	Ministry of Agriculture	Ministry of Agriculture
Date issued	23 May 2001	5 January 2002	5 January 2002	5 January 2002
Date revised	8 January 2011	1 July 2004	1 July 2004	1 July 2004
Level of authority	Administrative regulations	Departmental rules	Departmental rules	Departmental rules

GMO Biosafety Management Office (AGMOBMO). Risk assessment is conducted by the National Agricultural GMO Biosafety Committee (NAGMOBC). Currently, this committee meets twice each year to evaluate all risk assessment applications related to experimental research, field trials (small-scale trials), environmental release trials (medium-scale field trials), pre-production trials (large-scale field trials), commercialisation of agricultural GMOs and importation of GMOs (Xue, 2009). Normally the risk assessment will be assigned by the NAGMOBC to competent authorities (research institutes, ministerial departments or university labs, etc.). On the basis of the NAGMOBC's technical assessment results, the AGMOBMO prepares recommendations to the LGAGMOBM, which is tasked with taking decisions on approval or disapproval.

As the risk assessor, the NAGMOBC is composed of 56 experts allocated to several sub-committees that undertake different areas of work, and the assessment shall be undertaken twice a year (the application shall be made either before 31 March or before 30 September each year) under the guidelines of the MoA. The MoA shall provide a response within two months from the receipt of the application. Once the application has been accepted, the MoA has to make the final decision within three months after the expiration date of the application.

The guidelines regarding risk assessment are contained in MSEAAGMOs (see Table 18.1, third column). They provide specific principles on risk assessment and application procedures in a step-by-step manner and on a case-by-case basis. In general, the principle of substantial equivalence will apply for evaluation, which means that GMOs will be compared with their conventional counterparts for identification of the Safety Class.

An administration and evaluation system has been put in place to classify the agricultural GMOs into four classes, Safety Classes I, II, III and IV, depending on the degree of their potential risk to human beings, animals, plants, microorganisms and the environment, respectively. Accordingly, each class has its own requirements for safety assessment. The detailed classification standards are stipulated by the MSEAAGMOs: Safety Classes I, II, III and IV refer to GMOs 'with no known risk', 'with a low risk', 'with a medium risk' and 'with a high risk', respectively.

There are also detailed criteria for the determination of different classes according to the nature of the recipient organism as well as the type and degree of genetic operations which affect the recipient organism. Any recipient organisms that have no negative effect, or for which there is little chance of their causing any negative effect on human health or the environment, can be categorised as Class I, while the other three categories are for recipient organisms that may pose a risk to either human health or the environment, and the degree of the risk is classified as low risk, medium risk and high risk, respectively.

In terms of the genetic operations, three types have been listed in the MSEAAGMOs:

Type 1 refers to improving the safety of recipient organisms if the operation aims to remove genes known to be dangerous or inhibit the expression of such genes;
Type 2 refers to causing no effect on human health and the environment;
Type 3 refers to reducing the safety of recipient organisms if the operations cause a negative effect on human health and the environment.

According to the aforementioned criteria, all classes of recipient organisms and their GMO products obtained through a Type 1 operation shall fall under one of Safety Classes I, II, III and IV, according to their safety effects, whereas the operations are most likely to be classified as Class IV if the GM products are made from GMOs of Class IV through operations of Types 2 and 3.

One criticism is that the NAGMOBC, by virtue of its composition, is likely to pay less attention to the potential risks to the environment and human health (Wang, 2004). In addition, conflicts of interest exist in the assessment process as well. For example, some of the committee members who have participated in the assessment may personally have an interest in commercialising such products, or the committee members' personal connections could influence the result of an assessment. Another concern relates to the transparency of the whole process, since the list of members of the committee and the details of risk assessment have only been made publicly accessible after several NGOs and the media had consistently criticised the procedure.

In general, two initiatives could be considered in order to improve matters. Firstly, each committee member should declare any conflicts of interest before being selected as a member, and the committee members should be drawn from diverse sectors, including agriculture, environmental protection, health, etc. Secondly, guidelines and results of risk assessment should be made publicly accessible, and competent authorities should provide public consultations to elicit comments from the lay public. The recent draft Decision on Revision of MSEAAGMOs (Draft Decision, 2015) can be seen as a good example, since it was made available in order for stakeholders to make comments.

The Approval Process

Risk assessment provides preconditions for the risk management system (Kubasek and Silverman, 2007). Although risk management measures are responses to the result of risk assessment, this does not mean that risk management relies entirely on the result of risk assessment (Pollack and Shaffer, 2009). The responsible risk manager in China is the MoA. Once a risk assessment has been conducted and submitted to the LGAG-MOBM, the MoA is responsible for issuing the safety certificate to applicants in due course. Only when the safety certificate has been issued shall the applicants start to proceed to the next steps of application for a production licence and a business licence to commercialise GMO products.

Bt rice proves to be the most controversial case in the approval process. In December 2009 two Bt rice lines (Huahui 1 and Bt Shanyou 63) were issued safety certificates, and the term of validity of the certificates is five years. However, Bt rice was *de facto* suspended in the pipeline for commercialisation. The safety certificates were renewed only at the very last moment in December 2014, for another term of five years. Two points have been proposed to explain the unwarranted delay imposed by the MoA. Firstly, China is the world's largest producer of rice, which is one of the major staple foods for China. The public concern about the safety of Bt rice played an important role in the suspension. Secondly, it was found that unauthorised Bt rice had been cultivated illegally in several provinces and placed on the market in recent years even before the Safety Certificate had been issued. Moreover, some trade partners, such as the EU and Japan, have consistently reported finding a number of unauthorised GMO ingredients in products imported from China (Greenpeace, 2014).

It is noteworthy that the recent MoA press conference indicated that a roadmap for the pipeline of GMOs in the 13th Five-Year Plan of China is now ready. The priority for commercialisation will be given to non-food GMO plants first, followed by GM feeds, or processed GMOs, and finally GMO foods (e.g. Bt rice) (MoA, 2016a).

It is apparent that the process for the approval of GMOs in China is not clear and transparent, even though there are well-documented regulations. Several concerns have been raised, as follows.

Firstly, the state invests a huge amount of funding in research and development (Keeley, 2003). Consequently, there is pressure to commercialise new crops in order to benefit from this investment. According to the current regulations, those GMOs with positive risk assessment results will be issued a safety certificate, and then will be approved for commercialisation. However, it seems there are two concerns with regard to the regime. The first is that a socio-economic evaluation regime seems to be missing. However, besides the scientific risk assessment, risk management needs to take into account other justifiable considerations for decision-making. The second concern is about the unwarranted delay of the approval process for certain GM varieties, which undermines the legitimacy and effectiveness of risk regulation. It is noteworthy that the draft Revision of MSEAAGMOs in 2015 mentioned that the MoA should take socio-economic considerations into account during decision-making (Draft Revision, 2015). However, one of the major concerns regarding the Decision as reflected in the industry associations' comments is that there might be a lack of transparency about justification for a decision, and it might also be in violation of China's commitment to the World Trade Organization (WTO) (Comments, 2015). The issue regarding socio-economic considerations was eventually dropped from the final revision of the MSEAAGMOs in 2016 (Revision, 2016).

Secondly, the involvement of farmers and consumers in the research and development process is under question (Xue, 2009). China has a diverse smallholder farming system. On the one hand, it is very difficult to implement a uniform biosafety regime, and, on the other hand, whether and to what extent the farmers are really capable of benefiting from the new GM varieties is questionable. Decision-makers are aware of the divergent situation within China. Yet the excessively centralised decision-making mechanism cannot accommodate the diverse interests of farmers in different locations. Therefore, farmers are quite passive in this mechanism, and their interests cannot be sufficiently represented. It is vitally important to see whether relevant stakeholders could be more actively engaged in the policymaking process.

Labelling Regulation

The labelling regulations in China cover five GMOs, i.e., soybean, maize seeds, rape seeds, cotton seeds and tomato seeds, and any products made directly from those GMOs. The labelling regulation on GMOs covers the following aspects. Firstly, any agricultural GMOs listed in the labelling catalogue of China fall within the scope of the labelling regulation. Secondly, the carrying out of the labelling is the responsibility of the organisations or persons that produce or pack the agricultural GMOs. Relabelling is necessary if the original packages have been opened for sale. Thirdly, the detailed labelling methods have been listed in Articles 6 to 10 of the MALAGMOs (Table 18.1, fifth column). Any unlabelled or mislabelled GMOs should not be sold on the market or imported. Fourthly, the application procedures for labelling of domestic and imported GMOs are different. For domestic GMOs, the examination and approval authorities are the appropriate local agricultural administrative departments, whereas for the imported ones, the competent authority is the MoA, and copies of relevant documents shall be sent to the General Administration of Quality Supervision, Inspection and Quarantine (AQSIQ), the Ministry of Foreign Trade and Economic Cooperation (MFTEC) and other relevant competent authorities.

The labelling system was introduced in 2002, in order to regulate the huge amount of transgenic soybeans being imported from the USA, and very few technical considerations have been taken into account. Therefore, a well-functioning labelling system is needed for the management system. There are several problems regarding the current labelling regulation that need to be solved.

Firstly, the lack of any threshold in the labelling regulation results in *de facto* zero tolerance in the labelling of GM products. It is then very difficult to deal with contingent contamination with GMOs because of accidental factors and unavoidable technical factors. The threshold is 0.9% in the EU, 3% in South Korea and 5% in Japan. Therefore, a consistent national threshold needs to be introduced into the Chinese labelling system for effective implementation, and also to lower the cost of compliance for producers of GMOs (Xue, 2009).

Secondly, within the defining objectives, the labelling regulation applies only to agricultural GMOs and products made directly from the GMOs. There is no explanation of

Box 18.1 The Nestlé Labelling Case

In December 2003, Miss Zhu bought a Nestlé sweet (Qiao Banban) for her son, and found out from the internet that the product may contain GMO ingredients. She sued Nestlé Company at the Shanghai No. 2 Intermediate Court because the product was not expressly labelled with GM information and violated her right to know as a consumer. The key point of this case was whether or not the product fell within the category of products for which labelling is compulsory. It should be noted that the deciding factor strongly relied on the technical standard and method of detection. The court instructed the GM Food Laboratory of Shanghai Agricultural Academy to investigate the product in question. The laboratory initially detected GM ingredients with polymerase chain reaction methods, but then this was overruled by its second investigation with another national detection method, and it was ruled that no GMOs had been found in the sample product. Miss Zhu lost both the case and the appeal, but she raised a very interesting question regarding producers' labelling liability in non-mandatory cases. According to the Consumer Protection Law, business operators should disclose information regarding their products or services, and should not conduct false and misleading publicity. In this case, the court upheld two points to support the Nestlé company: one is that the Nestlé products may not be covered by the compulsory list and the other is that no GM ingredients were found in the second investigation conducted by an authorised authority. Therefore, Nestlé did not have an obligatory legal responsibility to disclose the requested information and did not bear any liability.

the term 'directly from the GMOs' in the relevant documents. Accordingly, indirectly processed or deep-processed products may be excluded from the labelling requirements and the marketing of such products without labelling may be exempt from any sanctions. Furthermore, the list of GM products that need to be labelled under the labelling regulation has not been updated since 2003. As the indirectly processed or deep-processed products are likely to contain GMO ingredients, the undefined norm will definitely cause confusion in the implementation of GMO regulation. Then the legitimacy of the labelling system would thereby be undermined (Xue, 2009). This concern is explained in the Nestlé labelling case (Box 18.1).

Public Participation

Public Awareness of GMOs

Several surveys have already been conducted by researchers to illustrate public awareness of GMOs in China (Peng and Huang, 2015; Zhang *et al.*, 2014). Peng and Huang carried out research in 2015 on the basis of the data collected from 2002 to 2012. One of the major findings is that 45% of consumers in 2012 believed that GMOs are having adverse effects on human health, compared with 13% in 2002. Similarly, 36% of consumers in 2012 believed that GMOs are having adverse effects on the environment, compared with 11% in 2002. It is important to note that almost 50% of consumers chose the option 'I don't know' during the survey both in 2002 and 2012.

Another survey (Zhang *et al.*, 2014) illustrated that over 50% of consumers hold a negative opinion of GM food, and 40% of them are against GM rice, while 23% are in favour of it. In addition, this survey indicated that over 75% of people would like to know more about GM food, and the proportion is even higher (80.5%) when they are told about GMOs by scientists. The surveys conducted by the MoA research projects and natural science projects in 2009 and 2010 also found that over 60% of interviewees believed that risk communication is necessary in order to increase consumers' confidence on GMOs.

Public Participation in GMO Regulation

As a contracting party to the Cartagena Protocol on Biosafety (CPB), China needs to include public participation in its biosafety regulations in order to fulfil its international obligations. Even though public participation is not explicitly mandated in the agricultural GMO regulations, some general laws stipulate that it is necessary. For instance, the Regulation on Disclosure of Government Information states that information concerning the vital interests of citizens, legal persons or other organisations should be disclosed and made widely known by affected stakeholders.

Several lawsuits in recent years have fuelled the debate regarding disclosure of information about agricultural GMOs in China. The case Huang vs. MoA failed to challenge the MoA in 2013, but did raise an issue regarding the scope of the information which must be made publicly available. More importantly, competent authorities are starting to become aware of this necessity for communication. The MoA stated in its response to the enquiry of Ms Cai *et al.* in 2015 that it had established an information-disclosure platform at its official website. For the first time, the MoA specified in detail the procedures and required documents for requests for information with respect to agricultural GMOs. The MoA concluded that it had finalised 27 replies in 2014 and 12 replies in 2015, in response to some 100 inquiries (MoA, 2016b).

It is arguable that the stalled commercialisation of Bt rice resulted from the existence of the traditional government-led decision-making procedures, in which public participation has been undervalued. The traditional regime's legitimacy has been widely challenged, and there needs to be more consideration of how to ensure broader public participation.

Firstly, both the government and scientists should improve public awareness and participation in the decision-making process regarding Bt rice commercialisation. Public attitudes and opinions not only play an important role in determining the final policies, but also shape the infrastructure of democracy in the governmental decision-making mechanism. This helps provide basic conditions for a well-functioning management framework for the development of Bt rice (Mao, 2011).

Secondly, it is necessary for the government and related sectors to disclose decisions and safety information about Bt rice to the general public, in order to build communication platforms and improve the transparency of the decision-making process. The communication platforms need to be formalised, such as public hearings, case discussions, a public forum, community discussions, etc. Bt rice, as a staple food, has a significant influence on the public's livelihood. In the face of uncertainties, a democratic decision-making mechanism that requires broad public participation helps to confer a higher degree of legitimacy on the final decisions (Mao, 2011).

Thirdly, scientists play an important role in shaping the public's awareness regarding Bt rice. They undertake their social responsibilities of disseminating scientific information, fairly evaluating the risk and benefits of GM crops, and helping the public to become aware of the uncertainties associated with such varieties, etc. The gap between scientists and the general public needs to be bridged by creating a better communication platform (Jiang, 2010; Wang, 2015).

Conclusion

It is noteworthy that China has shifted from a rather pro-GMO attitude to a more cautious approach since its adoption of the CPB and several national agricultural regulations in the early 2000s. The MoA has been playing a leading role in the risk regulation of GMOs. The risk assessment is conducted by the NAGMOBC. It is argued that the NAGMOBC will need to be more transparent and must engage more stakeholders in order to improve its accountability. The MoA is the *de facto* risk manager, and the approval process is expected to become more accountable. Since the MoA aims to introduce socio-economic considerations into the decision-making, a more sound regime will be needed in order to fulfil China's obligations under the international conventions, and to build a more comprehensive and enforceable domestic legal framework of biosafety. The labelling system for GMOs in China is not pragmatic, since it results in a *de facto* zero-tolerance regime insofar as the labelling of GM products is concerned, due to the lack of any thresholds. The establishment of a tolerance threshold for any minimal accidental or unintended admixture of GMOs is recommended in order to make the labelling system more practical.

Although the lay public is becoming increasingly aware of the issue of food safety and environmental protection, it is acknowledged that public awareness of the uncertainties associated with GMOs and of the benefits of GMOs is rather low. Therefore, the question of how to engage the broader public in order to make the risk regulation more accountable may pose a huge challenge to Chinese regulators. The regulatory challenges presented by biotechnology should not simply be seen as a scientific issue, but more importantly must be regarded as a social choice. GMO regulation should therefore ensure that the various stakeholders can participate in the process of regulation (Howse, 2000). It is recognised that the slow pace of commercialisation of GMOs is largely due to low public acceptance, among other considerations. Bt rice is such an example.

The conclusion to be drawn is that communication among governmental officials, scientists and the lay public should be enhanced (Li *et al.*, 2016). On the one hand, China should endeavour to increase public awareness by means of education and information-disclosure sessions. On the other hand, more stakeholders should be involved both in risk assessment and in risk management by means of consultation and public hearings, etc.

Acknowledgement

This work was supported by the University of Copenhagen's Excellency Programme KU2016.

References

Beck, U. (1992a). *Risk Society: Towards a New Modernity*. London: Sage Publications.

Beck, U. (1992b). From industrial society to the risk society: questions of survival, social structure and ecological enlightenment. *Theory, Culture and Society* **9**(1), S97–S123.

Brown, L. R. (1995). *Who Will Feed China? Wake-up Call for a Small Planet*. New York, NY: W. W. Norton & Company.

Comments (2015). Industrial Comments on Draft Revision of Measures on the Safety Evaluation Administration of Agricultural Genetically Modified Organisms, 2015. [Online]. Available from www.amcham-shanghai.org/ftpuploadfiles/insight/Comments%20to%20MOA%20on%20Draft%20Amendments_CN.pdf

Draft Revision (2015). Draft Revision of Measures on the Safety Evaluation Administration of Agricultural Genetically Modified Organisms 2015.

Greenpeace (2014). Investigation report: illegal cultivation of GM rice in 10 years. [Online]. Available from www.greenpeace.org/china/zh/publications/reports/food-agriculture/2014/illegal-rice-decade/

Howse, R. (2000). Democracy, science, and free trade: risk regulation on trial at the World Trade Organization. *Michigan Law Review* **98**, S2329–S2330.

Jiang, G. (2010). The commercialisation of Bt rice shall be cautious. *Journal of China Weekly* **2**, S60–S61.

Keeley, J. (2003). A biotech developmental state? The Chinese experience. [Online]. Available from www.ids.ac.uk/go/idspublication/a-biotech-developmental-state-the-chinese-experience

Kubasek, N. and Silverman, G. (2007). *Environmental Law*. Sixth edition. Upper Saddle River, NJ: Prentice Hall.

Li, Y. *et al.* (2016). The development and status of Bt rice in China. *Plant Biotechnology Journal* **14**, S839–S848.

Mao, X. (2011). On the industrialisation of GM rice in China: dilemmas and way out. *Journal of Nanjing Agricultural University (Social Sciences Edition)* **11**(9), S125–S130.

MoA (2016a). Press conference. [Online]. Available from http://news.sohu.com/20160420/n445153674.shtml

MoA (2016b). Official response. [Online]. Available from www.moa.gov.cn/zwllm/tzgg/tz/201508/t20150805_4776718.htm

Peng, B. and Huang, J. (2015). Chinese consumers' knowledge and acceptances of genetically modified food. *Agricultural Economics and Management* **1**, S33–S39.

Pollack, M. and Shaffer, G. (2009). *When Cooperation Fails. The International Law and Politics of Genetically Modified Foods*. Oxford: Oxford University Press.

Wang, Q. (2015). China's scientists must engage the public on GM. *Nature* **519**, S7.

Wang, X. (2004). Challenges and dilemmas in developing China's national biosafety framework. *Journal of World Trade* **38**(5), S899–S913.

Xue, D. (2009). *Biosafety and Regulation for Genetically Modified Organisms*. Beijing: Science Press.

Zhang, X. *et al*. (2014). GM Food: a study of Chinese public's recognition and attitude. *Journal of Anhui Agricultural Sciences* **42**(20), S6378–S6786.

Zhao, J. and Ho, P. (2005). A developmental risk society? The politics of genetically modified organisms (GMOs) in China. *International Journal of Environment and Sustainable Development* **4**(4), S370–S394.

19 Commercial Use and Governance of Bt cotton in China

Yunhe Li, Yanhui Lu, Eric M. Hallerman, Yufa Peng and Kongming Wu

Introduction

Cotton is one of the most important cash crops in China, which is the world's largest cotton-producing country, accounting for more than 20% of world production (United Nations Environment Programme (UNEP), undated). However, cotton production as traditionally practised is seriously affected by insect pests (Wu and Guo, 2005); a Ministry of Agriculture assessment (Zhang, 2011) indicated that increasing harm from pests and diseases is the major threat to the cotton sector. More than 300 species of herbivorous insects have been recorded in cotton fields in China, including cotton bollworm (*Helicoverpa armigera*), pink bollworm (*Pectinophora gossypiella*), beet armyworm (*Spodoptera exigua*), cotton worm (*Spodoptera litura*), corn borer (*Ostrinia furnacalis*), spiny bollworm, cotton leaf roller, aphids, spider mites, mirid bugs, thrips, whiteflies and leafhoppers (Wu and Guo, 2005). From the late 1980s to the mid 1990s, cotton bollworm outbreaks caused considerable yield losses throughout China's major cotton-planting regions (Guo, 1998). During that period, farmers applied large amounts of chemical insecticides to control cotton bollworm, which led to the evolution of insecticide resistance, pest resurgence, environmental pollution and poisoning of many cotton farmers (Guo, 1998). Hence, the cotton bollworm outbreak did not only pose agricultural and economic issues, but also brought environmental and health concerns.

Against this background, China devoted considerable resources to developing and utilising genetically modified (GM) cotton expressing δ-endotoxins from the bacterium *Bacillus thuringiensis* (Bt). These endotoxins have insecticidal action, especially against larval lepidoptera and coleoptera, which led to their use as insecticides, and more recently to the development of GM crops expressing the toxins (Kumar *et al.*, 1996). In order to efficiently control cotton bollworm and reduce environmental pollution associated with the application of chemical insecticides, transgenic Bt insect-resistant cotton was formally commercialised in 1997 in China. In this chapter, we discuss the development of Bt cotton in China, and the associated risks, risk management, governance and societal perspectives. China provides a useful case study because it commercialised Bt cotton relatively early and is one of the two largest producers of Bt cotton (James, 2013). Nationally representative data on inputs and yield of cotton production have been assembled and published annually.

Governance of GM Crops in China

The Chinese government has devoted considerable attention to the development of public policy for regulatory oversight of agricultural biotechnology (Li *et al.*, 2014). In the early 1990s, China started to establish its GMO biosafety regulations. The Ministry of Science and Technology (MST) issued the first biosafety regulation for agricultural biotechnology in 1993, stating general principles, safety issues, the need for risk assessments, application and approval procedures, safety control measures and legal responsibilities. Implementation Guidelines were issued by the Ministry of Agriculture (MoA) in 1996. The State Council of China (2001) decreed a new policy for administrative oversight of GMO biosafety in 2001, the Regulation on Safety Administration of Agricultural GMOs, providing detailed prescriptions for laboratory research, biosafety testing, production, processing, marketing and import and export activities with respect to agricultural GMOs. The MoA announced three implementing regulations in 2002. The Ministry of Public Health (MPH) promulgated the regulation on GMO food hygiene. These regulations collectively covered all aspects of development of GMOs, including laboratory research, food and environmental safety testing, labelling, production, processing, marketing, and import and export of GMOs and derived products. In addition, the food-safety law in China was revised in 2015; Article 69 specifies that foods must be clearly labelled as containing GM ingredients (Keller and Heckman LLP, 2015).

Against this background, from 2002 to 2012, the MoA approved 2,775 applications for pilot-scale testing of GM plants, 459 for environmental release for field testing and 317 for pre-production testing, and the MoA has issued 1,830 biosafety certificates. However, only seven GM crops, involving 10 integration events, have been approved by the MoA for commercial planting in China (Li *et al.*, 2014).

Development and Use of Bt Cotton in China

Research on transgenic Bt cotton started in the early 1990s. Bt gene transformation of cotton was first reported in China in 1991 (Zhao *et al.*, 2013). In 1994, a *cry1Ac* and *cry1Ab* fusion gene of *cry1A* was successfully transferred into cotton, and the resulting Bt cotton line exhibited high resistance to cotton bollworm, with a control efficacy of over 80% (Zhang *et al.*, 2007). This success made China the second country with the ability to develop Bt cotton plants, after the USA. To delay the evolution of resistance of target pests to Bt cotton, Chinese scientists attempted to transfer two insecticidal genes, *cry1A* and *CpTI* (cowpea trypin inhibitor) into cotton plants, and dual-gene transgenic cotton plants were successfully produced in 1996 (Zhang *et al.*, 2007). Both Bt and GNA (*Galanthus nivalis* agglutinin) genes were later transformed into cotton plants (Zhao *et al.*, 2013). Starting from the Bt cotton events, many Bt cotton hybrids have been developed (Zhang *et al.*, 2007). Thus far, 10 cotton events have been approved for commercial use in China (International Service for the Acquisition of Agri-biotech Applications (ISAAA), 2016), six of which express traits conferring insect resistance (Table 19.1). Among the approved cotton events, only the three Bt cotton varieties

Table 19.1 Insect-resistant cotton events approved in China

Name	Trade name	Developer	Insect-resistance transgene(s)	Stacked trait
GHB119	–	Bayer CropScience	*cry2Ae*	Herbicide tolerance
GK12	–	Chinese Academy of Agricultural Science	*cry1Ab–Ac*	–
MON15985	Bollgard II	Monsanto	*cry1Ac* + *cry2Ab2*	–
MON531	Bollgard, Ingard	Monsanto	*cry1Ac*	–
SGK321	–	Chinese Academy of Agricultural Science	*cry1A* + *CpTI*	–
T304–40	–	Bayer CropScience	*cry1Ab*	Herbicide resistance

Source of information: ISAAA (2016).

expressing *cry1Ac* (MON531 from Monsanto), *cry1A* (a *Cry1Ac* and *Cry1Ab* fusion gene) and *cry1A* + *CpTI* genes have been approved for cultivation in China, all of which express only the trait of insect resistance (Li *et al.*, 2014).

In 1997, the first year of commercial planting, less than 0.1 million ha of Bt cotton was planted. Since then, the area of Bt cotton planted has expanded rapidly (Figure 19.1). The total planting area reached 1.1 million ha in 2000 and 2.8 million ha in 2003 (Wu, 2007), peaked at 3.9 million ha in 2007, and has remained relatively stable since then. In 2014, 7.1 million Chinese farmers planted 3.9 million ha of Bt cotton, representing 93% of the total cotton area (James, 2014; Figure 19.1). In two of the three major cotton-planting regions – the humid Yangtze and Yellow River valleys – Bt

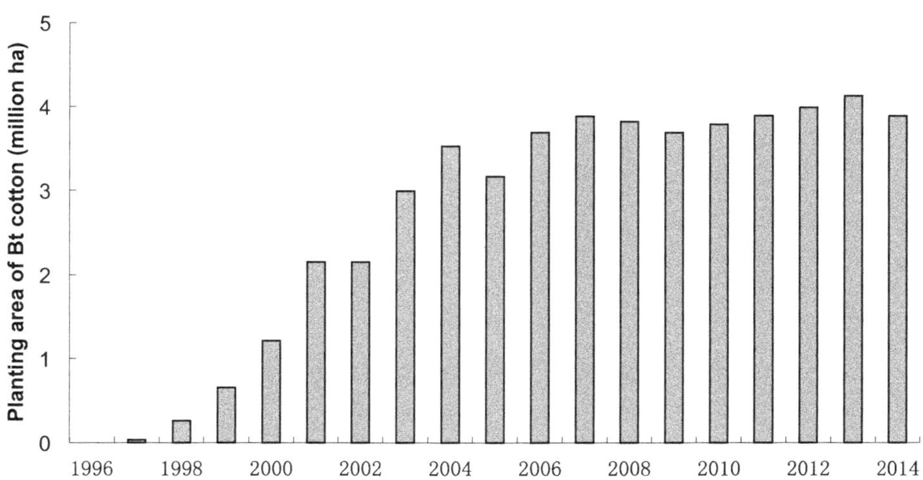

Figure 19.1. Yearly planting area of Bt cotton in China since 1997 (James, 2013).

cotton coverage is almost 100%. Published data on Bt cotton production aggregate all approved lines; we could not find any data comparing the plantings of the respective lines.

Economic Benefits Associated with the Use of Bt Cotton

Early work demonstrated that Chinese smallholder farmers benefited from Bt cotton. Farm budget analysis on data from surveys of 300–400 farmers in five provinces from 1999 to 2001 showed that complex adjustments of pesticide use and yield had occurred (Huang *et al.*, 2002; Pray *et al.*, 2002). For example, in Shandong Province during 1999, yields of Bt cotton were 9% higher and the amount of insecticide applied 71% lower. Farmers had a positive net income from Bt cotton, mainly through reduced input and labour costs (Huang *et al.*, 2003).

Whether the short-term benefits of Bt cotton adoption prove sustainable is at issue because ecological impacts and the evolution of resistance to Bt toxin may take years to emerge or become recognised. Qiao (2015) presented a descriptive analysis of the economic benefits of Bt cotton in China and its temporal dynamics from 1997 to 2012 in terms of pesticide expenditure, seed expenditure, labour use and cotton yield, and also developed econometric models to isolate the contribution of Bt cotton. His descriptive results showed that, with the spread of Bt cotton, pesticide use decreased significantly. This was especially true in the Yellow River valley, where pesticide expenditure in terms of RMB per hectare of land fell to less than 40% of that preceding the adoption of Bt cotton. Labour use was 383 days/ha in 2004–2012, down about 40% from 638 days/ha in 1990–1996. Much of the decrease in labour was attributed to a reduction in the amount of time spent spraying pesticide and manually removing large cotton bollworms. However, because Bt cotton seed is more expensive than conventional seed, the seed cost increased from 375 RMB/ha (US$60) to 828 RMB/ha (US$132.5) during the late adoption period. The cotton yield increased from 1,848 kg/ha during 1990–1996 to 2,547 kg/ha during 2004–2012. The increase in seed cost was similar to the decrease in pesticide cost, and did not offset the economic benefit of the increased yield. The economic benefit of Bt cotton did not diminish over the 15-year period studied. Results of econometric modelling supported those of the descriptive analysis. Because of the adoption of Bt cotton, Chinese farmers saved 4.12 billion RMB on pesticide use and 8.70 billion RMB on labour use. The benefit from the increase in yield was even higher. The total economic benefit of Bt cotton was more than 33 billion RMB over the 15 years since the commercialisation of Bt cotton, or 2.08 billion RMB per year.

Ecological Effects of Bt Cotton and Management Strategies

China has developed a comprehensive framework for assessing the biosafety of GMOs. The Ministry of Agriculture is the lead institution for nationwide supervision and inspection of agricultural GMO biosafety (State Council of China, 2001). The Office of Agricultural Genetic Engineering Biosafety Administration was created in 1996 to deal with routine operations with respect to GMO safety. In addition, the National Agricultural GMO Biosafety Committee was formed in 1997, with responsibility for the management of agricultural GMO research and risk assessment (Li *et al.*, 2014). Since Bt cotton was approved for planting, China has implemented a relatively well-developed monitoring programme. This monitoring has yielded a database that is useful for assessing strategies for managing ecological risks associated with the use of Bt cotton (Wu *et al.*, 2008; Lu *et al.*, 2010, 2012).

Target Pests

As noted above, the insecticidal Cry proteins produced by Bt cotton have specific activity against the larvae of various lepidopteran pests, including cotton bollworm, pink bollworm, spiny bollworm, corn borer and cotton leaf roller.

Cotton bollworm is a polyphagous insect pest upon more than 200 species of host plants, including cotton, maize, soybean, peanut, vegetables and many other main crops in China. With increasing adoption of Bt cotton, cotton bollworm populations have been not only effectively suppressed on this crop, but also greatly reduced on other non-Bt host crops (e.g., maize, soybean and peanut) in the same planting area (Wu *et al.*, 2008), a phenomenon termed the 'halo effect'. In China, cotton is the main host for cotton bollworm moths of the first generation of the growing season for laying eggs, thereby acting as the source for subsequent generations of moths on other host crops. Bt cotton then kills most of the emerging larvae of the second generation in cotton fields, reducing the breeding population for the following generations on other host crops; i.e., Bt cotton works as a 'trap' crop for the cotton bollworm population in the wider planting area.

Pink bollworm affects about 80 species of host plants, although cotton is almost the exclusive host crop in China. A similar 'halo effect' was found for this target pest; after the adoption of Bt cotton, the pink bollworm population greatly decreased both on Bt cotton and on non-Bt cotton (Wan *et al.*, 2012a; Huang *et al.*, 2013). Additionally, Bt cotton suppressed the populations of other target pests, such as spiny bollworm, corn borer and leaf roller (Wu and Guo, 2005).

Non-target Pests

Non-target pests are not susceptible to the Bt insecticidal proteins, which do not directly affect their individual development or population growth. However, under field conditions, the planting of Bt cotton may indirectly affect their population increase by altering the pattern of insecticide use. A typical case is mirid bugs. Mirid bugs were first recorded on cotton in China during the early 1930s, and were regarded as secondary pests on this crop during the last century (Lu and Wu, 2011). However, populations of mirid bugs have progressively increased and acquired pest status in cotton and multiple other crops, in association with a regional increase in adoption of Bt cotton. This multi-crop outbreak was correlated with their attraction to flowering plants and

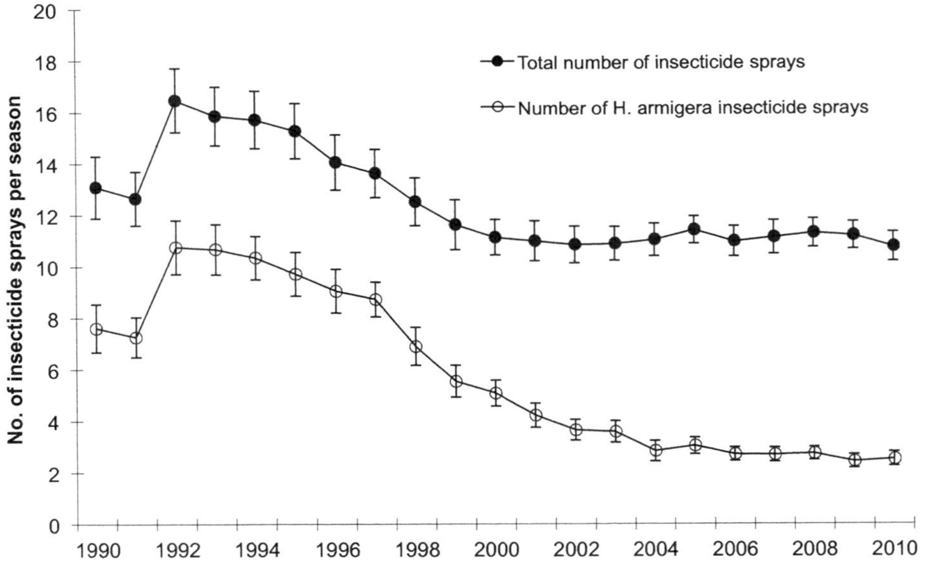

Figure 19.2. Numbers of insecticidal sprays on Bt cotton in China since 1997. Modified from Lu *et al.* (2012).

associated seasonal host alternation. As cotton is one of the few flowering host crops during mid-to-late June, mirid bugs largely move from early-season host plants to cotton fields, where their populations increase. Prior to the adoption of Bt cotton, broad-spectrum insecticide use targeting cotton bollworm also knocked down early mirid bug populations, with cotton acting as a trap crop (Figure 19.2). The absence of insecticide sprays on Bt cotton permits an unrestrained build-up of mirid bug population and a subsequent active spread or passive spill-over to other flowering crops. Hence, a reduction in insecticide use for cotton bollworm control in Bt cotton led to mirid bug outbreaks in cotton and various fruit crops in the broader agro-landscape (Lu *et al.*, 2010). With the increase of populations of mirid bugs or other secondary insect pests, pesticide had to be applied, raising the number of applications of insecticide in cotton fields in recent years (Figure 19.2). However, the total number of applications of insecticide is still smaller than that before commercialisation of Bt cotton (Figure 19.2). The secondary insect pest issue is the ecological consequence of planting Bt plants or utilising other insect pest control measures (Lu *et al.*, 2010), demonstrating that we cannot expect to solve all insect pest problems simply by planting Bt crops, which must be considered as but one component of an integrated pest management programme (Wu and Guo, 2005; Wu *et al.*, 2008). In China, scientists have been developing and applying various measures, such as cultural control, biological control, the use of trap crops, chemical control and other approaches to manage mirid bugs in Bt cotton. With the combination of multiple measures, mirid bug populations on cotton have been effectively suppressed (Lu *et al.*, 2012; Lu and Wu, 2011).

Natural Enemies

Bt insecticidal proteins do not have direct impacts on the survival, development or fecundity of natural enemies of crop pests (Li *et al.*, 2014, 2016). Impact on *Microplitis mediator*, a larval parasitoid wasp of cotton bollworm, has been reported. The impact is due to the indirect effect of poor host-quality of cotton bollworm larvae that had been fed Bt proteins rather than to direct toxicity of Bt proteins (Ding *et al.*, 2009). This phenomenon has been documented also with other Bt crops, such as Bt maize (Dutton *et al.*, 2003; Romeis *et al.*, 2006; Chen *et al.*, 2008; Li *et al.*, 2013).

Under field conditions, population levels of natural enemies of plant pests might change due to the change in insecticide use upon Bt cotton (Naranjo, 2009). With increasing adoption of Bt cotton and decreasing insecticide use, abundances of generalist predators (ladybirds, lacewings and spiders) gradually increased, resulting in decreasing abundance of aphids. Generalist predators usually have great dispersal ability, such that increasing predator abundance on Bt cotton ultimately promoted predator-mediated biocontrol services across the agro-ecosystem of China (Lu *et al.*, 2012).

Impact on Soil Ecosystems

In the natural environment, the insecticidal protein in Bt cotton degraded rapidly during the initial period following harvest (85% in the first month), more slowly during the winter season, and to undetectable levels by late April (Li *et al.*, 2005). Hence, after continuous use of Bt cotton, Bt protein residues were extremely low in field soil. The low level of Bt protein residue did not result in detectable biological activity, i.e., long-term cultivation of Bt cotton had no significant effect on the abundance of soil invertebrates (Collembola, Acarina, Araneae, Coleoptera, Dermaptera, Diplura, Isoptera and Diptera – Li *et al.* (2012)), numbers or diversity of microbial populations (bacteria, fungi, azotobacter, denitrifying bacteria and ammonia-oxidising bacteria – Li *et al.* (2011)), or densities and diversities of soil nematodes (Li and Liu, 2013).

Evolution of Resistance by Insect Pests

Insect pests targeted by Bt cotton may evolve resistance to Bt proteins. Experience with traditional pesticides shows that, if the Bt crop is used extensively without targeted management, the evolution of resistance is virtually inevitable. Indeed, resistance risk is arguably the main risk of Bt cotton production in China. Among the approaches that can be used to delay the evolution of resistance are (1) reducing selection pressure by maintaining refuge plants that do not express Bt; (2) using high-dose plants expressing multiple, 'stacked' Bt proteins; (3) reducing the selection differential between resistant and susceptible insects by applying other insecticides, agrotechnical methods or biological control; and (4) using chemical or environmental attractants to elicit the movement of susceptible males into Bt plantings.

Maintaining refuge plants is the most widely applied resistance management strategy in China. The maximum allowable distance from the refuge to Bt cotton depends on the frequency of movement and the distances over which adult crop pests disperse, i.e., males before mating, virgin females and mated females. Assuming that pest species may disperse 1–2 km from their site of emergence, and given the smallholder land ownership system in China, refuges are unstructured and rely on the presence of other crops. Cotton bollworm infests many non-Bt host crops (e.g., maize, soybean, peanut and vegetables), which collectively account for about 90% of the total agricultural planting area in China and serve as the refuge to produce cotton bollworm (Wu *et al.*, 2002, 2004; Wu 2007). After more than 18 years of production of Bt cotton with a single Bt gene, resistant populations of cotton bollworm have not developed and caused field control failure (Wu, 2007; Tabashnik *et al.*, 2012). Jin *et al.* (2015) confirmed the efficacy of the natural refuge strategy for delaying insect resistance to transgenic Bt cotton in China. Some studies, however, have indicated that cotton bollworm populations exhibit increasing resistance frequency and decreased susceptibility to Bt in places where there has been long-term, intensive planting of Bt cotton (Zhang *et al.*, 2011).

With the increasing adoption of Bt cotton, pink bollworm has gradually lost its refuge, i.e., non-Bt cotton. As a result, its LC_{50} values for Bt-Cry1Ac increased from 0.24 mg Cry1Ac per ml diet in 2005–2007 to 0.47 in 2008–2010, and the percentage of field populations yielding one or more survivors at a diagnostic concentration of Cry1Ac increased from 0% in 2005–2007 to 56% in 2008–2010. This result implies that the frequency of resistance to Cry1Ac has increased in field populations of pink bollworm. However, the median survival at the diagnostic concentration was only 1.6% from 2008 to 2010, and failure of Bt cotton to control pink bollworm has not been reported in China (Wan *et al.*, 2012b).

Looking ahead to future production conditions, should Bt maize and other crops be commercialised, then a programme of structured refuges would be needed in order to manage the evolution of Bt resistance in crop pests.

Gene Flow Risks

GM crop plants may pose ecological and evolutionary risks if their transgenes are transmitted to wild relatives of the crop plants. The most likely centres of origin of the two Old-World species of cotton, *Gossypium arboreum* and *G. herbaceum*, are the Indus Valley and Africa, respectively (Maestri and Hirst, 2014). Hence, production of Bt cotton in China does not pose risks of gene flow to wild ancestors of cotton. Only a few studies have investigated gene flow between Bt cotton and non-transformed cotton cultivars, and results showed that the maximal distance of Bt gene flow from Bt cotton to other cultivars is less than 100 m (Zhu *et al.*, 2013). Therefore, the recommended isolation distance of 150 m is effective to prevent gene flow between GM and conventional cotton plants.

Socio-Political Impacts and Implications for GM Crops in China

The economic, ecological and socio-political sustainability of GM crops has been a subject of heated debate in China. Those opposing GM technology (e.g., Wang *et al.*,

2008) argued that the economic benefit is only short-term due to the evolution of pest resistance and secondary pest outbreaks; such opposition was especially notable after the secondary pest outbreak in some Bt cotton fields in the early 2000s noted above. Bt technology was even described as a 'weapon' that developed countries use to attack developing countries (Jiang and Li, 2010). Scientists and researchers who supported GM crops have been labelled 'traitors' (*Economist*, 2013).

Results of a survey by Qu *et al.* (2011) showed that over 55% of people questioned believed that GM crops pose risks to human health and the environment. However, the results also suggested that Chinese people have very low cognition on GMOs, which may be attributed to (i) the technical complexity and novelty of GM technology and the low scientific literacy of the public; (ii) the existence of different opinions on GMOs among scientists, confusing the public; (iii) the fact that some scientists overstate the benefits of GM technology, hence the technology is mistrusted; and (iv) the fact that Chinese researchers have little interest in popularisation of science (Liu *et al.*, 2011). Commercialisation of Bt cotton has proven somewhat less controversial in China, as production, ecological and human health benefits have been realised, and cotton is a non-food GM crop.

Faced by increasing criticism of GM technology, the Chinese government has delayed the commercialisation of other Bt crops (Jiang and Li, 2010). For example, even though GM rice technology has been mature for years (Li *et al.*, 2014), the Chinese government has no plan for its commercialisation.

Empirical evidence that GM technology offers long-term economic benefits, that ecological benefits are realised, and that ecological risks can be managed might help resolve the public debate, which could prove helpful for Chinese policymakers managing GM crop adoption. The Chinese experience with Bt cotton and other GM crop lines might inform the development and implementation of agricultural biotechnology policies in other countries.

References

Chen, M. *et al.* (2008). A critical assessment of the effects of Bt transgenic plants on parasitoids. *PLoS ONE* **3**(5), e2284.

Ding, J. *et al.* (2009). The life parameters of a parasitoid *Microplitis mediator* (Hymenoptera: Braconidae), reared on cotton bollworm *Helicoverpa armigera* (Hübner) with *Cry1Ac* diet. *Biocontrol Science and Technology* **19**(9), 931–941.

Dutton, A. *et al.* (2003). Prey-mediated effects of *Bacillus thuringiensis* spray on the predator *Chrysoperla cannea* in maize. *Biological Control* **26**, 209–215.

Economist (2013). China's food fight: a fierce public debate over GM food exposes concerns about America. *The Economist*, 14 December 2013. [Online]. Available from www.economist .com/news/china/21591577-fierce-public-debate-over-gm-food-exposes-concerns-about-america-food-fight

Guo, Y. Y. (1998). *Researches on Cotton Bollworm*. Beijing: China Agriculture Press.

Huang, J. *et al.* (2002). Transgenic varieties and productivity of smallholder cotton farmers in China. *Australian Journal of Agricultural and Resource Economics* **46**, 367–387.

Huang, J. *et al.* (2003). Biotechnology as an alternative to chemical pesticides: a case study of Bt cotton in China. *Agricultural Economics* **29**, 55–67.
Huang, Y. X. *et al.* (2013). Diminishing returns from increased percent Bt cotton: the case of pink bollworm. *PLoS ONE* **8**(7), e68573.
ISAAA (2016). GM crop events approved in China. [Online]. Available from www.isaaa.org/gmapprovaldatabase/approvedeventsin/default.asp?CountryID=CN
James, C. (2013). Global status of commercialized biotech/GM crops: 2013. ISAAA Brief No. 46. Ithaca, NY: ISAAA.
James, C. (2014). Global status of commercialized biotech/GM crops: 2014. ISAAA Brief No. 49, Ithaca, NY: ISAAA.
Jiang, G. and Li, X. (2010). On the safety and risks of genetically modified crops industry. *Journal of Engineering Studies* **2**(2), 112–119.
Jin, L. *et al.* (2015). Large-scale test of the natural refuge strategy for delaying insect resistance to transgenic Bt crops. *Nature Biotechnology* **33**, 169–174.
Keller and Heckman LLP (2015). China passes sweeping amendment to food safety law: the most stringent to date. China Regulatory Matters, 28 April 2015. [Online]. Available from www.khlaw.com/China-Passes-Sweeping-Amendment-to-Food-Safety-Law-The-Most-Stringent-To-Date
Kumar, P. A. *et al.* (1996). Insecticidal proteins of *Bacillus thuringiensis*. *Advances in Applied Microbiology* **42**, 1–43.
Li, X. G. and Liu, B. (2013). A 2-year field study shows little evidence that the long-term planting of transgenic insect-resistant cotton affects the community structure of soil nematodes. *PLoS ONE* **8**(4), e61670.
Li, X. G. *et al.* (2011). No evidence of persistent effects of continuously planted transgenic insect-resistant cotton on soil microorganisms. *Plant Soil* **339**(1–2), 247–257.
Li, X. G. *et al.* (2012). Field trials to evaluate effects of continuously planted transgenic insect-resistant cottons on soil invertebrates. *Journal of Environmental Monitoring* **14**(3), 1055–1063.
Li, Y. H. *et al.* (2013). Bt rice expressing Cry2Aa does not cause direct detrimental effects on larvae of *Chrysoperla sinica*. *Ecotoxicology* **22**, 1413–1421.
Li, Y. H. *et al.* (2005). Degradation dynamics of *Cry1Ac* insecticidal protein in leaves of Bt cotton under different environments. *Science Agriculture Sinica* **38**(4), 714–718.
Li, Y. H. *et al.* (2014). Biosafety management and commercial use of genetically modified crops in China. *Plant Cell Reports* **33**, 565–573.
Li, Y. H. *et al.* (2016). The development and status of Bt rice in China. *Plant Biotechnology Journal* **14**, 839–848.
Liu, P. L. *et al.* (2011). Analysis of risk communication for transgenic biotechnology in China. *Journal of China Biotechnology* **31**, 145–149.
Lu, Y. H. and Wu, K. M. (2011). Mirid bugs in China: pest status and management strategies. *Outlooks on Pest Management* **22**(6), 248–252.
Lu, Y. H. *et al.* (2010). Mirid bug outbreaks in multiple crops correlated with wide-scale adoption of Bt cotton in China. *Science* **328**(5982), 1151–1154.
Lu, Y. H. *et al.* (2012). Widespread adoption of Bt cotton and insecticide decrease promotes biocontrol services. *Nature* **487**(7407), 362–365.
Maestri, N. and Hirst, K. K. (2014). The domestication history of cotton (Gossypium): the origins of cotton. [Online]. Available from http://archaeology.about.com/od/cterms/a/Cotton.htm
Naranjo, S. E. (2009). Impacts of Bt crops on non-target invertebrates and insecticide use patterns. *Perspectives in Agriculture, Veterinary Science, Nutrition and Natural Resources* **4**, 1–23.

Pray, C. E. *et al.* (2002). Five years of Bt cotton in China – the benefits continue. *The Plant Journal* **31**, 423–430.

Qiao, F. (2015). Fifteen years of Bt cotton in China: the economic impact and its dynamics. *World Development* **70**, 177–185.

Qu, Y. D. *et al.* (2011). Survey analysis of the cognition of GMO risk and safety among Chinese public. *Journal of China Agricultural University* **16**, 1–10.

Romeis, J. *et al.* (2006). Transgenic crops expressing *Bacillus thuringiensis* toxins and biological control. *Nature Biotechnology* **24**, 63–71.

State Council of China (2001). Regulations on safety of agricultural genetically modified organisms. [Online]. Available from www.gov.cn/flfg/2005-08/06/content_21003.htm

Tabashnik, B. E. *et al.* (2012). Early detection of field-evolved resistance to Bt cotton in China: cotton bollworm and pink bollworm. *Journal of Invertebrate Pathology* **110**(3), 301–306.

UNEP (Undated). The cotton sector in China. [Online]. Available from www.unep.ch/etu/publications/Synth_China.PDF

Wan, P. *et al.* (2012a). The halo effect: suppression of pink bollworm on non-Bt cotton by Bt cotton in China. *PLoS ONE* **7**(7), e42004.

Wan, P. *et al.* (2012b). Increased frequency of pink bollworm resistance to Bt toxin *Cry1Ac* in China. *PLoS ONE* **7**(1), e29975.

Wang, S. *et al.* (2008). Bt cotton and secondary pests. *International Journal of Biotechnology* **10**, 113–121.

Wu, K. M. (2007). Monitoring and management strategy for *Helicoverpa armigera* resistance to Bt cotton in China. *Journal of Invertebrate Pathology* **95**(3), 220–223.

Wu, K. M. and Guo, Y. Y. (2005). The evolution of cotton pest management practices in China. *Annual Review of Entomology* **50**, 31–52.

Wu, K. M. *et al.* (2002). Evaluation of the natural refuge function for *Helicoverpa armigera* (Lepidoptera: Noctuidae) within *Bacillus thuringiensis* transgenic cotton growing areas in north China. *Journal of Economic Entomology* **95**(4), 832–837.

Wu, K. M. *et al.* (2004). Evaluation of maize as a refuge for management of resistance to Bt cotton by *Helicoverpa armigera* (Hübner) in the Yellow River cotton-farming region of China. *Crop Protection* **23**(6), 523–530.

Wu, K. M. *et al.* (2008). Suppression of cotton bollworm in multiple crops in China in areas with Bt toxin-containing cotton. *Science* **321**(5896), 1676–1678.

Zhang, H. N. *et al.* (2011). Early warning of cotton bollworm resistance associated with intensive planting of Bt cotton in China. *PLoS ONE* **6**(8), e22874.

Zhang, R. *et al.* (2007). Retrospect and prospect of research on Chinese transgenic insecticidal cotton. *Journal of Agricultural Science and Technology* **9**(4), 32–42.

Zhang, T. (2011). Country sector overview: the cotton sector in China. The Sustainable Trade Initiative. [Online]. Available from http://issuu.com/idhsustainabletradeinitiative/docs/cotton_in_china_-_country_sector_an

Zhao, D. *et al.* (2013). Research of transgenic cotton and the prospect in China. *Liaoning Agricultural Sciences* **2013**(1), 41–44.

Zhu, J. L. *et al.* (2013). The influence of wind direction on pollen-mediated gene flow in transgenic insect-resistant cotton. *Acta Ecologica Sinica* **33**(21), 6803–6812.

20 The Regulatory Regime for Genetically Modified Crops in India

Krishna Ravi Srinivas

Introduction

India was one of the few developing countries to initiate policies for the development of biotechnology even when the biotechnology revolution had not developed fully. Initially the Government of India established in 1983 a Board on Biotechnology, which was converted into the Department of Biotechnology (DBT) in 1985. Since then biotechnology has been supported by the Government of India through various programmes, including efforts in capacity building and human resources development. The opening up of the economy and liberalisation since 1991 provided many opportunities for the biotechnology sector in India. India became one of the favourite destinations for biotechnology research and development (R&D) and at the same time attracted foreign direct investments. With India becoming a global player in information technology and enabled services, bioinformatics and bioservices became important sectors in biotechnology. According to a recent estimate, India's bioeconomy is valued at about US$35 billion and the share of agribiotech is US$9 billion (ABLE, 2016).

The DBT published its latest biotechnology strategy in December 2015. Many state governments have their own policies and programmes to promote biotechnology. Thus the policy milieu in India for biotechnology has been favourable and, with the increased emphasis on 'Make in India' and new opportunities in sectors like biopharmaceuticals and post-genome life sciences, governments have become all the more supportive of biotechnology. Regarding GM crops though, Bt (*Bacillus thuringiensis*) cotton is the only crop that has been allowed for commercialisation. Other GM crops in the pipeline and their respective traits are very important for India to meet its needs in agriculture when threats like climate change, constraints on using water resources and soil degradation loom large. Among scientists in India there is almost a consensus that GM crops have a key role to play and that they pose no significant risks to the environment and human health (Giri and Tyagi, 2016). While the policies for promoting and incentivising biotechnology are important, equally important is a biotechnology regulatory regime that would ensure that benefits from biotechnology are not achieved at the cost of environmental safety, human health and conservation of biodiversity.

This chapter discusses the regulatory regime in India for GM crops and the challenges faced by it. It analyses the experience with regulating Bt cotton, the controversies and moratorium on Bt brinjal (aubergine) and the ongoing evaluation of GM chickpea and

Table 20.1 Key departments involved in regulating GM crops

Department/ministry	Role
Ministry of Environment, Forest and Climate Change (MoEF&CC)	Nodal agency for implementing the Rules 1989 Nodal ministry for the Cartagena Protocol and CBD Responsible for conservation and protection of the environment
Department of Biotechnology (DBT) (under the Ministry of Science and Technology)	Promotion of biotechnology Support in implementing biosafety guidelines
Ministry of Agriculture	Indian Council of Agricultural Research (ICAR) is responsible for monitoring agronomic performance and evaluation of GM crops Monitoring post-release performance of GM crops
Ministry of Health and Family Welfare	Food Safety and Standards Authority of India is responsible for regulating GM foods

GM mustard. Further, it proposes a new regulatory regime, the features of which are outlined.

Regulation of GM Crops in India

There is no specific Act to regulate GM crops in India. Rather the rules framed and orders issued under the Environment (Protection) Act 1986 (EPA) form the core of the regulatory framework (Table 20.1). The EPA is implemented by the Ministry of Environment, Forests and Climate Change (MoEF&CC). The Rules for the Manufacture/Use/Import/Export and Storage of Hazardous Microorganisms, Genetically Engineered Organisms or Cells 1989 (henceforth abbreviated to Rules 1989), pursuant to Sections 6, 8 and 25 of the EPA, form the basis for regulation of activities related to genetically modified organisms (GMOs) and products thereof. In addition to these, the Food Safety and Standards Act 2006, Drugs and Cosmetics Rule 1988, Biological Diversity Act, Plant Quarantine Order 2003 and Directorate General of Foreign Trade (DGFT) Notification relating to inclusion of GM trade policy in foreign trade policy (2006–2009) are also applicable for activities relating to GMOs including GM crops. India is a Party to the Cartagena Protocol on Biosafety and the Convention on Biological Diversity. The key authorities with respect to approvals, regulation and monitoring are listed in Table 20.2.

The regulatory power is shared among ministries, and a balance has evolved over the years between the two key ministries, i.e., the DBT and MoEF&CC. According to Gupta, 'the regulatory system has sought a balance of authority between concerned governmental actors, with the central axis of conflict being between a proactive Department of Biotechnology under the Ministry of Science and Technology, which aggressively promotes development and adoption of transgenic crops, and a more precautionary Ministry of Environment and Forests' (Gupta, 2011, p. 737).

Table 20.2 Authorities and their roles

Authority	Role
Recombinant DNA Advisory Committee (RDAC)	Advisory
Institutional Biosafety Committee (IBSC)	First-level approval for experiments and trials
Review Committee on Genetic Manipulation (RCGM)	Second-level approval on the basis of rules and review of proposals from the applicants (institutions, companies, universities etc.)
Genetic Engineering Appraisal Committee (GEAC)	Appraisal and recommendation for commercial release. The recommendations are not binding and the MoEF&CC will take the final decision.
State Biotechnology Coordination Committee (SBCC)	Monitoring (the SBCC cannot give approval or permission for trials)
District Level Coordination Committee (DLCC)	Monitoring (the DLCC cannot give approval or permission for trials)

The RCGM and GEAC can approve field experiments and controlled trials, but permission to conduct them is required from the state government of the region in which the site of the trial is situated. Thus, while the federal government has the power to permit trials, that cannot override the power of the state government to approve them or not.

While the RCGM is housed within the DBT, the GEAC, which is the apex regulatory committee, is within the MoEF&CC. The RDAC is formed by the DBT, and its role is to frame guidelines and review developments in biotechnology at the national and international level and recommend suitable safety regulations. Its role is purely advisory. The IBSC forms the first-level authority for biosafety and regulation, and all organisations involved in rDNA activities should have one such committee. The RCGM is responsible for specifying guidelines for conducting GMO research covering the uses indicated in the schedule of Rules 1989.

The GEAC is the final authority to review data/results from field trials and make recommendations to the government on commercial release. The GEAC supervises the implementation of rules and guidelines. It carries out supervision through the SBCC or DLCC, or by authorised persons. The approvals are for a period of four years and are renewable for a further two years. The GEAC can revoke approvals on account of non-compliance with conditions imposed by the GEAC. The GEAC can call for additional data, can seek new information and can request that applicants share data with third parties, if it considers that such sharing is necessary. The Monitoring and Evaluation Committee (MEC) is formed by the GEAC and is empowered to visit sites, inspect facilities, analyse data and make recommendations to the RCGM/GEAC regarding transgenics.

Until 2010 the GEAC was known as the Genetic Engineering Approval Committee, but in 2010, through a notification, it was ordered that the GEAC would henceforth be the Genetic Engineering Appraisal Committee, which meant that it could only appraise and recommend and has no power to approve release of GM crops. Technically this had

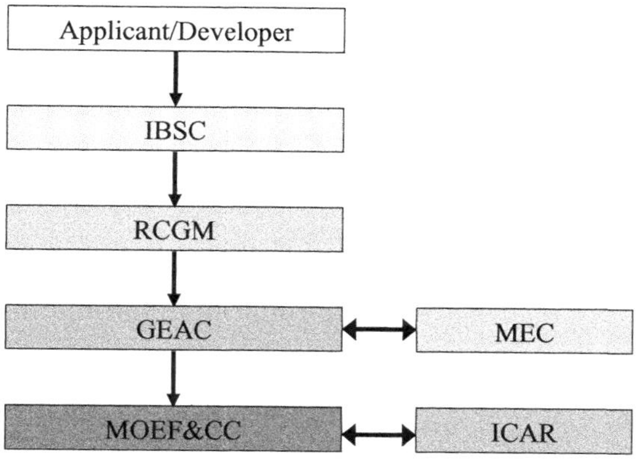

Figure 20.1. The approval process for GM crops in India.

not resulted in any major change in regulation, but it was obvious that it meant that the GEAC would have a limited but important role in the approval process.

The SBCC formed by the respective state government plays a role in coordination of activities relating to GMOs in that state and is the nodal agency at the state level to assess what damage, if any, has occurred on account of release of GMOs and to pursue control measures. It has powers to investigate and inspect sites where trials are held. The DLCC has powers to monitor at the district level. Its role is similar to that of the SBCC.

Figure 20.1 summarises the regulatory process with respect to applications and approvals. India's GM regulatory regime is based on a case-by-case review and approval process. Risk assessment is based on scientific risk assessment, and procedures and requirements for data collection are based on international norms. In 2008 the GEAC adopted an 'event-based' approval system. An event is defined as 'a genotype produced from the transformation of a single plant species using a specific genetic construct' (MoEF&CC, 2015). Thus, technically, a trait can be incorporated in more than one event. The GM event has to undergo extensive agronomic evaluation through field trials, under the supervision of an ICAR institution or a state agricultural university (SAU), for a minimum of two crop seasons. Product developers can conduct agronomic trials either after environmental clearance from the GEAC or in conjunction with biosafety trials. The final approval of commercial use of an event is given by the Government of India. After such an approval, the developer or applicant can register and market the seeds in states as per the provisions of the 2002 National Seed Policy and other applicable rules, which are specific to each state. The field performance of approved GM crops is monitored for three to five years, by the Ministry of Agriculture with the Departments of Agriculture in the various states. In the case of stacked events, even a stacked event consisting of already-approved events will be considered as a new event (Global Agricultural Information Network (GAIN), 2015).

There are no specific regulations on coexistence of GM crops and non-GM crops, and there is no liability and/or redress regime in this regard. Given the commercial export value of basmati rice and in view of the potential loss of export markets on account

Table 20.3 Approved events in Bt cotton

Gene/event and source	Developer	Usage
Cry1Ac (Mon 531) – Monsanto	Mahyco Monsanto Biotech Ltd	Fibre/seed/feed
Cry1Ac (Mon 15985) – stacked gene event provided by Monsanto	Mahyco Monsanto Biotech Ltd	Fibre/seed/feed
Cry1Ac (Event 1) – gene provided by IIT, Kharagpur	JK Agrigenetics	Fibre/seed/feed
Cry1Ab and Cry1Ac (GFM event) – gene provided by China	Nath Seeds	Fibre/seed/feed
Cry1Ac (BNLA1)	Central Institute of Cotton Research, Nagpur	Fibre/seed/feed
Cry1Ac (Event MLS 9124)	Metahelix Life Sciences Pvt Ltd	Fibre/seed/feed

Source: GAIN (2015).

of contamination and other issues, in 2007 the GEAC decided that no multi-location GM rice trials will be permitted in the basmati rice-growing regions, in farmers' fields particularly in the appropriate areas in Punjab, Uttarakhand and Haryana.

Regarding labelling for GM foods, the Food Safety & Standards Authority of India (FSSAI) is the regulatory authority. It is mandated to perform food safety assessment of GM foods (viable, i.e., living modified organisms (LMOs), and processed) and to decide on approval for commercial release of GM foods (processed). Only recently has the FSSAI published a document outlining the proposed regulatory structure for evaluating and approving GM foods (FSSAI, 2016). Under the proposed regulations, the FSSAI will be responsible for food safety assessment of GM foods (viable and processed) and for approval for commercial release of processed GM foods, while the GEAC will be responsible for approval of viable GM foods. The proposed GM Foods Safety Assessment Unit (GMFSAU) will be located at the National Institute of Nutrition, Hyderabad.

The whole system is based on a set of rules, guidelines and orders issued under EPA 1986. While this has given flexibility in developing a regulatory regime, it has hampered the development of a coherent regime with a single institution/authority responsible and accountable as a regulator.

India's Experience with Its Regulatory Regime

Regulation and Release of Bt Cotton

So far Bt cotton is the only crop to have been approved for commercial cultivation in India. Six events relating to Bt cotton have been approved, as indicated in Table 20.3.

In 1995, with permission from the RCGM, Mahyco, a developer, imported 100 grams of seeds containing Cry1Ac gene from Monsanto, USA. Later this trait was introgressed into cotton parent materials. Between 1966 and 2002 a series of risk assessments for Bt cotton was performed, followed by controlled multi-location field trials in important cotton-growing states during 1998–2000. During 2000–2002, large-scale field trials of Bt cotton were done to assess its performance, particularly in terms of gene flow and impact on non-target beneficial organisms and soil microorganisms, and in comparison with non-Bt cotton varieties. The GEAC requested the ICAR to validate the safety, efficacy and performance of Bt cotton hybrids under the All India Coordinated Cotton Improvement Project in 2001 (Choudhary *et al.*, 2014).

Finally, in 2002, after review of data and results from the trials and reports of the MEC, the GEAC approved commercial release of Bt cotton. This led to a Bt cotton boom in India, and Bt cotton from Monsanto–Mahyco captured a significant share of the market within a decade. The regulatory regime worked well in the case of Bt cotton, and Bt cotton hybrids have a market share of about 95% in India. About 1,000 Bt cotton hybrids were available by 2014, from various seed companies. Switching over to Bt cotton resulted in significant reductions in pesticide use (Qaim, 2016, pp. 75–76). Events in recent months, however, namely decreases in yield in some states on account of whitefly and pink bollworm and controversies over royalties charged by the technology provider and the government's attempts to fix the price of Bt cotton seeds through an administrative order (which has since been withdrawn), have kindled debates on the affordability of Bt cotton seeds and the long-term sustainability of Bt cotton.

Regulatory Hurdles for Bt Brinjal

The case of Bt brinjal, however, indicates that challenges to the regulatory regime from various quarters could impact on the approval process and raise questions about the credibility of the regulatory regime. Brinjal is consumed widely in India. Bt brinjal is resistant to 'fruit and shoot borer', *Leucinodes orbonalis.* Mahyco submitted an eight-volume dossier with data on safety and potential risks from Bt brinjal. On 14 October 2009 the GEAC declared that Bt brinjal event EE-1 was safe for environmental release and forwarded the report of the Expert Committee of the GEAC on safety and efficacy of Bt brinjal with its recommendation to the MoEF&CC for a final decision. Meanwhile, many civil society groups such as Greenpeace had been opposing the decision of the GEAC and urged the MoEF&CC not to sanction the commercial release of Bt brinjal. They ran a campaign against its release and also convinced some state governments not to approve/permit cultivation of Bt brinjal in their respective states.

The polarisation on this issue was total, and the civil society groups exerted pressure on the MoEF&CC. The then Minister for Environment and Forests, Jairam Ramesh, decided that public consultations would be organised to arrive at a decision. The MoEF&CC also published the GEAC report and sought comments. Civil society groups and a few scientists pointed out what they considered to be deficiencies in data and contended that the tests done were inadequate to assess the toxicity of Bt brinjal and its safety (Millstone, 2014). Concerns about gene flow to brinjal relatives and reservations about the assessment by the expert committee were other factors that were considered. As brinjal is used in Indian systems of medicine as an ingredient, fears that Bt brinjal would destroy the medicinal properties were also expressed (Kudlu and Stone, 2013, p. 33).

They called for more extensive tests and questioned the credibility of the risk assessment which had been accepted by the GEAC. After holding a series of public consultations and considering the opinions submitted on this issue, on 9 February 2010 the MoEF&CC decided to impose a moratorium on release of Bt brinjal. The moratorium has continued ever since, and no approval has been granted for commercial release of any GM crop, although the then Minister has clarified that the moratorium was only for Bt brinjal. While declaring the moratorium, the Minister announced that further studies and more data would be needed before any regulatory decision on Bt brinjal could be made. But what exactly would be needed in terms of data was not made explicit. In this case the opponents of Bt brinjal have won, at least temporarily. While this experience showed that, in the case of GM crops used as food, public sensitivity to safety and health impacts could have a considerable impact on decision-making, it also revealed that, even if the central government had approved Bt brinjal, as many as ten state governments were unwilling to permit its commercial cultivation. This divide among various stakeholders, including environmental activists, scientists, regulators and policymakers at different levels of government, has the potential to affect cultivation of GM crops, even after approval by the central government.

Evaluation of GM Mustard and GM Chickpea

While Bt brinjal has not been cleared for commercialisation, GM mustard and GM chickpea have yet to be approved, although both have undergone multi-location field trials. In recent months the GEAC has deliberated on approval for GM mustard and GM chickpea. The transgenic mustard hybrid DMH-11 has been developed by the team led by Professor Deepak Pental at the Centre for Genetic Manipulation and Crop Plants at Delhi University. By using genes sourced from soil bacteria, the Centre has developed a technology that facilitates developing hybrid varieties in mustard, which is generally a self-pollinating plant. It is claimed that the yield from the transgenic mustard hybrid is about 25%–30% more than that from the available high-yielding varieties, but civil society groups have challenged this claim. The GM chickpea was developed by Assam Agricultural University, and the technology has been licensed to Sungro Seeds Ltd. This transgenic crop has two genes (Cry1Ac and Cry2Aa) isolated from Bt, and is claimed to be resistant to the pod borer insect pest. The expected yield increase is about 25%–30%. As India imports huge quantities of pulses, their cultivation, if it were approved, could decrease the importation of pulses.

It is likely that the issues relating to reliability of test data, inadequacies in trial design and experimentation, and increase in yield will be raised again and again by civil society groups in the months to come, irrespective of a granting of approval for commercialisation. A GM pigeon pea developed by the International Crop Research Institute for Semi Arid Tropics is also in line for approval. According to a survey conducted by the MoEF&CC, experimental work was being pursued on 85 different plant species and about 20 applications for commercial approval were likely to be made by the end of 2016. Insect-resistance, fungus-resistance and drought-tolerance are the three most important traits that are being studied (MoEF&CC, 2016).

Thus, in the next few years both the number of GM crops with different traits and/or ready for commercialisation and the number of field trials are likely to increase. But the question is whether the current regulatory regime can successfully handle all the issues ranging from granting permission for trials to giving final approval for commercialisation. The regulatory costs in India for approval have been estimated to be in the range US$1,500,000–2,000,000 for a food crop like rice (Pray *et al.*, 2005). Besides the high costs, the delay in approval for trials and the time taken for decision-making act as a disincentive for small companies to develop, test and commercialise GM crops. The delay in approvals affects public-sector research and testing too. In revamping the regulatory regime, the negative consequences of the current high regulatory costs and lengthy durations of processes should be evaluated, with the intention of improving the regime's effectiveness.

Litigation and the BRAI Bill

Since 2005 cases on regulation of GM crops and biotechnology in India have been pending before the Supreme Court of India, with Aruna Rodrigues vs. Union Of India W.P Civil 260 (2005) being the first instance of litigation. In May 2012 the Supreme Court heard the cases and appointed a six-member Technical Experts Committee (TEC) to review and recommend risk assessment studies for all GM crops before conducting open field trials. In the interim report submitted in October 2012, the TEC recommended that a ban be imposed on trials currently under way until gaps and deficiencies in the existing biosafety regulation system had been addressed. The Government of India contested the suggestions made in the interim report. In July 2013 the TEC submitted its final report and reiterated the earlier suggestion of a ban on ongoing field trials, with five members concurring with this and one member dissenting. In responding to this report, the Government of India opposed and rejected the recommendations made by the majority of the members of the TEC. The TEC report called for a revamp of the current regulatory regime and opposed the introduction of herbicide-tolerant crops in India. It also took into account the data available in the dossiers on Bt cotton and Bt brinjal. After examining them, the TEC in its report questioned the quality and relevance of data provided by developers/applicants and accepted by the regulators, and indicated that it had serious issues with the risk assessment methodology and the type of tests sought to be undertaken under the rules (TEC, 2013). Instead of a single committee (the GEAC or RCGM) doing all the evaluation, the TEC suggested that the regulatory body should have area-wise sub-committees/expert groups in disciplines such as toxicology. The last hearing was in May 2014, and the Supreme Court has yet to give its verdict on the case.

In 2011 the DBT proposed setting up a regulatory authority – the Biotechnology Regulatory Authority of India (BRAI). A bill for that purpose was drafted by the DBT. The bill proposed that the regulatory regime be replaced with a single authority with the above name that would regulate all aspects of biotechnology in India. The bill envisaged setting up a fully fledged regulatory body, with units for risk assessment and evaluation of trial data, that would be empowered to grant approval for trials and recommend commercialisation. The last version of the bill proposed a Risk Assessment Unit that would have experts from different disciplines, an economic analysis unit, three regulatory and policy units under the three divisions, an environmental appraisal unit and a product review committee, in addition to a legal unit, with scientific advisory panels under each division (Rao, 2014). The bill lapsed with the term of Lok Sabha ending in May 2014. The DBT has set up a Risk Assessment Unit, and is planning to strengthen it by hiring experts in relevant disciplines.

In the case of Bt cotton, commercialisation and fast adoption indicated that farmers benefited and the gains were visible. But in the case of Bt brinjal and Bt mustard, civil society has questioned the claims of increases in yield and the adequacy of risk assessment. The controversies over risk assessment, safety and increases in yield are not likely to be resolved soon, given the polarisation of views on GM crops. As the TEC has criticised the risk assessment models used and tests done in the cases of Bt cotton and Bt brinjal, its views are likely to be used to question the approval of GM mustard and GM chickpea, if their commercialisation is approved. According to Herring 'Uncertainty about regulatory criteria also depresses investment in both public and private sector research and development. If there is no definable hazard, and no rational way to set limits of acceptable risk, scarce public resources are diverted to surveillance and control rather than agricultural innovation' (Herring, 2014, p. 208).

The challenges facing the system are as follows.

(1) Legal issues. The case before the Supreme Court could have a significant impact on the regulatory regime. The Court might not fully reject the recommendations of the TEC and might order that the regime should address concerns expressed by the TEC, or might recommend a series of actions for revising/strengthening the regime. It is unlikely that it will take the view that the regulatory regime is fine and need not be changed.
(2) Technological developments. In the USA and elsewhere regulators have to take into account developments such as gene editing and new plant breeding techniques. Gene editing actually raises questions about the definition of a GMO, and it is debatable whether the current definition is adequate. The rules under the EPA in India were framed in an earlier era of genetic engineering, and how adequate they are in the context of new developments in technology and regulation is an important question. Whether the case-by-case approach is adequate for meeting future regulatory needs is a key issue.
(3) Credibility and acceptability. The regime's credibility has been questioned by sections of civil society, while the TEC has found fault with the regime on many issues. The acceptability of its recommendations and findings hinges on its credibility. Despite its approval by the GEAC, some state governments indicated that they would not permit cultivation of Bt brinjal. The moratorium continues, and controversies over GM mustard and GM chickpea show that the gaining of approval could be a protracted process. While the GEAC has been listening to the critics and complying with their requests for data, the gap in credibility remains. It is impossible for the GEAC or the regime to convince the section of civil society that is totally opposed to GM crops, but it is important to convince the state governments and other stakeholders such as farmers, state governments and consumers. It has

been suggested that regulation of biotechnology should take into account lessons learnt, future needs and advances in science and technology, and incorporate socio-economic impact assessment as part of the regulation (Chaturvedi and Srinivas, 2013).

Towards a New Regulatory Regime for GM Crops?

The BRAI bill proposed a new regulatory regime, and this should be seen as an acknowledgement by the policymakers that the system needed a drastic change. The moratorium has not been followed up with a clear signal that the government would go ahead with accepting recommendations from the GEAC as and when they are made. In the previous section I have listed new and emerging challenges to the regulatory regime for GM crops in India. These challenges could be addressed either by making changes in the current regime or by establishing a new regime in the form of a regulatory body. If such a body were to be established, it would be better for it to be housed in the MoEF&CC rather than in the DBT, since the DBT's mandate is to promote biotechnology and the MoEF&CC is more appropriate for housing a regulator focusing on environmental risk assessment and on the health and safety aspects of biotechnology.

Given the nature of biotechnology and rapid developments in biotechnology, it would be better to develop a regime that is more suited to biotechnology. The current regime is flexible and adaptable to circumstances, but this also makes the process amenable to too many and frequent changes, and having too many orders, guidelines and rules makes the system less coherent. With more than one committee/body dealing with these, the result is that the regulatory system functions without having a stable structure with well-defined roles and responsibilities coupled with accountability. Hence I suggest the following for establishing a regulatory regime for GM crops.

(1) Establish through an Act approved by the Parliament a regime under which there should be a single regulator, i.e., a single authority for all biotechnology-related activities. The guiding principles and structure for the authority can be developed from the points made by the critics of the current regime, the TEC report and debates on the BRAI bill, and a comparative study of regulatory regimes for biotechnology in different countries. The gaps in the current framework should be addressed, and future needs for regulatory capacity should be taken into account in developing a new regulatory regime.

(2) India should opt for neither the precautionary-principle-based model in Europe nor the substantial-equivalence-principle-based model in the USA, as the effectiveness of both models has been questioned. Instead it should try to evolve a model that is suited to India's needs without compromising either on biosafety or on its commitments under the CBD and CPB.

(3) To effectively regulate second- and third-generation GM crops, stellar competence, emphatic engagement and utmost integrity must constitute the three core areas of regulatory excellence (Howlett and Mukherjee, 2016). India should enhance its capacity to regulate the next generation of GM crops and the emerging technological choices in biotechnology, such as gene editing and gene drives.

References

ABLE (2016). *Make in India: Biotech Handbook*. Bangalore: ABLE.

Chaturvedi, S. and Srinivas, K. R. (2013). Genetically modified crops: policy logjam. *Economic & Political Weekly* **48**(14), 19–23.

Choudhary, B. *et al.* (2014). Regulatory options for genetically modified crops in India. *Plant Biotechnology Journal* **12**, 135–146.

FSSAI (2016). Operationalizing the Regulation of Genetically Modified Foods in India. New Delhi: FSSAI.

GAIN (2015). *India: Agricultural Biotechnology Annual GAIN Report IN5088*. Washington, D.C.: GAIN (USDA).

Giri, J. and Tyagi, A. K. (2016). Genetically engineered crops: India's path ahead. *Nature India*, 4 March 2016.

Gupta, A. (2011). An evolving science-society contract in India. *Food Policy* **36**, 736–741.

Herring, R. J. (2014). On risk and regulation: Bt crops in India. *GM Crops & Food* **5**, 204–209.

Howlett, M. and Mukherjee, I. (2016). Achieving regulatory excellence in the agri-food biotechnology sector: building policy capacity. *Asian Biotechnology and Development Review* **18**, 35–46.

Kudlu, C. and Stone, G. D. (2013). The trials of genetically modified food: Bt eggplants and Ayurveda in India. *Food, Culture and Society* **16**, 21–42.

Millstone, E. (2014). Science and politics in Indian GM crop regulation: a u-turn down a blind alley. In *India's Risks: Democratizing the Management of Threats to Environment, Health and Values*, ed. R. Moor and M. V. R. Gowda. New Delhi: Oxford University Press, pp. 205–226.

MoEF&CC (2015). Confined field trials of GE plants. New Delhi: BCIL and MoEF&CC.

MoEF&CC (2016). Genetically engineered plants in the product development pipeline in India. New Delhi: MoEF&CC.

Pray, C. E. *et al.* (2005). The cost of biosafety regulations: the Indian experience. *Quarterly Journal of International Agriculture* **44**, 267–289.

Qaim, M. (2016). *Genetically Modified Crops and Agricultural Development*. New York, NY: Palgrave.

Rao, S. R. (2014). Indian GMO regulations and implications of new plant breeding molecular technologies. International Workshop on Regulation of Animal Biotechnology, Jaipur.

TEC (2013). Final Report of the Technical Expert Committee (TEC) set up by the Supreme Court in a Public Interest Litigation on Genetically Modified Organisms (GMOs). New Delhi. [Online]. Available from www.indiaenvironmentportal.org.in/content/380047/final-report-of-the-technical-expert-committee-tec-set-up-by-the-supreme-court-in-a-public-interest-litigation-on-genetically-modified-organisms-gmos/

21 The Argentinian GMO Biosafety System: An Evolving Perspective

Carmen Vicién and Eduardo Trigo

Introduction

Argentina is the third-largest grower of genetically modified (GM) crops. Using data from growing seasons 1996/97–2010/11, Trigo and Cap (2011) estimated that the gross profit from the cultivation of GM soybean in Argentina amounted to 65 billion dollars. The distribution of the benefits accrued from GM soybean is such that about 79% is absorbed by the complex seed production agro-industry, while the remaining 21% belongs to the State.

The high level of adoption of this technology is the result of a complex combination of circumstances, which included political support from the agricultural area and the ability to satisfy needs prevalent among farmers (Burachik, 2010). The development and evolution of GM crops would not have been possible without the existence of an adequate regulatory framework for biosafety; its establishment in the early 1990s was crucial for the development of agricultural biotechnology.

This chapter presents an overview of the regulatory framework in Argentina considering some aspects of its evolution since its creation at the beginning of the 1990s. We consider trade issues that had a significant impact on regulatory aspects and strategies adopted in order to improve the system over the years, ending with comments on asynchronicity and some prospects for the future.

The Regulatory System

In practice, regulatory systems around the world have evolved on a 'piece-by-piece' basis, frequently in response to demands or needs of the moment, like assembling the pieces of a jigsaw puzzle (McLean *et al.*, 2002). The different levels of institutional development, in particular of the systems governing innovation, plus the diverse trading positions of the countries and societal perceptions about biotechnology, resulted in national strategies for construction of regulatory systems that, with rare exceptions, were isolated and lacked mechanisms for international coordination (Vicién and Álvarez, 2013).

The unprecedented nature of the matter under regulation and the dynamics of the processes of technological development resulted in the fact that in many cases the needed items of know-how from different areas, such as seeds, plant health, agrochemicals, environmentalism, food safety and even anthropological and economic studies, were put together, in the newly born regulatory institutions and their norms, without any clear and logical order (Vicién and Álvarez, 2013).

In the case of Argentina, the first applications for confined field trials with GM crops were evaluated in 1991 by a new committee (the National Advisory Commission on Agricultural Biotechnology, known by the Spanish acronym CONABIA) at the Secretariat of Agriculture, Livestock and Fisheries, which was the competent authority on the matter. The regulatory body included government personnel, as well as representatives both from the public sector and from the private sector. This combination of resources resulted in a balanced and productive amalgam of science, societal interests and field experience, and crucially, in addition to providing knowledge, brought greater transparency to the regulatory process in comparison with other cases in which decisions are made solely by the technical bodies of political organisations. This has been very important for the later support that the committee has received from different sectors (Burachik, 2010).

The formation of CONABIA was triggered by the necessity of a framework to assess this technology, due to the fact that the government was aware that there were several companies planning to perform experimental trials with GM soybean, corn and cotton in Argentina, and there was a strong feeling that GM crops should be evaluated on the basis of science-based considerations. The membership of CONABIA includes researchers on seeds, the environment and health from national government, as well as representatives from academia and private-sector organisations (Ministry of Agroindustry, 2016).

Nowadays, full responsibility for the authorisation of any release of GMOs into the agro-ecosystem continues to lie with the Ministry of Agro-industry (the former Ministry of Agriculture, Livestock and Fisheries), which both houses the consultative body with public and private representation which is responsible for the risk evaluation and performs the administrative acts for the release once the scientific and technical issues have been resolved. The primary responsibility for this process lies with the Directorate of Biotechnology; the Agri-Food Health and Quality National Service (SENASA) participates in matters having to do with human and animal health and the National Seed Institute (INASE) advises on matters related to supervision of field trials and issues pertinent to seeds. Risk assessment, the design of biosafety measures and risk management (at various stages of evaluation) are the responsibility of CONABIA. The Directorate of Biotechnology is responsible for the Executive Secretariat of CONABIA. Food and feed safety assessment is up to the Directorate of Food Quality from SENASA, supported by the Advisory Technical Committee on Food Use of GMOs (CTAUOGM is the Spanish abbreviation). The analysis of impacts on production and marketing that may result from authorisation of the commercialisation of GMOs is the responsibility of the Directorate of Agricultural Markets (see below). The different assessments (environmental, food/feed and commercialisation) can in practice be initiated either simultaneously or in any order (SENASA, 2002).

The fact that the basic framework of the regulatory system involves ministerial resolution, rather than legislative action, has provided the basis for a system that is able to evolve quickly in line with changing institutional aspects as well as with respect to advances in science and technology and in assessment methodologies. The authorisations for commercial release of GM plants are granted by the political authority of the Ministry of Agro-industry on the basis of a three-step procedure: an assessment of the impact on agro-ecosystems; a food and feed safety evaluation; and an analysis of the production and marketing of a particular GM crop. The assessments are prepared by the two consultant agencies (CONABIA and the CTAUOGM) and the Directorate of Agricultural Markets. Although the assessments leading to the formation of these three expert opinions constitute stages that ought to be completed in order for a commercialisation permit to be granted to a given GMO, in all cases they are recommendations and, as such, not binding for the Secretary's final decision (Burachik, 2010).

A significant aspect in the case of the Argentinean system is that it is widely recognised as being a structure 'uncontaminated' by a bureaucratic history, where scientific and technical credibility and enforceability prevail, which is critical when one is dealing with a sensitive issue for society such as GM crops. CONABIA – the cornerstone of the system – has shown over the years a strong *esprit de corps*, even among participants from the public and private sectors, with the authorities, in turn, mostly respecting its autonomy. All this has resulted in a high level of institutional prestige. In understanding this situation it is useful to give a little historical perspective.

Trade Issues

Beyond regulations and regulated matters, there is the need to consider another important aspect of the Argentinian system: Argentina's position as a net exporting country, and the related internalisation of potential trade problems. This evolved, eventually, into one of the main causes of delays in granting approvals, since it was the government itself which subordinated the final approval decision to the existence (or not) of market risks.

By the end of the 1990s, Argentina had implemented an export impact assessment stage, which became an integral part of the approval procedure for GM crops. The purpose was to determine the potential consequences which the commercial release of the GM crop concerned might have for the international trade of Argentina. It considered the situation of trade competitors, potential markets and the relative importance of the crop in the existing trade with the countries to which Argentina exported it and the share of these imports in their supply (Vicién, 2012).

From that time onwards, the so-called 'mirror policy' was put in place and, therefore, the final decision on GMO assessment was based on a technical analysis and a political decision. The implementation of this 'mirror policy' generally followed the approval patterns of the European Union (EU) as the main benchmark, in order to minimise impacts on future exports. This was important because of the role of the EU as a main trading partner of Argentina in agricultural commodities. In the particular case of soybean events, the importance of soybeans and soybean oil sales also required that close attention be paid to approvals granted in India and China (Vicién and Álvarez, 2010).

The elicitation of commercial opinion, unlike other assessments, usually led to a 'waiting period', prior to the final approval, since the nature of export-market analysis may implicitly define the duration of postponement of the approval decision. This is so because, when the cause of the risk associated with trading in a GM material disappeared since the latter had been approved in the country of destination for exports, the opinion changed automatically from unfavourable to favourable. The nature of this opinion is different from the others because this is not a toxicity or environmental impact problem. Clearly, if it were, no one would allow such 'mobility' in decision-making (Vicién and Álvarez, 2010).

From the follow-up analysis on the procedures within the Argentinian regulatory system, it can be concluded that, in almost every case, there was a favourable technical opinion as to the approval of certain events concerning the assessments of food safety and environmental risk, with final approval being subject to the commercial opinion, which became the third stage of the authorisation process. Two significant conclusions can be drawn. The first is that the speed of event approval was greater in the Argentinian system than it was in most of the importing markets, since a prior technical opinion was issued first, and then there was a wait for the commercial approval. The second is that agility in the regulatory system, counteracted by caution in order not to incur a commercial problem, generally delayed the final decision (Vicién, 2012). In fact, the time spent at the stage of commercial analysis was 'virtual' since the preliminary studies required in order for an opinion to be formed did not usually take more than one or two months. However, no final opinion was generally issued until it could be verified that any potential trading restriction had been removed.

Considering approval times in Argentina, it can be estimated that there was a period of approximately one or two years between the stages of the food safety or environmental impact assessment, as the case may be, and the final regulatory approval. Therefore, such a period would be equivalent to the virtual time attributable to the commercial analysis, and corresponds to the time needed for resolution of the opposing interests of growers, seed companies, exporters and the government itself. In some circumstances, exporters' interests prevail because it is a priority for exporters not to alter the essential characteristics of their business, which could undoubtedly be complicated by the introduction of an unapproved new material in their destination markets (Vicién and Álvarez, 2010).

In this context, for several years, Argentina closely followed the approvals granted by the EU and, in recent years, those given by India and China, as reference models for its own approvals. But recently, albeit informally, Argentina has been paying attention to progress made by Brazil, which is a competitor country in many sectors.

The next step could be to reinforce monitoring in relation to markets and their future evolution, as well as monitoring market performance vis-à-vis approvals for new events. This will allow the prediction of scenarios, rather than merely a recommendation of caution when faced with a potential problem; i.e., the ability to act strategically in the case of a trade policy involving local technological developments will be acquired. Another role of the observation on trade and regulatory aspects consists of seeking enhanced diversification of export destinations, by adding promotion policies that may make them attractive for exporters, through specific mechanisms such as tax or financing incentives. Reflecting this and without discussing the importance of trade information or the necessary caution vis-à-vis potential problems, in the sphere of administrative acts a clear separation of food/feed safety and environmental risk evaluations from market-related analysis was introduced, stating clearly that commercial authorisations are given on the basis of three technical and independent assessments (environmental, food/feed and commercialisation).

'Fit for Purpose' Strategies

The biosafety regulatory system in Argentina had an early development and there is a certain consensus about it, as it has adequately provided the basis for monitoring developments that have been going on for the last few decades (Trigo *et al.*, 2002).

Biosafety regulatory systems are perfectible and need operational simplifications. The requirements and procedures of regulatory systems dealing with GMOs do not always include proven advances in the field of know-how, and, faced with the latest developments, it is necessary to undertake periodic reviews of regulations, without necessarily completely amending such rules. Instead, there should be a specific review of cases in point. The risk assessments (the 'heart' of the regulatory system) must comply with every requirement outlined in the country's regulatory framework and provide clearly laid out robust scientific information to facilitate this decision-making process; in other words, the risk assessment must be 'fit for purpose' (García-Alonso, 2013). Strategies to streamline risk assessment processes include responding to protection goals, establishing clear criteria for data quality, employing proportionate risk assessment, developing alternative methodologies and the effective use of simple regulatory tools.

As an example of decision-making by the Argentinian regulatory body, the case of GM canola can be mentioned. Canola is a minor crop in Argentina, and indigenous crucifers closely related to canola exist in vast regions of Argentina. These populations belong to several genera (Brassica, Diplotaxis, Hischfledia, Raphanus and Sinapsis) and grow mainly in places where the soil has been disturbed by agriculture and cattle raising, as well as in urban areas and their peripheries. During 1996, confined field tests of canola tolerant to glyphosate were authorised by CONABIA under stringent conditions. When the applicant requested authorisation for a large planting (two 250-hectare lots) in 1997, the review was intensive and a long debate preceded the final decision to deny the request. CONABIA's members believed that, on the basis of current knowledge on the problem of plant invasions, it could be asserted that gene flow would occur and that its effects in terms of the impact on the ecosystems would be hard to predict. Also, as in the case of populations of rapeseed (*Brassica napus* L.) already cultivated in Argentina, it could be stated that some GM canola seeds could enter the seed pool, providing propagules for the indigenous flora. The evaluation indicated that there was no reason to consider *a priori* that the introduction of herbicide-tolerant GM canola, or any increase in the *Crucifera* populations, would lead to impacts on the current properties of the ecosystem or cause negative effects on the flora and fauna of the relevant regions. The higher predictable risk was agricultural: within the areas of cultivation of canola, gene flow and seed dispersion processes would lead to a rapid evolution of herbicide-tolerant cruciferous weeds, and in the end this would result in a significant reduction in efficacy of the herbicide most commonly employed by farmers, leading to a need for substitution of that herbicide for weed control.

During the mid 2000s delays were observed in the approval process. These delays were not linked to the implementation of the so-called mirror policy but due to issues that needed to be clarified in risk assessment procedures. This fact was evident because several countries, such as the EU states, Japan, South Korea and China, had approved GM materials for which analysis was still pending in Argentina.

Because of this, during 2011 an in-depth study employing a participatory process to assess, and propose an updating of, legal and regulatory aspects was developed, in order to give a renewed momentum to the Argentinian regulatory system. The review included several aspects for which adjustments were required, such as introducing legal definitions specifying the areas of intervention and coordination of the institutions involved in the processes; developing regulations for contained activities (*vis-à-vis* confined ones); reviewing the application forms for the various phases of evaluation, giving some definitions in the case of risk assessments for stacked events; reviewing the analysis on production and commercialisation for the commercial approval of GM crops; and revising the procedures for counter-season applications.

A brief description is as follows (Ministry of Agro-industry, 2016). Specific rules have been designed and introduced to take into account the distinction between 'contained' and 'confined'. 'Contained' means enclosed in such a facility as a laboratory or greenhouse. 'Confined' means that genes and plant materials are confined within a specific area (as is the case in confined field trials).

With respect to confined field trials, particular aspects were considered within the different stages in the process of obtaining the GMO as well as the various techniques for evaluating materials, according to the nature of the character introduced and the data requirement when preparing evaluations for authorisations for large-scale releases or food safety assessments. Finally, the process of computerisation of the applications was improved.

For the evaluation stage, a crucial aspect was the necessity to clarify the criteria for assessments of stacked events, i.e., events in which several transgenes are involved. Such stacked events now constitute a significant proportion of the GM varieties, both in development and at the marketing stage, and by virtue of their ability to give rise to multiple combinations they generate a potential workload in the regulatory system. Fine-tuning the criteria for evaluating the combined events would also make clear to the institutions and companies concerned the nature of the information to be presented.

At present, applicants are offered a mechanism for direct interaction with the evaluators, which serves to clarify any doubts about the information to be included on the form and to define the general criteria to be used in the application. This is an option, but its existence does not imply that there is a forum for debate and setting of normative standards, and consequently it does not replace CONABIA's intervention in the evaluation. It consists of a formal demand, from the applicants, about the authorities' opinion. During the meeting the applicant can present orally the case submitted for consideration. Depending on the subject, there is the possibility of summoning experts on specific topics. Finally, CONABIA will present a written report with opinions and other considerations that could have come up during the analysis of the presentation.

Recent reviews have included, amongst other things, approaches to evaluate various new breeding techniques (NBTs) (Lema and Whelan, 2015) and the development of a construct-based risk assessment framework (Beker *et al.*, 2016). Regarding NBTs, by means of an early consultation, applicants have to submit information on the NBT-derived crop to establish whether the result of the breeding process is carrying a novel combination of genetic material and hence has to be regulated. A criterion included is that a genetic change shall be considered as a new combination of genetic material when one or more genes or DNA sequences that are a part of a defined genetic construct have been introduced permanently into the plant genome by a stable and joint insertion (Lema and Whelan, 2015).

Comments on Asychronicity

The differences in ways of thinking, dissimilar institutional developments and political and trade issues result in different time frameworks for approval of the same event in different countries, a fact that results in frequent asynchronicity, and inevitably ends up affecting the smooth flow of international markets. It is inevitable that the country that developed a particular GM crop is the first to give approval and it is only after a certain amount of time – usually on the initiative of the seed companies involved, or other local interested stakeholders – that it will be put up for approval by other countries for planting or food/feed consumption. The delay in approvals can be considered one of the most significant costs deriving from the regulatory process, with verifiable impacts at the level of both seed companies/research institutions and countries.

Despite the differences in philosophical and political approaches that define the characteristics of regulatory systems in different countries, in general, their scientific and technical legislations require broadly similar information in order to perform environmental and food safety assessments of GMOs. However, this is not reflected in the regulatory decision-making process, the rate of approvals or the final decision to authorise or prohibit commercialisation of a certain GMO (MacKenzie, 2000).

Between the required information and the decision to approve a certain GM crop there are matters related to protection values, according to environmental or food security goals, and institutional and political aspects, which contribute to those differences. There are countries whose regulatory systems contain exclusionary criteria for protection of the environment and/or health, which have generated partial or total bans on the planting and marketing of GMOs, and others whose strategies might be characterised as dilatory, which have instituted regulatory systems with many levels of consultation and long periods for arriving at a final decision. On the other hand, there are cases of countries with high productive potential that do not have a high level of consistency in the development of their regulatory system but have achieved rather fast growth in terms of the area planted with transgenic crops. Finally, some countries do not yet have regulatory systems in place, and in some cases are net food importers and are facing complex decisions (Vicién and Álvarez, 2013).

In this context it can be recalled that the conflict generated by the '*de facto* moratorium' of the EU made clear the lack of definition about the treatment of GMOs at multilateral level. Argentina, Canada and the USA took the EU to the WTO Dispute Settlement Body because of the '*de facto* moratorium' imposed by the EU with regard to all approvals of biotechnology products since 1998, as well as the non-consideration of applications for the approval of certain agricultural products derived from biotechnology of interest for Argentina and, finally, the prohibition by some EU states of certain biotechnological products of particular interest for Argentina that had previously been approved at the EU level. It has to be noted that Argentina did not question the rules of the EU on the approval of biotech products, but objected to their lack of implementation since 1998; in other words, the EU had imposed a delay regarding the implementation of its own procedures. It should be noted that, while EU legislation specifies that the analysis of biotech products has to be carried out on a 'case by case' basis, this was in contradiction with the 'moratorium', which is a measure of a general nature that had been applied to the EU's system of approvals and, therefore, affected all biotech products without distinction. By 2006, the WTO had ruled in favour of Argentina, Canada and the USA (Burachik *et al.*, 2006).

In relation to these issues, over the last couple of decades there has been a lot of discussion about the 'underlying need for confidence in the risk assessments undertaken in the country (countries) of export' and for 'governments to take a pro-active approach by ensuring that risk assessments are made publically accessible in a timely manner' (Center for Environmental Risk Assessment (CERA), 2014). International differences relating to country profiles, regulatory systems, histories of adoption and public perception make it difficult to synchronise safety assessments and, therefore, approvals for the cultivation and importation of GM crops, resulting in possible disruptions of trade and potential problems with the food supply. These differences create a growing need for dialogue among countries in order for the competent authorities in each country to increase their understanding of the approaches to the safety assessment of GM crops in other countries, build trust and share experiences (Bartholomaeus *et al.*, 2015).

In that context, it is worth mentioning an initiative that led to an international meeting of GM food safety assessors from the main importing and exporting countries in Asia and the Americas that was held in Buenos Aires, Argentina, in June 2013. Participants shared details of their approaches to evaluation, identified similarities and challenges, and used their experience to propose areas for future work. Recommendations for improving risk assessment procedures and avenues for future collaboration were also discussed. The importance of making all possible efforts to move towards more integrated and harmonised regulatory oversight for GMOs was strongly emphasised. This exercise showed that such harmonisation is a feasible goal that would help to sustain a fluid trade of commodities and ultimately enhance food security (Bartholomaeus *et al.*, 2015).

To achieve harmonisation there is a need for more common ground: this concerns economic or commercial considerations, technological joint initiatives, the avoidance of disputes amongst countries or with regard to third parties, and the establishment of common values and concerns. Experience indicates that what matters is adequate functioning of regulatory systems on the basis of technical instruments obtained by consensus, including the preparation and publication of documents summarising, in a clear manner, the decision process employed by a country when approving a certain GMO (Vicién and Álvarez, 2013). Finally, generating instruments with public–private participation to anticipate scenarios, by tracking and monitoring global markets and their future evolution, as well as the expected approvals of new events, will be crucial in order to articulate trade policies and the promotion of technological innovation at the regional level.

Final Considerations

The Argentinian biosafety system is one of the oldest in operation and one that has had to surf the long and complex evolution of GMO development and markets. It was set up very early, when many of the controversial issues that emerged later around this technology were not significant, and in that sense it has benefited from the possibility of being able to mature without interference from the political discussions of more recent times. At the same time, it has also had to deal with the fact that Argentina is an agricultural exporting country, meaning that other countries' policies in this field directly impinge on its market access conditions and have direct and important economic consequences. The experience of how it has evolved since the early 1990s highlights the complexities of the matter at hand but also proves that it is possible to evolve in a positive way.

The system has worked effectively ever since 1991, its ability to evolve with accompanying technological development being perhaps one of its main virtues. Its effectiveness is evident from the fact that it has been able to intervene in the processing of a large number of permits for releases into the environment. Furthermore, its members have been active participants in the international discussion on the issue of biosafety and associated regulatory processes.

The organisation of the system, through advisory bodies, allowed the inclusion of human resources with scientific capacity, which has contributed to the development of comprehensive regulations with flexibility in their implementation, because the fundamentals of the system rely on the knowledge of the experts. This form of organisation has also provided transparency in the decision-making process.

Given that biosafety systems deal with evolving concepts, they inherently need to anticipate in their procedures and requirements advances in scientific knowledge and technological development, for which a periodic review of their regulations and operation is always necessary. Therefore, in an international context in which some systems are more predictable than others, it is seen as desirable that regulatory systems evolve simultaneously in such a way as to simplify their tasks; have a clear definition of values to be protected and of the risk hypothesis to be analysed; and prepare documents on issues on which there is scientific consensus, thus eliminating the need for recurrent consultation and discussion in the daily process of evaluation of GM materials.

References

Bartholomaeus, A. *et al.* (2015). Recommendations from the workshop on Comparative Approaches to Safety Assessment of GM Plant Materials: a road toward harmonized criteria? *GM Crops & Food* **6**(2), 69–79.

Beker, M. P. *et al.* (2016). Development of a construct-based risk assessment framework for genetic engineered crops. *Transgenic Research* **25**(5), 597–607.

Burachik, M. (2010). Experience from use of GMOs in Argentinian agriculture, economy and environment. *New Biotechnology* **27**(5), 588–592.

Burachik, M. *et al.* (2006). Introducción a las Negociaciones Internacionales en Agrobiotecnología. Secretaría de Agricultura, Ganadería, Pesca y Alimentos. Oficina de Biotecnología, 52 pp.

CERA (2014). *Low-Level Presence in Seed: A Science Based Approach to Expedited Environmental Risk Assessment – Workshop Proceedings*. Washington, D.C.: CERA (ILSI Research Foundation).

García-Alonso, M. (2013). Safety assessment of food and feed derived from GM crops: using problem formulation to ensure "fit for purpose" risk assessments. In *Collection of Biosafety Reviews*, volume 8. Trieste: International Centre for Genetic Engineering and Biotechnology, pp. 72–101.

Lema, M. and Whelan, A. (2015). Regulatory framework for gene editing and other new breeding techniques (NBTs) in Argentina. *GM Crops & Food: Biotechnology in Agriculture and the Food Chain* **6**(4), 253–265.

MacKenzie, D. (2000). International comparison of regulatory frameworks for food products of biotechnology. The Canadian Biotechnology Advisory Committee Project Steering Committee on the Regulation of Genetically Modified Foods, 62 pp.

McLean, M. *et al.* (2002). A conceptual framework for implementing biosafety: linking policy, capacity and regulation. ISNAR Briefing Paper 47. The Hague: ISNAR.

Ministry of Agro-industry (2016). Biotechnology Directorate. [Online]. Available from www.agroindustria.gob.ar/sitio/areas/biotecnologia/ogm/

SENASA (2002). Resolution 412/2002. [Online]. Available from www.senasa.gov.ar/cadena-vegetal/oleaginosas/industria/biotecnologia

Trigo, E. and Cap, E. (2011). Quince años de cultivos genéticamente modificados en la agricultura argentina. ArgenBio. [Online]. Available from http://docplayer.es/15679839-Quince-anos-de-cultivos-geneticamente-modificados-en-la-agricultura-argentina-eduardo-j-trigo.html

Trigo, E. *et al.* (2002). *Los transgénicos en la agricultura argentina. Una historia con final abierto*. Buenos Aires: Libros del Zorzal.

Vicién, C. (2012). Socioeconomic assessment – a requirement for authorisation for cultivation of GM crops in Argentina. In *Workshop Proceedings: International Workshop on Socio-economic Impacts of Genetically Modified Crops (Co-organised by JRC-IPTS and FAO)*, ed. M. Lusser *et al.* Office of the European Union and Food and Agriculture Organization of the United Nations.

Vicién, C. and Álvarez, G. (with the collaboration of G. Petri *et al.*) (2010). La asincronicidad en las aprobaciones comerciales de los materiales genéticamente modificados. ArgenBio, 78 pp.

Vicién, C. and Álvarez, G. (2013). La asincronicidad de las aprobaciones comerciales de los cultivos genéticamente modificados. Conferencia internacional "Comercio agrícola y América Latina: cuestiones, controversias y perspectivas". Cátedras OMC de la FLACSO – Argentina, del ITAM – México y del Instituto de Estudios Internacionales – Chile. [Online]. Available from www.researchgate.net/profile/Gustavo_Alvarez7/publications

22 The Brazilian GMO Regulatory System: A Historical View and Perspective

Edivaldo Domingues Velini, Maria Lúcia Zaidan Dagli, Gutemberg Delfino de Souza, Rubens José Nascimento, Tassiana Fronza Pinho, Paulo Paes de Andrade and Helaine Carrer

Introduction

Brazil is a leading exporter of crops, which represent an important element in its trade balance. At the same time, the country has a solid infrastructure for research and development in agriculture, health, fuel production and innovation, and in many other important areas, which are also relevant for the trade balance and for Brazilian social and economic development. In this favourable scenario it is not surprising that products derived from modern biotechnology were welcome from the very beginning, as were the new technologies derived from genetic engineering, both in research and in innovation.

As a consequence of the international consensus on regulating products derived from biotechnology and/or the technology itself, a long national discussion started in 1989 on the uses and regulation of biotechnology-derived products and how to assess risks associated with the intentional use of genetically modified organisms (GMOs) or arising from their accidental release. A plethora of different stakeholders contributed to the establishment of the first regulatory framework in 1995 and its refinement in 2005. Moreover, Brazil joined the Cartagena Protocol on Biosafety and has been a party to it since 2003, which has profound implications in its trade relations with other parties and countries, in general.

The establishment of a comprehensive regulatory framework enabled the country to quickly adopt biotechnology in agriculture and other productive sectors and, as a consequence, Brazil is now one of the leading crop exporters and has the second-largest cultivated area of genetically modified (GM) crops, 44.2 million hectares (ha), just behind the USA with 70.9 million ha in 2015 (James, 2015).

This chapter describes the history of GMO regulation in Brazil, the biosafety regulatory framework and the contribution of agro-biotechnology to the Brazilian economy.

A Brief History of GMO Regulation in Brazil

The establishment of a biosafety law in Brazil began with discussions in the Lower House in 1989. However, only after the Convention on Biological Diversity in 1992 did the pressure to approve biosafety rules for the use of genetic engineering technologies increase. Scientists, government agencies, environmentalists and private companies urged that such rules should be adopted. This resulted in the approval of a Legal Biosafety Regulatory System, Law 8974, on 5 January 1995, with the objective of setting up rules for controlling the development, cultivation, manipulation, transportation, marketing, consumption, release and disposal of GMOs, while protecting human health and the environment in compliance with the precautionary principle (Mendonça-Hagler *et al.*, 2008). A few months later, the National Technical Commission on Biosafety (CTNBio) was created by Decree 1520 on 12 June 1995, to establish biosafety policies and regulation for activities and projects related to GMOs. During the period 1996–2002, CTNBio published 20 normative instructions concerning biosafety issues, such as the importation of GM plants and animals for research, GMO transportation, regulation of genetic manipulation of human cells and activities related to GM derivatives. Moreover, public lectures on specific matters were held when necessary. A problem emerged in 1998 when CTNBio issued its first technical opinion to approve the commercial release of a GM glyphosate-tolerant soybean (soybean Roundup Ready). The completion of an Environmental Impact Report (EIA) was not required by CTNBio because of previous risk assessment results and accumulated experience in other countries, where no evidence of harm to the environment and health had been reported. CTNBio's decision was challenged by environmentalists and other stakeholders, resulting in a public civil action being filed in court by the Office of Consumer Affairs (IDEC), which further developed into an extensive and argumentative trial that lasted for several years. The judges accepted the case due to an existing conflict of authority in the first biosafety law: although CTNBio was responsible for GMO risk assessment, its technical decision needed the referendum of other agencies of the federal government, which had technical opinions and ideological alignments different from those of the Ministry of Science and Technology. Legal concerns resulted in the enactment of new laws and decrees, creating a complex regulatory framework, which was not effective. The technology was being discussed by civil society, which had repercussions for the government.

The confusing regulatory scenario and successive civil actions in court led to the blocking of the commercial release of all GM plants. Meanwhile, the pressure to improve productivity triggered the smuggling of a lot of GM soybean seeds to be planted in southern Brazil between 2001 and 2004. The existence of extensive plantations of illegal GM soybeans forced the federal government to issue a decree provisionally approving the planting and harvesting of this GM variety, and encouraged discussions about the need for a new law. There ensued a contest among the conflicting interests of the various stakeholders in the agricultural scenario: agribusiness, multinational seed companies, environmental activists, organisations supporting small farmers, peasant movements, scientists, regulators and other groups strongly influenced the discussion of the new law. Ultimately, the forces pushing agribusiness prevailed, supported by the scientific consistency of GMO risk assessment and the certainty that the risks to the environment and health are low. Nevertheless, it took more than a year and a half of discussions in Congress for the approval of the new legislation, Law 11105/05, which was promulgated on 24 March 2005. This law was slightly modified by Decree 5591/05 on 22 November 2005 (Legislação Brasileira, 2005).

The new law followed essentially the same normative path as the previous one, but without conflicts of authority: the new CTNBio was now to decide alone on risks associated with GMOs and on their commercial release, delegating to the oversight and registry agencies the task of registering the product without additional considerations as to its biosafety. Other relevant provisions of the new law included the creation of the Information System on Biosafety (SIB) to provide information on the biosafety of GMOs in Brazil and the decision-making related to it, improving the transparency of all of the processes involved. Moreover, it also nominated the National Biosafety Council (CNBS), which is composed of representatives of the ministries responsible for proposing biosafety policies, taking into account socio-economic issues. Certainly, this new biosafety law effectively contributed to harmonisation among legal instruments, ending controversies surrounding GMOs and initiating a new era for biotechnology in the country.

Brazil is a party to the Cartagena Protocol on Biosafety. In 2009 the Ministry of External Relations was established as the National Focal Point. As a consequence of its entrance into the Protocol, Brazil has a tightly controlled GM export procedure in order to comply with the many rules set up by the Protocol and agreed to by its parties.

The Brazilian GMO Regulatory Framework

The Brazilian GMO regulatory system is built on four pillars, represented by the Internal Biosafety Committees (CIBios), the National Biosafety Technical Commission (CTNBio), the National Biosafety Council (CNBS) and the Surveillance and Registration Bodies. CTNBio has the leading role, but risk assessment and biosafety surveillance are also done at the institutional level by the CIBios. Moreover, CTNBio has the duty to help the CNBS determine and improve the National Biosafety Policy. Once a new GM product has been approved as being safe for humans, animals, plants and the environment, the Surveillance and Registration Bodies still have to register it before commercialisation.

Any new GM product or GM activity, at the laboratory research level, in controlled field experiments or for commercial release, needs to have its risks assessed. In summary, the CIBios can authorise research in containment and GMO imports, CTNBio assesses risks, authorises field and commercial releases and follows the development and results of field experiments and post-release monitoring activities, and the CNBS assesses the possible negative impacts of every new product at the social and economic levels and defines the National Biosafety Policy.

The CIBios

The chain of responsibility starts at the CIBios: all public and private institutions must get a Biosafety Quality Certificate (CQB) from CTNBio before starting any activities

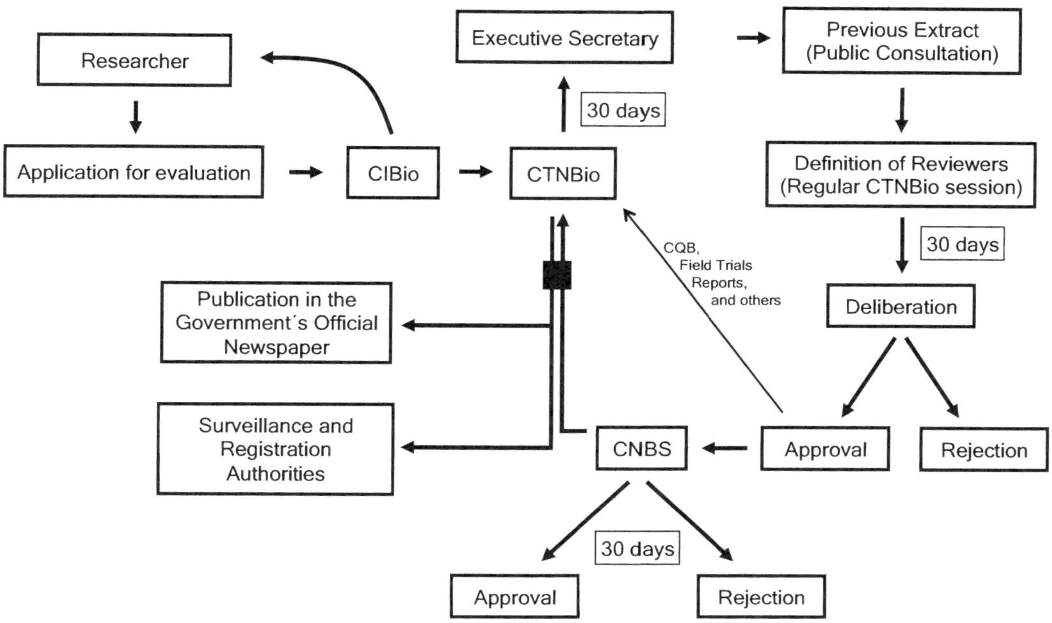

Figure 22.1. A flowchart representing the application and decision process.

involving GMOs and at the same time must create an Internal Committee for Biosafety (CIBio), usually composed of technicians with expertise in biosafety. The CIBios assess risks of Class I GMO confined research at the institutional level and annually report results to CTNBio.

CTNBio

The new biosafety law restructured CTNBio. It became a multidisciplinary consultative and deliberative committee established under the Ministry of Science, Technology and Innovation to provide the federal government with technical and advisory support for the implementation of national biosafety policies and the elaboration of a law regulating all activities that use technologies involving recombinant DNA and derivative products. CTNBio is composed of 27 members and their respective deputies, all of whom are required to have a PhD and to be professionally active in some area of biosafety, biotechnology, biology, human and animal health, and environmentalism.

CTNBio is the sole Brazilian authority that evaluates, and allows the development of, experimental activities with GMOs in laboratories, under containment and also in field trials. For that purpose, CTNBio has released general rules, normative resolutions and memos, concerning specific criteria for different species. All information obtained from confined experiments and field trials is used to support a robust risk assessment for the commercial release of a GM product.

All activities involving GMOs must have their risks assessed either by one of the CIBios or by CTNBio, according to the flowchart shown in Figure 22.1. The CIBio of a public or private institution is the first committee to evaluate a request, submitted by a researcher in that institution. If the request is approved, the CIBio submits it to CTNBio through the Executive Secretary, who verifies the documents. The process, from receiving the documents to nominating reviewers, may take up to 30 days. Projects related to human and animal health are analysed by a sub-committee related to these subjects, and projects on agriculture are analysed by a plant and environment sub-committee. Reviewers representing the four areas (i.e., specialists in human and animal health, plants and the environment) are required to consider the commercial release processes. Their evaluations are read and discussed by the members in a plenary meeting, at which they decide whether or not to approve release.

It is important to emphasise that CTNBio only deliberates on risks to the environment and health of GMOs and does not evaluate socio-economic issues related to their commercial release, which are analysed by the CNBS.

The CNBS

The CNBS is responsible for all socio-economic aspects and national interests relating to commercial releases of GMOs. The Chief of Staff of the Presidency of the Republic is the President of the CNBS and the members are the Ministries of Science, Technology and Innovation, Health, Environment, Agriculture, Cattle-raising and Supply, Agrarian Development, Defence, Justice, Development of Industry and Foreign Trade, International Relations, and Aquaculture and Fisheries.

As depicted in Figure 22.1, after a favourable technical opinion has been received from CTNBio regarding the commercial release of a GM product, the decision is held in abeyance for 30 days, pending the evaluation and decision of the CNBS. If no formal report is issued by the Council, the process is forwarded to the Surveillance and Registration Bodies and published in the government´s official newspaper. CTNBio and the National Surveillance and Registration Bodies can also directly require the CNBS to analyse the social and economic aspects related to the commercial release of a product, but this has happened only twice: for the approval of Liberty Link maize (event T25) and Bt11 maize.

The CNBS does not have defined rules for socio-economic assessments of the adoption of new biotechnologies. This board will arbitrate questions and complaints from sectors affected by new biotechnologies; however, it will not interfere in the technical decision on biosafety that has been made by CTNBio. There has never been a case in which the decision made by CTNBio regarding commercial release has been changed.

Surveillance and Registration Bodies

The Health Surveillance Agency (ANVISA), the Ministry of Agriculture, Livestock and Supply (MAPA) and the Brazilian Institute of Environment and Renewable Natural Resources (IBAMA) are not directly involved in risk assessment, except through their representatives in the Commission. They do, however, have the duty to assess the product and to ascertain its usefulness and adequacy for the Brazilian market, but no further

Table 22.1 The rate of adoption of GM commodity crops in Brazil in 2015

	GM adoption rate 2015		
Crop	Cultivated area (millions of ha)	Percentage of total production	GM growth rate (%)
Soybean	31.4	93.2	5.4
Maize	13.5	86.5	6.3
Cotton	0.75	77.5	9.9

Source: Céleres (2016).

questioning on risks directly arising from the genetic modification of the product is expected. Without proper registration, the product cannot be marketed in Brazil.

The Adoption of Biotechnology in Brazil

Brazil is one of the world's largest countries by area (8.5 million km^2) and population (206 million); it is currently a key player in the international market and has great power among emerging countries as the world's seventh-largest economy. Agriculture plays an important role in Brazil's economy, being responsible for approximately 40% of all exports and employing 20% of the active workforce. About one-third of Brazil's Gross Domestic Product (GDP) comes from agribusiness (Momagri, 2012; IBGE, 2016).

In order to assess the positive impacts of Law 11105 regarding applications of biotechnology on agriculture, it is essential to assess information and statistics about the relevance and evolution of this sector of the Brazilian economy. Brazil is the second-largest producer of biotech crops in the world (the USA being in first place), planting 44.2 million ha of GM crops in 2015, an increase of 5% from 2014 (James, 2015). The adoption rate in 2015 reached 92.3% of the total number of seeds planted. For the 2017 harvest, the expectation is 49.1 million ha (Céleres, 2016). For the 10-year period from 2005 to 2015, grain production increased by 5% per year, possibly due to the adoption of biotechnology crops. The rate of adoption of GM soybean, maize and cotton has dramatically increased in the last few years (Table 22.1).

Between 2006 and 2014, Brazil's agriculture exported US$708 billion, resulting in a positive trade balance of US$590 billion. Excluding agriculture, all other sectors of Brazil's economy exported US$1,109 billion, but with an accumulated deficit of US$386 billion. The sum of the surplus produced by agriculture and the deficit originating from other sectors of Brazil's economy resulted in a surplus of US$204 billion. Given the scenario established since 2006, one can conclude that Brazil's main agricultural and livestock products have ensured the maintenance of a positive trade balance and, consequently, improved the economic and social stability of the country.

Also, modern technology has allowed the development of more effective products for preventing or treating human and animal diseases. In fact, Brazil has improved the production of healthier animals, resulting in higher production and better distribution

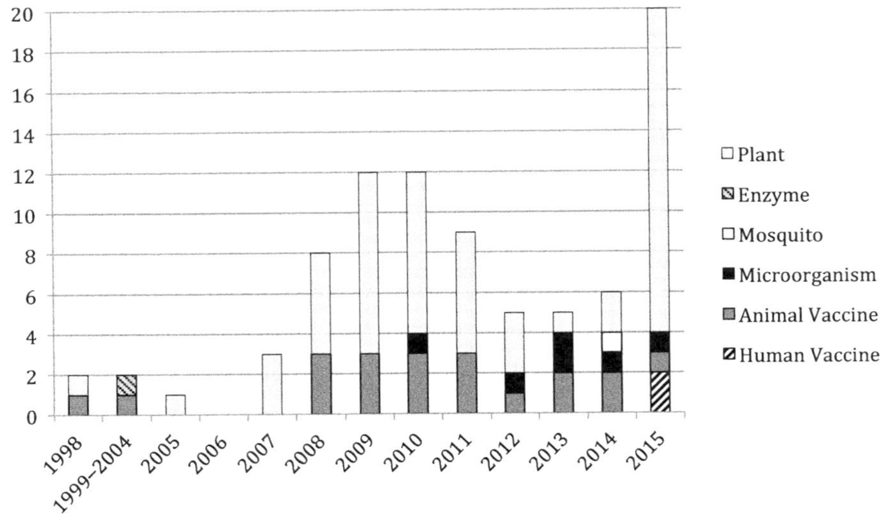

Figure 22.2. Commercial release of GMOs from 1998 to 2015.

in the domestic market, as well as increasing the amounts of chicken and pork exported to more than 150 countries. Vaccines for horses and dogs have also been analysed for biosafety and commercially released. Several laboratories are accredited by CTNBio for the production of vaccines against infectious diseases such as influenza, dengue, etc. Research on the development of genetically modified animals as bioreactors has been approved. For example, research on transgenic goats expressing lysozyme in milk, which can be used to treat infant diarrhoea, is being conducted by a consortium of universities in Ceará State, with promising results (Bertolini *et al.*, 2016; Righetti, 2010).

In the 10 years since the promulgation of the new biosafety law, Law 11105/2005, CTNBio has approved 82 GMOs, namely 52 events in plants, 20 animal vaccines, two human vaccines for dengue virus, seven microorganisms, and one mosquito, *Aedes aegypti*, genetically modified for population control. Figure 22.2 shows the commercial approvals of GMOs up to 2015.

Driving Forces in the Brazilian GM Scenario

The demand for biotech products tends to grow rapidly, and a favourable regulatory framework supports the adoption of new products. Brazil has widely employed transgenic seeds in agriculture, with great success. Likewise, vaccines and enzymes produced by GMOs or GM viruses are also in common use. The country produces alcohol and oils from GMOs, and a GM mosquito has been approved by CTNBio for disease-vector control. However, hardly anything has been developed within the country, with the notable exceptions of the virus-resistant bean, developed by EMBRAPA; Eucalyptus for rapid growth, developed by Futuragene; a yeast for second-generation ethanol production, developed by Granbio, which is already in use; and a soybean variety developed by EMBRAPA in collaboration with BASF for tolerance to herbicide. What prevents the country from bringing to market products coming from its universities and research centres or institutions?

The adoption of biotechnology in Brazil has not come without protests from stakeholders representing various sectors of society, in line with the global trend of anti-biotechnology activism. Protests began in the late 1980s and grew in intensity, particularly from 2005, when the new biotechnology law allowed the entry of transgenic plants onto the market. Has this movement against agricultural biotechnology played a role in the country's inability to bring its own products to market? What other driving forces have kept local innovations out of the market?

Research and Development in Biotechnology

Basic and applied research in biotechnology in Brazil essentially follows the same path as in developed countries (ABDI and CGEE, 2009): there is a concentration of scientific articles in the areas of genomics, proteomics and gene expression, as well as on cloning and heterologous protein function and ultimately on the generation and use of GMOs. Other areas of biotechnology (biomass, pharmacogenetics, etc.) also depend on the frequent use of GMOs in the generation of experimental data. It is clear that a strict regulatory framework did not prevent the development of academic research in biotechnology. On the contrary, despite restrictions on research and production (Soares, 2014), and public opinion sometimes having been unfavourable to GMOs, the adoption of biotechnology has been rapid and remains so.

Almost all of the basic and applied research has been, and continues to be, done in public and private academic or research institutions, with support from public funding agencies. The production of scientific work specifically in biotechnology by the private sector in Brazil, however, is negligible. When patents in biotechnology are scored for the same period in Brazil, only foreign or multinational private companies appear as applicants (ABDI and CGEE, 2009).

The main reason for the lack of patents and commercial products is the high cost determined by the Brazilian regulatory requirements: although there is no estimate for Brazil, the regulatory costs should be similar to those in the USA, which have been estimated at about $35 million (Phillips McDougall, 2011). After all, many of the innovative Brazilian companies are start-ups and, even with the possible support of venture capital, they could hardly amass the resources needed for the commercial release of a GM plant. Public institutions also have serious financial constraints, and funding agencies are not prepared to cover regulatory costs.

Risk Perception and Opposition to GMOs

Public perception of GMO risks, particularly of transgenic plants, is generally positive in Brazil, but, as in other parts of the world (Porth and El-Kassaby, 2014), concerns regarding the dominating role of large biotech companies and its impact on food security and agricultural independence have been raised by environmental groups and movements. The adoption of biotechnology in Brazil has thus faced, ever since its inception, intense opposition from groups with varied strategies of action. In the first regulatory framework, this opposition could effectively foist a moratorium on biotechnology on the country, but the opposition was unable to stop the adoption of biotechnology from 2005 on, owing to the force of the new biotechnology law. The presence of a constant and very active opposition in the media, however, played an important role in the formation of public opinion regarding GMOs in Brazil.

According to a long retrospective study (CGEE, 2015), Brazilians are very interested in science and technology (S&T), levels of interest being even higher than those in Europe, but they do not seem to have knowledge of biotechnology commensurate with their interest in the subject, which is also a common finding in other countries. The vast majority of Brazilians believe that S&T bring more benefits than harm to humanity. This opinion prevails in all age groups and has been growing in strength since 1987. This positive view extends to scientists, who are seen as agents of social transformation, which contributes to the growth of the country and the welfare of society.

The high level of concern about the impact of GMOs on the environment and health declared in interviews would seem to be paradoxical: although science has shown that there is no risk from GMOs, Brazilians perceive various risks and express their concern regarding agricultural biotechnology almost as strongly as they care about deforestation of the Amazonian region and global warming. Here there is clearly an influence of media bias, which is a powerful agent in moulding public concern regarding risk.

The opinion poll results published by Capalbo *et al.* (2015) also show the conceptual bias in public perception: while a large majority of Brazilian respondents have a positive perception of biotechnology, biosafety and even genetic engineering, most have a neutral or even negative position on transgenic plants and GMOs in general. The influence of the media, especially alternative media, determines that GMOs for food production are the main source of concern and also of disagreement among respondents, as opposed to biotechnology. Groups with positive attitudes towards transgenic crops are predominantly composed of senior citizens who trust experts as sources of information. In contrast, those with negative attitudes towards GM crops tend to be young people, particularly students, who rely on NGOs as an information source about GMOs and biotechnology. The NGOs in Brazil align with peasant movements, just as in other parts of the world (Brazil's Landless Rural Workers' Movement (MST), Family Farming and Agroecology (AS PTA), Via Campesina and Greenpeace, among others). The flags raised by these groups have the colour of environmentalism, but also those of movements against capitalism and against big business. The scientific and technical content of this opposition is generally very limited.

The opposition to GMOs, although limited to a few groups, has promoted invasions of experimental stations and factories in collaboration with peasant movements, causing the destruction of research material and the halting of activities. Occasionally opposition groups, in association with peasant movements, have also invaded CTNBio, preventing voting and the taking of decisions and creating a climate of fear and discontent among the majority of its members. These events, however, have not been frequent, and the

Brazilian social environment has been mild for agricultural biotechnology, and much more so for industrial biotech.

Labelling and Food Safety

The concern about GMOs is reduced significantly when it comes to buying food products or animal feed, and choice on the basis of the GM label – which is mandatory in Brazil – seems to be rare. With the increasing adoption of agricultural biotechnology, virtually all food or feed products containing corn or soybeans are labelled. The consumer, however, does not seem interested in the label (a capital T in a yellow triangle) and generally does not know the meaning of that label on the packaging (Furnival and Pinheiro, 2009). The opposition to GMOs, on the other hand, seeks to alert the public to the importance of labelling as a civil right, but, in the absence of clear scientific arguments that there are risks, the population does not exclude GMOs from the shopping list and cares more about other elements on the label, for example the expiry date and, naturally, price.

Risk Communication: The Missing Part

Although the opposition to agricultural biotechnology is real and sometimes very strong, an organised and effective way to disclose the advantages and safety of biotech products does not exist in Brazil, as in other countries of the world. Universities have little interest in the question and there is no funding for large projects relating to this matter. Initiatives such as the University of Cornell Alliance for Science (http://allianceforscience.cornell.edu/) could be propagated in the country, to the benefit of the public and private sectors and consumers in general.

Conclusion and Perspectives

In Brazil, all stages of the development of a GMO, be it a plant, a microorganism or derivatives, are evaluated. It is important to state that the approval for use of a commercial GMO does not stop the surveillance carried out by CTNBio. After commercial approval, the applicant must provide a post-release commercial monitoring plan that will be approved by CTNBio and accompanied by periodic reports. Furthermore, the Commission will review its decisions if new scientific data concerning products or technologies come to light.

Our experience in risk communication is distorted. The perception that CTNBio ‘approves all’ and therefore would not have correct criteria for evaluation, serving only as an agency seal that companies have, is totally wrong. Our legal system establishes rigorous assessment criteria, and companies adhere to these criteria and standards for the preparation of their proposals. When an item is not clear or has methodological failures, the reporters of the claims request rectifications and corrections necessary to adapt the claims.

The approval and subsequent adoption of GM herbicide-tolerant crops, achieved thanks to a robust and functional regulatory system, played a key role in transforming the way the country produces grain. Brazil is now a world leader in crop management with minimum- and no-tillage systems, which helps decrease erosion and improve general soil quality and recharging of groundwater (Lopes, 2016). As a consequence, a significant decrease in the amount of chemical fertilisers applied to crops was achieved, as well as reduced environmental impacts such as water resource contamination with nitrates or other harmful elements. These and many other innovations incorporated by Brazilian agriculture allowed an increase in efficiency of resource use, higher productivity and a more efficient use of land, greatly reducing the environmental costs of farming.

The Brazilian life-science industry, as well as agriculture, has profited from the comprehensive, science-based regulatory framework (Rampelotto, 2016), which, in turn, encouraged the funding agencies to spend on research and the federal government to plan and put into practice a set of bold, successful programmes to support biotech development. Research and development in the academic and research institutions therefore immensely profited from a reliable system to assess risks, while at the same time clearly defining the responsibilities shared among the different stakeholders, including an extended funding programme. However, the very high costs of producing the required data for GMO regulatory purposes have discouraged the transformation of discovery into innovation, and today there is no product on the Brazilian market or in CTNBio's approval pipeline derived from a public or private native university.

The way CTNBio assesses GMO risks is essentially the same as recommended by the Ad Hoc Technical Expert Group of the Convention on Biodiversity (CBD, 2012) and in line with the procedures adopted by many other official GMO risk assessment agencies in the world, such as the EFSA (EU), the OGTR (Australia), CONABIA (Argentina), etc. It is based on science, as advocated by experts (Conko *et al.*, 2016), and assesses only the new biological risks directly produced by GMOs. The versatility of the Brazilian risk assessment approach, proceeding case by case, step by step, allows the assessment of GMOs obtained by any technology, not only by conventional genetic engineering. The centralisation of the assessment in a single organ facilitates the updating and training of evaluators and streamlines decision-making. The agility with which CTNBio has acted, while at the same time being solidly supported by science, has brought great respect for the country and resulted in a high level of confidence in the technical decisions involving GMOs, both domestically and internationally.

In addition to the approved GM crops, other plants are being tested in controlled field releases supervised by CTNBio and subject to inspection by the Ministry of Agriculture and Livestock: rice, passion fruit, citrus, Eucalyptus, cowpea and sugarcane. They express different traits such as yield improvement, drought tolerance, fungus resistance, oil quality and wood density. New products, derived from new technologies such as gene editing, will soon arrive in CTNBio's assessment pipeline. Also in industrial biotechnology, it is expected that gene editing will strongly contribute to a net gain in efficiency, as well as resulting in the introduction of novel products and technologies into the market: bioplants for drugs, special oils, insecticides and polymer production, etc., and new ways of producing energy.

In the public-health sector, GM products, such as vaccines and recombinant proteins for diagnosis and therapy, are proving to be of great help. A new recombinant dengue vaccine developed by the Brazilian Butantan Institute, a public research institution, was recently approved by CTNBio to enter a Phase III trial. GM insects will be possibly the basis of a radical transformation of vector-borne disease control. Indeed, the use of CRISPR/Cas9 gene editing and its association with gene drive systems allows the modification/suppression of entire insect populations, with relatively small budgets and with a high benefit/cost ratio. Other genetic modifications can be used for disease-vector control, as in the case of the RIDL *Aedes aegypti*, which is currently in use in Piracicaba, in the State of São Paulo, with very promising results and no safety concerns (Andrade *et al.*, 2016).

Biotechnology evolves very quickly, and legislation related to biosafety has to monitor these developments, in order to evaluate their biosecurity, without interfering with the scientific progress and the consequent trade. Thus, CTNBio has created working groups to analyse, for example, the biosecurity of organisms generated by new techniques such as genome editing, and to deliberate how Brazil will deal with these GMOs while harmonising with global perspectives.

All the aforementioned technologies, which fall under CTNBio's regulatory scope, are of paramount importance for Brazil's development. The major goal and challenge has been to allow this development and guarantee human and animal health, while respecting the environment. All products regulated by CTNBio to date have not caused any adverse effect on human or animal health or any environmental damage. This indicates that the risk assessment procedures and management measures adopted in Brazil have been sufficient and have successfully ensured the biosafety of GMOs without precluding their use.

References

ABDI and CGEE (2009). *Panorama da biotecnologia.* Brasília: Agência Brasileira de Desenvolvimento Industrial (ABDI) and Centro de Gestão e Estudos Estratégicos (CGEE), 170 pp. [Online]. Available from www.abdi.com.br/Estudo/Panorama%20Setorial%20Biotecnologia.pdf

Andrade, P. P. *et al.* (2016). Use of transgenic *Aedes aegypti* in Brazil: risk perception and assessment. *Bulletin of the World Health Organization* **94**, 766–771.

Bertolini, L. R. *et al.* (2016). The transgenic animal platform for biopharmaceutical production. *Transgenic Research* **25**, 329–343.

Capalbo, D. M. F. *et al.* (2015). A study of stakeholder views to shape a communication strategy for GMO in Brazil. *Frontiers in Bioengineering and Biotechnology* **3**, 179.

CBD (2012). Guidance on risk assessment of living modified organisms, 64 pp. [Online]. Available from www.cbd.int/doc/meetings/bs/mop-06/official/mop-06-13-add1-en.pdf

Céleres (2016). 2° levantamento da biotecnologia agrícola no Brasil, safra 2015/16. [Online]. Available from www.celeres.com.br/2o-levantamento-de-adocao-da-biotecnologia-agricola-no-brasil-safra-201516/

CGEE (2015). Percepção pública da ciência e tecnologia 2015 – Ciência e tecnologia no olhar dos brasileiros. Sumário executivo. Brasília: Centro de Gestão e Estudos Estratégicos. [Online]. Available from www.cgee.org.br

Conko, G. *et al.* (2016). A risk-based approach to the regulation of genetically engineered organisms. *Nature Biotechnology* **34**(5), 493–503.

Furnival, A. C. and Pinheiro, S. M. (2009). O público e a compreensão dos rótulos de alimentos: o caso dos transgênicos. *Revista Digital de Biblioteconomia e Ciência da Informação (Campinas)* **7**(1), 1–19.

IBGE (2016). Projeções e estimativas da população do Brasil e das Unidades da Federação. [Online]. Available from www.ibge.gov.br

James, C. (2015). Global Commercialization of Biotech/GM Crops: 2015. ISAAA Brief 2015.

Legislação Brasileira (2005). [Online]. Available from www.planalto.gov.br

Lopes, M. A. (2016). Biotechnologies in action in Brazil. In *FAO International Symposium on The Role of Agricultural Biotechnologies in Sustainable Food Systems and Nutrition*, pp. 12–13. [Online]. Available from www.fao.org/3/a-bc562e.pdf

Mendonça-Hagler, L. *et al.* (2008). Trends in biotechnology and biosafety in Brazil. *Environmental Biosafety Research* **7**(3), 115–121.

Momagri (2012). Agriculture, a strategic sector for Brasil's economy. [Online]. Available from www.momagri.org

Phillips McDougall (2011). The cost and time involved in the discovery, development and authorisation of a new plant biotechnology derived trait. A Consultancy Study for Crop Life International, 24 pp. [Online]. Available from https://croplife.org/wp-content/uploads/2014/04/Getting-a-Biotech-Crop-to-Market-Phillips-McDougall-Study.pdf

Porth, I. and El-Kassaby, Y. A. (2014). Current status of the development of genetically modified (GM) forest trees worldwide: a comparison with the development of other GM plants in agriculture. *CAB Reviews* **9**(8), 12.

Rampelotto, P. H. (2016). The Brazilian life science industry: advances and challenges. *Industrial Biotechnology* **12**(1), 3–10.

Righetti, S. (2010). Ceará fará leite com proteína humana. *Jornal Folha de São Paulo* 28 August 2010. [Online]. Available from www.recodisa.ufc.br/news/Ceara-fara-leite-com-proteina-humana.php

Soares, E. (2014). Restrictions on Genetically Modified Organisms: Brazil. Law Library of Congress, USA. [Online]. Available from www.loc.gov/law/help/restrictions-on-gmos/brazil.php

23 An Effective Regulatory Regime Supported by Research and Development Is Key to Adoption of GM Technology in West Africa: Burkina Faso and Nigeria as Case Studies

Olalekan Akinbo, Ademola A. Adenle and Diran Makinde

Introduction

Genetic modification (GM) technology has great potential to improve agricultural productivity and contribute to food security challenges in Africa when properly regulated. The adoption of GM technology continues to increase in the developing world, particularly among smallholder farmers. The 2015 report of the International Service for the Acquisition of Agri-Biotech Applications (ISAAA) showed that, for Africa, this is the 18th year of successful commercialisation of biotech crops. Only three African countries planted GM crops in 2015, Burkina Faso, South Africa and Sudan (James, 2015). The lack of an enabling policy and legal environment for biosafety regulation, among other factors, currently serves as a hindrance to the adoption of GM crops in Africa. In West Africa, only three countries (Burkina Faso, Ghana and Nigeria) have functional biosafety laws that allow laboratory and field experimentation on genetically modified organisms (GMOs) to take place, while three countries (Mali, Senegal and Togo) have biosafety laws without field trials (Figure 23.1). Other factors that have contributed to the slow adoption of this technology in West Africa include the limited critical mass of regulators with enhanced competencies in biosafety and the seeming lack of biosafety communication and cooperation. Biosafety activities by the New Partnership for Africa's Development (NEPAD) Agency African Biosafety Network of Expertise (ABNE; www.nepadbiosafety.net) and other biosafety service providers are key to enhancing the knowledge and skills of regulators, scientists, lawyers and policy/decision-makers in adopting best practices for performing their mandated functions and also broadening their understanding of core issues in biosafety decision-making, thereby increasing confidence in the regulatory system. In the same vein, we cannot underestimate the impact of the constantly expressed concerns of the environmentalists and activists that have led to the slow adoption or non-adoption of the technology in many African countries, including West Africa.

Figure 23.1. Regulatory status of agricultural biotechnology and biosafety in West Africa.

In this chapter we discuss achievements made by Burkina Faso with insect-resistant (Bt) cotton, especially how it has benefited small-scale farmers; how Burkina Faso became a showcase for GM technology in Africa; the current scenario of the technology in West Africa with a focus on Burkina Faso's experience; and developments in other West African countries, specifically Ghana and Nigeria. The chapter describes the challenges of adoption of GM technology, citing the example of Bt cotton qualities, as recently revealed in Burkina Faso, and how this may affect other countries' uptake of GM technology in the sub-region. Finally, we discuss the role of the regional economic communities and public–private partnerships, and provide suggestions on how to make progress towards the attainment of food security in Africa.

Background: Current Status of Agricultural Biotechnology and Biosafety Policy in West Africa

In the Economic Community of West African States (ECOWAS), of 15 member states, six countries have biosafety laws, namely Burkina Faso, Ghana, Mali, Nigeria, Senegal and Togo, with only three currently conducting confined field trials on commodities that are important to them. The other three countries have not been able to conduct experiments on GM crops because the laws are prohibitive to technology developers.

Burkina Faso

Burkina Faso is a good example of a country whose biosafety law allows both confined field trials (CFTs) and subsequent commercialisation of GM crops. The commercialisation of Bt cotton started in 2009; its biosafety law was revised and enacted by the national assembly in 2013 to comply with international law. Within this legal framework, approvals were given for CFTs for a new herbicide-tolerant cotton variety, Maruca-resistant cowpea and biofortified sorghum. Despite the fact that Burkina Faso has a functional regulatory regime, cultivation of GM crops faced stiff opposition from anti-GM activists, especially the Francophone media. The government therefore developed programmes geared towards effective biosafety communication and public awareness to ensure that Burkina Faso continues to harness benefits from the new technology and maintain its leading role in biotech production in the sub-region. In addition, Burkina Faso is one of the six West African countries to have signed up to the Comprehensive Africa Agriculture Development Programme (CAADP) compact and allocates more than 10% of the national budget annually to agriculture. Other West African countries that have attained this target include Ghana, Guinea, Mali, Niger and Senegal. These countries have allocated more resources to agriculture and have made progress towards raising agricultural productivity by at least 6% per annum and increasing public investment in agriculture to 10% of their national budgets annually by signing the CAADP compact. However, a lack of trained scientists, weak infrastructure and institutional challenges are currently hindering the development of agricultural biotechnology programmes in many of these countries (Traoré *et al.*, 2014a).

Nigeria

Nigeria's influence in agricultural output in West Africa is enormous (Nin-Pratt *et al.*, 2011), and it will have a greater chance of resuming its role as the breadbasket of the West Africa sub-region if appropriate technologies, including GM technology, are harnessed. For decades, cotton production (lint and cotton seed) has been a driving force for economic development in Nigeria. The neglect of the agricultural sector during the oil-boom years (the 1970s and 1980s) had a direct impact on the cotton sector. In recent years, due to poor management and reduced production of both lint and seed, the cotton sector has slackened. The main feature of Nigerian cotton cultivation is that 80% of the total production is by small-scale farmers under rain-fed conditions. Cotton production in Nigeria was 30,000 metric tonnes (MT) in 1954, which increased to 60,000 MT in 1979; then it picked up to its current production level of 95,000 MT. However, cotton production suffers losses from insects and diseases (Organisation for Economic Co-operation and Developments (OECD), 2006), and poor-quality seed also remains a challenge affecting yield, thereby affecting farmers' income.

Nigeria has been conducting CFTs for a number of commodities, including biofortified cassava, biofortified sorghum and Maruca-resistant cowpea, but has yet to commercialise GM crops, partly due to the delay in granting assent to the biosafety bill by the Presidency. The Nigerian investment in public agricultural biotechnology research and development (R&D) is low (0.2% as a percentage of gross domestic product (GDP) (National Bureau of Statistics (NBS), 2015)), despite the fact that Nigeria has the largest manpower for biotechnology in the sub-region. This continues to raise questions about the country's ability to execute the government's agricultural biotechnology policy. As part of the process of ownership, validation and subsequent finalisation of biosafety documents, the National Biosafety Management Agency (NBMA), which was created shortly after the Biosafety Law was assented to by the Presidency in April 2015, identified the need to subject these documents to review by stakeholder ministries and other national institutions.

Ghana

Agriculture is Ghana's most important economic sector, employing more than 50% of the population on a formal and informal basis and accounting for almost half of GDP and export earnings from grains, cocoa, oil palms, kola nuts and timber. The country produces a variety of crops in various climatic zones, which range from dry savanna to wet forest and run in east–west bands across the country. Cotton production expanded rapidly in the early and mid 1970s, reaching 24,000 tons in 1977, but it fell back to one-third of this figure in 1989. Ghana's biosafety regulatory system is maturing following the passage of the Biosafety Act in 2011, and it adopted an interim measure by passing an implementing regulation to the CSIR Act, Biosafety Regulations (Management of Biotechnology; LI 1887), which served to regulate research in modern biotechnology while the substantive act was being worked on. Currently work on passing regulations to implement the Biosafety Law, 2011 (Act 831) is in progress. The biosafety system of Ghana exhibited considerable resilience in 2015 following a legal suit that challenged the set-up, mandate and operations of the National Biosafety Authority (NBA), which previously was the National Biosafety Committee (NBC). A civil society organisation brought a legal suit against the NBC, NBA and other state actors seeking an interlocutory injunction on the development and release of GMOs in Ghana. Ultimately, the group wanted a ban on the release of GMOs in Ghana for any purpose. The high court of the Republic of Ghana dismissed the application for an injunction on the grounds that it lacked merit. This case notwithstanding, the Biosafety Authority continues to function with confidence. This is an important indication that Ghana has a robust and resilient regulatory system.

Crops currently being tested in CFTs include insect-resistant and herbicide-tolerant cotton (stacked traits); Maruca-resistant cowpea; nitrogen-use-efficient, water-use-efficient and salt-tolerant (NEWEST) rice (triple stack). CFTs with a high-protein sweet potato were approved but have yet to commence. Biosafety service providers in Africa, including NEPAD Agency ABNE, conducted a campaign to educate Ghana's news editors and farmers' leaders; this was to equip the public with factual information on biosafety, facilitate dialogue and contribute to the process of mending the divided front (regarding controversial GMO issues) of the farmers' leadership, as farmers are the end users of the technology.

Mali, Senegal and Togo

The Francophone countries in West Africa which have biosafety laws are Mali, Senegal and Togo; however, the laws have not been investor-friendly to technology developers, and efforts to revise the laws and get them passed by the various countries' national assemblies are being made. Despite the political crisis in Mali, which did not allow much intervention, the farmers are demanding Bt cotton, having seen the benefits which accrued to their counterparts in Burkina Faso. It is worth noting that, despite this, Mali is actively involved with the technologies in the *Target Malaria* programme with Burkina Faso and Uganda. This is a programme that aims at reducing malaria transmission using transgenesis to control the population of *Anopheles gambiae.* Mali ratified the Kuala Lumpur Supplementary Protocol on Liability and Redress in 2015. This ratification was considered a key milestone that will provide justification for the revision of Mali's biosafety law. Senegal has shown interest in developing a workable biosafety system because Senegalese farmers are interested in Bt cotton. Togo prides itself on having the best cotton cultivar in the sub-region, but the challenge is for the country to establish an enabling legal environment. Togo was among the first African countries to sign the Nagoya–Kuala Lumpur Supplementary Protocol on Liability and Redress in 2011.

The Adoption of Bt Cotton in Burkina Faso – a Learning Experience for the Sub-region

The cotton sector plays an important role in the national economy as it contributes 8% of GDP, but suffers losses due to pests and diseases and farmers' inability to afford costly pesticides (Savadogo, 2009). The application of GM technology has been demonstrated to be effective against biotic and abiotic stresses in order to boost crop productivity in many developing countries, including African countries. Bt cotton production started in 2009 following three generations of backcrossing of the Bt trait into three widely grown cotton cultivars in Burkina Faso. The country has since then increased its adoption of Bt cotton cultivation to as much as 73% of total cotton hectares in 2015 (James, 2014). Literature evidence indicates that Burkina Faso farmers have benefited from the adoption of Bt cotton in several ways, including higher income per hectare cultivated; less use of insecticides; reduced cost of production; and reduced risk to human and animal health from pesticide usage, thereby leading to human health and environmental gains (Bennett *et al.*, 2004; Brookes and Barfoot, 2012; James, 2013, 2014). As a matter of fact, Burkina Faso was the top cotton producer in Africa in 2015, with over 700,000 MT of cotton seed produced (Dowd-Uribe and Schnurr, 2016). The 2009 field surveys indicated that farmers (who adhered to recommended pesticide spraying) recorded 37% yields higher than for conventional cotton and a 67% reduction in pesticide usage (Vitale *et al.*, 2011). These benefits followed the global trend of a 37% reduction in the use of pesticides, 22% increase in crop yield and 68% increase in farmers' profits from biotech crops (James, 2014). A total income gain of US$137 million has been recorded for the country in the past seven years. In fact, in 2013 alone, the country earned US$37 million from Bt cotton production. The country has a functional regulatory system and efficient cotton industry managed by *Sofitex*, *Socoma* and *Fasocotton*. Aside from Burkina Faso being the highest producer of cotton in Africa, Burkina Faso cotton was known for its high quality, with fibres that are long, strong and uniform, and therefore used to command a premium price because of these quality traits. Following the adoption of Bt cotton, a decline in staple length was observed. The resultant effect has been a reduction in its value on the international market. In the interim, farmers and cotton companies agreed to drastically reduce the cultivation of Bt cotton to prevent the short-fibre trait from becoming more prevalent. This was not understood as a failure of the technology, since the fibre quality depends on several parameters, including genetics, environmental conditions, soil fertility and the timing of harvesting. The variation in cotton fibre quality is a function both of the environmental conditions and of the genetic background of the varieties (Dowd-Uribe and Schnurr, 2016). This variation exists between all cotton varieties (conventional or genetically engineered) and it is independent of the Bollgard II trait (Traoré, 2016, personal communication). Efforts to restore the fibre quality within the foreseeable future are being made. Unfortunately, the propaganda generated from this made it look as if the government wanted to abandon the technology, which is completely untrue. This was an attempt by some activists to dissuade other countries in the region from adopting the technology. In view of the benefits associated with the Bt cotton technology, a new programme (a partnership of women with the association of Bt cotton growers) has been created to facilitate access to extension services and strengthen the cotton sector in the country.

Of cotton-producing countries (including Ghana, Mali and Senegal) in the subregion, Nigeria has the largest textile industry, but many factories are not operational, partly due to low cotton production (OECD, 2006). The lack of improved cotton varieties contributes to low cotton production, so adoption of Bt cotton can play an important role in the textile industry, particularly in creating job opportunities. Nigeria is one of many African countries that initially exhibited a precautionary approach towards the adoption of GM agriculture. This was obvious with the biosafety bill, which was delayed for several years before finally being approved by the last-but-one President in April 2015. With Presidential assent to the bill and positive results from CFTs on some food commodity crops, the country is now poised for commercialisation of GM crops, including Bt cotton. Bt cotton is likely to be adopted first as the research team working on CFT has made a lot of progress. The regulation of GMOs is one of the major constraints on adopting GM products in Nigeria. The efficient regulation and adoption of Bt cotton in Burkina Faso is a good example to other countries in Africa. Part of this is due to the fact that limiting the number of government institutions involved in regulatory frameworks concerning agricultural biotechnology has the advantage of requiring less interaction among ministries, leading to a more efficient and timely decision-making procedure for processing applications for permits. Nigeria's National Biosafety Management Agency (NBMA) was created shortly after the passing of the biosafety bill which had been delayed for five years (2011–2015). In contrast, in Burkina Faso, rapid adoption of GM technology was facilitated by political will and enabling legal frameworks, in the form of decrees, to authorise the establishment of CFTs, the creation of the agency for regulation of the technology and a ready market for the products (Traoré *et al.*, 2014b).

Burkina Faso also has a well-coordinated approach to value-chain management for introducing Bt cotton to the farmers. The aforementioned three national cotton seed companies in Burkina Faso are officially engaged in commercial production and distribution of Bt cotton seeds to the farmers. The human capacity of the local seed companies has therefore been enhanced, with concomitantly less dependence on foreign seed companies for seed multiplication. Nigeria and others can learn from Burkina Faso how to form a strong partnership and strengthen their cotton sector when Bt cotton becomes available in the future. Royalty negotiation or sharing of the benefits of the technology was based on mutual agreement among the cotton industries, the technology developer and the government to facilitate better access and improve the livelihood of farmers in Burkina Faso (James, 2014). Moreover, the standard operating procedures for seed production and distribution in Burkina Faso play an important role in the negotiation involved in buying back the seeds from the farmers, serving as great incentives for the rapid adoption of GM technology (Traoré *et al.*, 2008, 2014a). However, there is a need for continuous stewardship for the Bt cotton production by the developer and local seed companies to ensure that unadulterated products are delivered to the farmers and to ensure the sustainability of the benefits of the technology to all stakeholders. It is also important for the Bt cotton produced in Burkina Faso to remain competitive; this can be achieved by strengthening the value chains for sustainable markets locally, regionally and globally so as to maintain the competitiveness of the cotton sector. While the adoption of Bt cotton has benefited from good policy, more support is required from the government to facilitate the development of new GM crop varieties.

Challenges Facing the Adoption of GM Technology in the Sub-region

Burkina Faso

The initial hurdles preventing the early adoption of GM technology were partly due to lack of public participation and the inclusion in the biosafety law of some clauses that were not investor-friendly. This was cited as one of the main reasons why parliament had to return it for further public consultation (Savadogo, 2014). Although there were some indications of the impending danger of a collapse in the Burkina Faso cotton industry, stringent conditions were applied during the CFTs, resulting in extended trial durations of over five years (2003–2008). However, the initiation of the adoption process was farmer-driven, with the support of the cotton industry and the political will of the government (Vitale, 2010). Aside from setbacks caused by overly stringent regulation, some cotton growers' associations identified variation in inputs and seed prices as factors that had affected cotton prices (Traoré *et al.*, 2014b). Coexistence of the three cotton crops (Bt cotton, conventional cotton and organic cotton) in the same environment was another challenge that the adoption of the GM technology faced at the onset, followed by the issue of provision of 'refugia' of approximately 20% of the total land cultivated, which was somewhat impractical with smallholdings of area less than one acre. The reported complete phasing out of GM cotton in Burkina Faso, on account of inferior lint quality (Dowd-Uribe and Schnurr, 2016), is not a failure of the technology but due to other, extraneous factors. It is expected that the technology developer, in this case Monsanto, will find a solution soon.

Nigeria

Fear of loss of biodiversity in Nigeria has been identified as one of the constraints on the adoption of GM technology (Adenle, 2014). One of the major concerns expressed by a Nigerian environmental group and researchers was the fear of gene flow from GM crops into indigenous wild relatives. In fact, some Nigerian scientists are still concerned about the potential environmental impact of GM crops when they become commercialised in the future (Adenle, 2014). Other scientists have argued that it may be too early to talk about a negative impact of GM crops as there is no evidence from other countries that have adopted the technology to justify this argument, and conventional crops may pose similar risks. The poor understanding of GM technology and the perceived potential risks to human/animal health and the environment have also contributed to slow adoption of GM technology in other African countries. For example, Kikulwe *et al.* (2011), writing on the banana in Uganda, and Bett *et al.* (2010), writing on the food industry in Kenya, expressed concerns about the potential negative impacts on health and the environment. However, evidence gathered over the past 15 years suggests that biotech crop cultivation has helped to conserve biodiversity by saving 132 million hectares of land from degradation (Carpenter, 2011; James, 2014).

It is our considered opinion that fibre quality and ginning ratios may not be of that much concern to other African countries such as Nigeria whose sole aims in the adoption of Bt cotton technology are to increase yield, reduce pesticide usage and reduce the deleterious effects of pests and diseases on the crop. Despite all the concerns from Burkina Faso, the Nigerian NBMA recently granted approvals to Monsanto Agriculture Nigeria Limited for the commercial release/placement on the market of Bt cotton (MON 15985) and for multi-locational trials of maize (NK603 and MON 89034 × NK603), which has been genetically modified for insect resistance and herbicide tolerance.

Investment in Research and Development Is the Key to Adopting GM Technology

West African countries currently conducting research on GM crops should emulate other advanced developing countries such as Brazil and China that have been cultivating GM crops for almost two decades (see Chapters 18, 19 and 22). Those two countries have adopted a clear agricultural research and development (R&D) policy that integrates modern biotechnology such as GM technology in the agricultural sector. Strong agricultural biotechnology programmes were set up through increased investment in R&D systems that enhance the capacity (both human and physical) of the scientific community, particularly at the national level. Public funding in agricultural R&D programmes is very limited in many African countries, including Burkina Faso and Nigeria, whereas, in Brazil, public funding in the field of modern biotechnology continues to increase and is directed towards solving the problems faced by the agricultural and health sectors.

West African countries keen on GM technologies must increase investment in modern biotechnology R&D to facilitate the development of different types of local GM varieties, particularly in addressing fundamental challenges that are associated with the

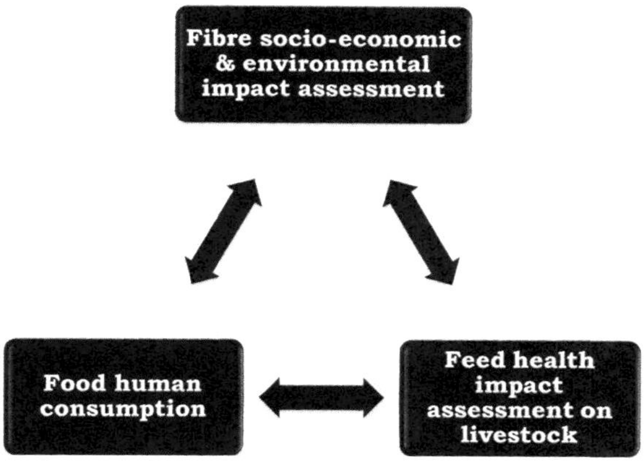

Figure 23.2. Introduction of GM technology through the Food–Feed–Fibre (F3) approach.

problems of low agricultural productivity and food security. Institutional and regulatory mechanisms that affect agricultural R&D must be put in place to enhance the application of modern biotechnology in agricultural production.

In a nutshell, Burkina Faso, Nigeria and other African countries must prioritise agricultural biotechnology policy in order to target and strengthen key research institutions and universities and provide adequate support to encourage the application of modern biotechnological techniques. Specifically, building human capacity, such as by the training of young scientists, provision of a conducive working environment, increasing agricultural R&D and carrying out infrastructure development can put the continent on the same path as Brazil and China.

The Way Forward

Burkina Faso and Nigeria, like many any other African countries, have yet to commercialise GM food crops. This is partly due to unsubstantiated claims regarding the so-called potential risks associated with GM foods and the lack of scientific expertise to regulate these GM crops. In view of potential environmental and health risk concerns, the Food–Feed–Fibre (F3) approach (Figure 23.2) has been proposed as the way to introduce GM products in Africa (Adenle *et al.*, 2013). The F3 approach encourages the adoption of Bt cotton first, followed by GM livestock feed that must undergo rigorous risk assessment before production of GM foods for human consumption, as stated earlier. That is, 'the approach is essentially designed to familiarize farmers and public with the new technology and to allay concerns about potential risks of GMOs' (Adenle *et al.*, 2013).

The experience and benefits associated with the adoption of Bt cotton in Burkina Faso indicate that the application of GM technology is very important in tackling some of the fundamental challenges associated with low agricultural productivity and improving the socio-economic status of resource-poor smallholder farmers in Africa. To overcome some of these challenges, African national governments need to increase the level of expertise in biosafety areas and strengthen the scientific community by investing in modern biotechnology R&D programmes. Moreover, stringent international regulation, particularly relating to the Cartagena Protocol on Biosafety (CPB), presents a significant challenge to countries in sub-Saharan Africa. Also, the current interpretation of the CPB, especially with regard to the precautionary principle, hinders the application and release of GM products in the continent. Therefore, regulatory provision that encourages less expensive and quicker assessment and safe evaluation of local GM varieties is needed in order to address these challenges. An agreed and harmonised regulatory regime among the ECOWAS member states is required in order to support the development and delivery of new technologies to African countries.

Some stakeholder groups, as described by Adenle *et al.* (2013), recommended a centralised approach to risk assessment modelled on that of the European Union as it would help reduce the cost of risk assessment in GMO approvals for each country. The regional economic bodies of West Africa, namely the West Africa Economic and Monetary Union (WAEMU, eight member states) and ECOWAS (15 member states, including all of the WAEMU states), together with the Permanent Interstates Committee for Drought Control in the Sahel (Comité Permanent Inter-États de Lutte contre la Sécheresse dans le Sahel (CILSS, 13 member states) are engaged in jointly developing a regional biosafety framework. The process started in 2007 but has not yet been concluded, and the framework is now being profoundly amended. WAEMU is leading the process. If the framework is adopted, it will apply to all member states and may supersede the national laws. This is unlike the regional biosafety policy framework of the Common Market for Eastern and Southern Africa (COMESA) which shows flexibility by focusing on policies and guidelines while providing opinion to national biosafety bodies. In focusing on regulations, the West African biosafety framework holds a high potential to influence national decisions and, if properly harmonised, will be of great benefit to the sub-region. The challenge to regional biosafety harmonisation is that member states in the regional economic communities, such as ECOWAS, take different positions on GMO issues. The idea of a centralised or harmonised regulatory approach is not impossible if there is political will and there exist good institution(s) to coordinate it. When a good system is put in place, it will encourage many countries in the West African sub-region that are discussing a roadmap for a new biosafety regulatory system or revising current biosafety policy. For example, West African countries such as Togo and Senegal are revising their biosafety laws to facilitate the introduction of new technology in view of the successes with Bt cotton recorded in Burkina Faso (Savadogo, 2014). This is an indication that a better and well-organised biosafety regulatory framework can pave the way for many more countries in Africa to adopt GM technology. In addition to addressing biosafety challenges, education and public awareness regarding the benefits and potential risks of GMOs and more participation of local scientists are fundamental to encouraging the adoption of GM crops (Adenle, 2014).

The role of public–private partnerships (PPPs) in the development of GM crops and their introduction to the farmers and consumers in Africa cannot be overemphasised. The evidence from Brazil has demonstrated that the successful adoption of GM crops during the past two decades occurred due to the increasing role of PPPs in that country. Both the public sector and the private sector can play crucial roles in facilitating the rapid development and diffusion of new technologies, particularly among farmers. The involvement of other key stakeholders, especially regional economic communities, is also important. The NEPAD Agency ABNE, which works with the national focal points in at least 12 member states in Africa and brings together key actors, including farmers, technology developers and local seed industries, to ensure that GM crops under CFTs are successfully brought to commercialisation, is a good example. This will help the PPPs to achieve their goals and sustain product integrity through stewardship within the countries. Nigeria, being the largest economic power in Africa and the most populous African country, needs to get its food security right. Then the whole continent of Africa will be close to achieving food security.

References

Adenle, A. A. (2014). Stakeholders' perception of GM technology in West Africa: assessing the responses of policymakers and scientists in Ghana and Nigeria. *Journal of Agricultural and Environmental Ethics* **27**(2), 241–263.

Adenle, A. A. *et al.* (2013). Status of development, regulation and adoption of GM agriculture in Africa: views and positions of stakeholder groups. *Food Policy* **43**, 159–166.

Bennett *et al.* (2004). Economic impact of genetically modified cotton in India. *AgBioForum* **7**(3), 1–5.

Bett, C. *et al.* (2010). Perspectives of gatekeepers in the Kenyan food industry towards genetically modified food. *Food Policy* **35**, 332–340.

Brookes, G. and Barfoot, P. (2012). *GM Crops: Global Socio-economic and Environmental Impacts 1996–2010*. Dorchester: PG Economics Ltd.

Carpenter, J. E. (2011). Impacts of GM crops on biodiversity. *Landes Bioscience* **2**, 1–17.

Dowd-Uribe, B. and Schnurr, M. A. (2016). Briefing: Burkina Faso's Reversal on Genetically Modified Cotton and the Implications for Africa. *African Affairs* 1–12. [Online]. Available from http://afraf.oxfordjournals.org

James, C. (2013). Global Status of Commercialized Biotech/GM Crops: 2013. ISAAA Brief No. 46. Ithaca, NY: ISAAA.

James, C. (2014). Global Status of Commercialized Biotech/GM Crops: 2014. ISAAA Brief No. 49. Ithaca, NY: ISAAA.

James, C. (2015). Global Status of Commercialized Biotech/GM Crops: 2015. ISAAA Brief No. 51. Ithaca, NY: ISAAA.

Kikulwe, E. M. *et al.* (2011). Attitude, perceptions, and trust. Insights from a consumer survey regarding genetically modified banana in Uganda. *Appetite* **57**(2), 401–413.

NBS (2015). Nigeria in 2014: Economic Review and 2015–2017 Outlook. [Online]. Available from http://sparc-nigeria.com/GREAT/files/5-Economic-Update-source-materials/NBS-Economic-Review-2014-and%20Outlook-2015.pdf

Nin-Pratt, A. *et al.* (2011). Yield gaps and potential agricultural growth in West Africa and Central Africa. Washington, D.C.: International Food Policy Research Institute.

OECD (2006). Cotton – West Africa in the international market. [Online]. Available from www.oecd.org/swac/publications/38409410.pdf

Savadogo, K. (2009). Le contexte macroéconomique. Background paper for the Burkina Faso Country Economic Memorandum. Washington, D.C.: World Bank.

Savadogo, M. (2014). Case study – Burkina Faso. In *Biosafety in Africa: Experiences and Best Practices*, ed. D. Keetch *et al.* Michigan, MI: Michigan State University Press, pp. 82–88.

Traoré, H. *et al.* (2014a). Agricultural R&D in Burkina Faso. An assessment of the Environment and Agricultural Research Institute. [Online]. Available from www.asti.cgiar.org/pdf/coraf/BurkinaFaso-SummaryNote.pdf

Traoré, H. *et al.* (2014b). Bt cotton in Burkina Faso demonstrates that political will is key for biotechnology to benefit commercial agriculture in Africa. In *Biotechnology in Africa: Emergence, Initiatives and Future*, ed. F. Wambugu and D. Kamanga. Basel: Springer International Publishing Switzerland, pp. 15–36.

Traoré, O. *et al.* (2008). Testing the efficacy and economic potential of Bollgard II under Burkina Faso cropping conditions. *The Journal of Cotton Science* **12**, 87–98.

Vitale, J. D. (2010). The commercial application of GMO crops in Africa: Burkina Faso's decade of experience with Bt cotton. *AgBioForum* **13**(4), 320–332.

Vitale, J. D. *et al.* (2011). Enhancing sustainability of cotton production systems in West Africa: a summary of empirical evidence from Burkina Faso. *Sustainability* **3**, 1136–1169.

24 A Pathway for Biosafety Regulation of GMOs in Sub-Saharan Africa

Julius Ecuru

Introduction

Scholars argue that, with the right policy and regulatory environment, Africa is likely to benefit more from the 'gene' revolution of the twenty-first century than from the 'green' revolution of the early twentieth century, which it missed (Koch, 2014). The Green Revolution was characterised by increased use of agricultural inputs, especially fertilisers, mechanisation, irrigation and new high-yielding crop varieties. Africa could not tap into these opportunities because of the poor nature of its subsistence farmers, and the lack of infrastructure, capital and skill to deploy new farming techniques (Pingali, 2012). The 'gene' revolution is more knowledge-intensive, and encompasses a broad mix of modern bioscience tools and the use of living organisms as a means of production (Pingali and Raney, 2005). Keese *et al.* (2002) noted that, in Africa, 'molecular biotechnology could address much more significant concerns such as food security, poverty, malnutrition and disease'. Similarly, Juma (2011) observed that 'Africa's economic transformation is starting in the new age of biology'. This optimism is expressed in different forms in a number of African policies and strategies (Juma, 2011). Within the African Union, for example, frameworks such as the Comprehensive African Agriculture Development Programme (CAADP) and the Science, Technology and Innovation (STI) Strategy for Africa 2024 point to the potential gains from the gene revolution, which includes the development and use of genetically modified organisms (GMOs) (New Partnership for Africa's Development, 2003; African Union Commission, 2014). Despite providing the overarching policy framework, these strategies have yet to make a reasonable impact on introducing genetic modification (GM) technology in African countries.

It is with this growth and development perspective that this chapter discusses the pathway of biosafety regulation in Africa. Most sub-Saharan African countries have a goal of transitioning from peasant to modern and more commercially orientated farming. Additionally, the pressing challenge of ensuring that there is enough safe food for everyone in the continent, considering the demands of a rapidly rising population, is compelling most African countries to look for alternatives, including GMOs for specific needs. It has been estimated that Africa's population by 2050 will be more than two billion people (United Nations Department of Economic and Social Affairs Population Division, 2015); thus the ability to produce more food from less land, and ensure that everyone has a proper meal, is a challenge for the continent. In all fairness, therefore, the gene revolution, whose main products are GMOs, is expected to play an important part in Africa's agricultural transformation and contribute to realising the aspiration of most Africans for an integrated, prosperous and peaceful continent by 2063 (African Union Commission, 2016). GMOs are defined elsewhere in this book, but, broadly, the term refers to living organisms (plants or animals) whose genomes have been artificially changed to express certain desired traits (Alexandrova *et al.*, 2005). Biosafety regulations are the measures which governments put in place to prevent or manage risks that GMOs may pose to human health and the environment.

The irony, however, is that, for most of the continent, biosafety regulation is at variance with continental aspirations of gaining value from the development and use of GMOs. Often the biosafety regulatory systems are not in sync with advances in GMO applications. This apparent contradiction may be explained by the limited willingness and ability of the relevant stakeholders in the continent to take biosafety regulation as an integral part of the wider gene technology innovation system, which should connect with the continent's goals for economic growth and environmental sustainability. Arguably, GMOs are not the only, but merely the most prominent, products of a functioning gene technology innovation system.

This chapter, therefore, emphasises the need for biosafety regulation to be viewed as an integral element of the gene technology innovation system, if sub-Saharan Africa is to benefit from the gene revolution. The chapter addresses the development of biosafety regulation in the context of the continent's shared vision for growth and environmental sustainability. It suffices to say that countries in Africa such as Sudan, Burkina Faso and South Africa, which have commercialised GMOs, are applying the right framework of positioning biosafety regulation as a functional element of their emerging gene technology innovation systems. The role of biosafety regulation in the gene technology innovation system is to ensure that GMOs pose negligible risk to human health and the environment. This implies that the effectiveness of the biosafety regulatory process should be measured by its level of impartiality, transparency and efficiency.

The next section in this chapter introduces the concept of innovation systems in relation to biosafety regulation. Subsequent sections discuss various dimensions of biosafety regulation and its co-evolution in the continent. The last section gives some suggestions on the way forward.

Biosafety in the Context of the Gene Technology Innovation System

The concept of an innovation system has been applied mainly in discussions on technical change, economic growth and industrialisation. The concept was first proposed by Lundvall, Nelson and Freeman in the late 1980s and early 1990s (Lundvall, 2010). Essentially, what it means is that, for innovation to happen, actors should interact with and learn from each other. 'Actors' here may refer to individuals or organisations involved in the process of knowledge production, use and diffusion. The relationship between the actors is guided by certain traditions, policies, laws or standards, i.e., what some scholars commonly refer to as the 'rules of the game' (Edquist and Johnson,

1997). In this regard, biosafety regulation would be considered part of the 'rules of the game' that developers or users of GMOs should abide by. Some scholars, such as Etzkowitz, take the concept a little further by emphasising that the triple helix of university, industry and government relations is absolutely essential for a meaningful translation of knowledge into economic benefit (Etzkowitz, 2002). Therefore, the concept of innovation systems can be used as a focusing device to explain technological changes, such as advances in GMO development and use. But innovation systems also exist in reality, for example, the interactive learning and relationships that individuals and organisations engage in in the development and use of GMOs. It is worth noting here that, with respect to GMO development, the multiplicity of actors usually extends beyond national boundaries.

Biosafety regulation is normally exclusively the role of government, but the task of ensuring biosafety regulation in the laboratory or the field involves the practitioners of the technology. Therefore, in the triple-helix dispensation, the responsibility for biosafety regulation is shared between the government and technology developers to ensure appropriate use and diffusion of GM technology to meet the countries' aspirations of growth and development. This is important because GMOs should neither harm people nor endanger the environment. This becomes the social contract that government has with GMO developers on behalf of the people to protect human health and the environment. Biosafety regulation, conceived this way, thus becomes an important aspect of legitimising GMOs. In other words, a biosafety regulatory system provides assurance both to developers and to consumers that a GMO has been evaluated to have negligible risk to human health and the environment. Taken this way, biosafety regulation, therefore, should ideally not be used as an impediment to the progress of GMO development and use; rather it should be a neatly woven measure that defines the 'rules of the game' in the gene technology innovation system.

Dimensions of GMO Development in Africa

GMO development and use in Africa have two dimensions. The first is the promise of GMOs to provide solutions to key challenges of food and nutrition security, and help build resilience to climate change. Though in fact it is better understood that GMOs are not necessarily a panacea to the aforementioned challenges, but rather one of the technologies in the 'tool box' offering only part of the solution. So far, most of the developers of GMOs in Africa have been scientists in the public agricultural research organisations, often working in partnership with donors abroad, including international agricultural research organisations and private life-science companies or foundations. The tendency in most sub-Saharan African countries has been to promote GMOs for the public good. This is evident, for example, in the type of GM crops being developed in Uganda, such as sweet potato, bananas, cassava and maize. These crops are grown mainly by subsistence farmers, and therefore any promising interventions with GMOs, e.g., pest- and disease-resistance or drought- and herbicide-tolerance traits, are being tailored to the needs of the smallholder farmers.

The second dimension is the potential for GMOs to support and enhance the competitiveness of the local industry. Here, the insect-resistant maize (Bt maize) grown in South Africa and the insect-resistant cotton (Bt cotton) grown in Burkina Faso could be good examples, where the private sector and farming communities alike could gain economic value through the commercial cultivation of these GM crops. Often this commercial advantage, when managed well, becomes the driving force for investment in and adoption of GMOs, with the goal of having less inputs, more yield and a more profitable enterprise. *Bacillus thuringiensis* (Bt) is a common naturally occurring soil bacterium, which produces crystalline proteins that are toxic to certain types of insect. Hence the GM maize and GM cotton traits carry a Bt gene, which produces a toxic protein that kills the common pests on these crops, namely the maize stalk borer and cotton bollworm.

From the two aforementioned scenarios, it appears that the pathway for biosafety regulation in Africa is influenced by the inclination towards either the public good or the higher commercial value of GMOs. Biosafety regulatory systems seem to develop faster in countries with a stronger commercial or economic focus in the utilisation of GMOs than in countries that promote GMOs purely for the public good. This is probably due to the financial and economic incentives involved. However, regardless of where the inclination lies, the common factor in both scenarios is the importance of concentrating efforts on proper risk analysis to ensure that GMOs of interest pose negligible harm to human health and the environment. Another consideration is the need to allow biosafety regulation to co-evolve with progress in the development of GMOs, e.g., through research and experimental testing. These aspects are explained further below. First, a brief look at the history of biosafety regulation in Africa is provided in the next section.

Africa, a Latecomer in Biosafety Regulation

Africa, with the exception of South Africa, came to biosafety regulation somewhat late. Biosafety regulation of GMOs became prominent in the rest of Africa only in the early 1990s, presumably because of the negotiations leading to the Convention on Biological Diversity, and it became even more so in the late 1990s, with field experiments starting in some countries, such as Kenya and Uganda, and approval for commercial releases of GM cotton and GM maize in South Africa in 1997. Elsewhere in the developed world, development of GMOs started way back in the 1970s and 1980s (Cohen, 2013). By 1980, recombinant human insulin (humulin) was being produced (Ladisch and Kohlmann, 1992). In the developed world, e.g., in Australia, GMOs were already being regulated to some extent by the Recombinant DNA Committee of the Academy of Sciences. By the 1990s successful experiments were being done to produce GM crops, e.g., potatoes, tomatoes, etc. The very first commercial releases of GM maize in Africa were in 1998 in South Africa (Alexandrova *et al.*, 2005). It was only then that governments in other African countries began to put in place explicit biosafety frameworks. In South Africa, for example, work to develop the GMO Act started in 1994, and it was passed into law in 1997 (Jaffe, 2004). Even after the passage of several decades, biosafety regulation in most of sub-Saharan Africa is still lagging somewhat behind advances in GMO development. In most cases, biosafety regulation has been developed as a direct response to controversies over GMOs. The downside to this is that most sub-Saharan African governments end up being reactive, instead of proactive in developing their biosafety regulation as an integral part of the gene technology innovation system. The Zambian government, for example, in 2002, banned importation of genetically modified food aid even though 2.5 million people in Zambia faced starvation due to a famine (Knight, 2002). At about the same time, Tanzania and Ethiopia adopted strict liability regulations for dealing with GMOs.

Therefore, for the most part, biosafety regulation in Africa became more of a deterrent to the advancement of GMOs. This may be partly explained by the fact that, within sub-Saharan Africa, biosafety regulation did not initially co-evolve with GMO development as appears to have been the case in the developed world. The exception to this is South Africa, where it was felt that there was no legislation at the time that would cover a field trial of GM strawberries that had been developed within South Africa; this, to a large extent, triggered the development of the South African GMO Act. Subsequently, a number of African countries rushed into making biosafety regulations without real-life experience with GMOs, and often without due regard to the science behind GMOs (this was especially true of the countries which developed GMO legislation in the absence of any ongoing research work on GMOs and of any pipeline of GM products). Some countries simply adopted the biosafety regulations of other countries without sufficient consideration of contextual differences. Additionally, and perhaps more importantly, biosafety regulation was seldom viewed as an integral part of the gene technology innovation system, save for in a few countries as discussed below.

Therefore, by the time biosafety regulation was being developed in a number of African countries, the debate on GMOs had already become polarised on two fronts. On the one front, the promoters of GMOs, supported by industry, portrayed GMOs as the solution for Africa's food crop problems. Some anti-GMO activists, on the other extreme end, painted GMOs as a monstrous and evil human endeavour, which should be fiercely fought and banned. So, the reality is that the pro- vs. anti-GMO debates, which were mainly between Europe and the USA and largely fuelled by trade rivalry and moral arguments about the need to preserve nature in its intrinsic form, have had significant influence on the development of biosafety regulation in sub-Saharan Africa. Instead of focusing on the real concerns of protecting human health and the environment from potential risks posed by GMOs, most biosafety regulations emphasised socio-economic issues and stringent legal requirements for dealing with GMOs. This situation was escalated by the position of the African chief negotiators of the Cartagena Protocol on Biosafety (CPB) at the time, and the ensuing African Union Model Law on Biosafety, which vehemently emphasised the precautionary approach (Chambers, 2013). Consequently, most sub-Saharan African countries ended up with biosafety legislation (or no legislation at all) that contradicted their countries' aspirations for economic growth and environmental sustainability. Fortunately, some of these countries, such as Ethiopia and Tanzania, have recently embarked on reviewing and amending their biosafety regulatory frameworks, making these frameworks a vital part of their gene technology innovation system.

Allowing Biosafety Regulation to Co-evolve with GMO Development: The Ugandan Example

Early in 1992, a researcher from Makerere University, Uganda's biggest public university, made an application to the Uganda National Council for Science and Technology (UNCST) for a study of recombinant bovine somatotropin (rBST) in milk production in cows. This hormone is produced naturally in the pituitary glands in cattle. However, larger volumes of it could now be produced more cheaply through genetic modification, and artificially given to cows to enhance milk production.

The UNCST is a government agency that, among other things, oversees the conduct of research and development in the country. In Uganda, researchers obtain a permit from the UNCST to conduct research and experimental development. Therefore, the aforesaid application was made as a regular research application to the UNCST. At the time, the UNCST did not have specific guidelines for dealing with GMOs or GMO products. This submission of a GM-derived BST triggered the biosafety regulatory process in Uganda. It became necessary for the UNCST to develop guidelines for review and approval of experiments and studies involving GMOs. However, the situation was complicated by the debate in Europe over the use of GMOs. The European Union had decided to place a moratorium on the marketing of rBST in 1990 (Brinckman, 2000). Around the world, the United Nations Convention on Biological Diversity (CBD) had been concluded and preparations for the negotiation of the CPB were under way. This posed a huge challenge to the leadership of the UNCST. They had to decide either to take the extremely cautious route, probably with a ban on GMOs, or to be more pragmatic and establish how the risks posed by GMOs could be reasonably and adequately managed. The risks of GMOs, about which there was a heated debate (which continues to this day), included the possibility that GMOs may turn into 'superweeds' and become invasive, and that some GMO products may cause allergies. Some argued that antibiotic resistance may spread, since antibiotics such as kanamycin were being used as selectable markers in GMO experimentation and production processes. There were also moral questions about humans tampering with nature, and big multinationals enslaving African farmers by requiring the latter to keep on buying fresh seed, and so on and so forth. The leadership of the UNCST, at that time, however, decided to take a pragmatic route, which allowed biosafety regulation to co-evolve alongside GMO development. National biosafety guidelines were subsequently developed, and a national biosafety committee was established in 1996. By 2000, Uganda, with the support of the United Nations Environment Programme (UNEP) working under the Global Environment Facility (GEF), had a functional biosafety framework, which allowed experimentation with GMOs (Sengooba *et al.*, 2005). The existing legislation, i.e., the UNCST Act of 1990, and the interpretations of the sections thereof, provided the legal basis for the experiments. To date, Uganda has approved more than 15 events of field experiments

Table 24.1 Authorised field trials of genetically modified crop events in Uganda

Crop	Trait	Year authorised
Banana	Resistance to black Sigatoka disease	2006
Cotton	Resistance to insect (bollworm) and resistance to herbicide (Roundup)	2008
Banana	Nutritional enhancement (provitamin A and iron)	2009
Cassava	Resistance to cassava brown streak disease	2010
Maize	Tolerance to drought	2010
Banana	Resistance to banana bacterial (Xanthomonas) wilt disease	2010
Cassava	Resistance to East African cassava mosaic virus and African cassava mosaic virus	2011
Cassava	Resistance to cassava brown streak disease (multi-site trials under confinement)	2012
Banana	Resistance to nematodes	2012
Maize	Resistance to insects (maize stem borer)	2012
Rice	Tolerance of low soil nitrogen and tolerance of drought	2012
Sweet potato	Resistance to sweet potato virus disease	2013
Potato	Resistance to potato late blight disease	2015

Source: UNCST database, 2016.

with GM crops, most of them with promising results (see Table 24.1). This is besides clinical trials of recombinant human HIV and Ebola virus vaccines, and laboratory experiments in contained facilities. However, even with this seemingly impressive progress, Uganda has yet to commercialise GMOs. Nonetheless, importation of humanitarian food aid that may contain GM products has been allowed since 2003 using the existing biosafety mechanism. These products, which have mainly been cooking oil, corn and soybean, must be in milled or processed form to be allowed into the country, i.e., they should contain no viable seeds or materials.

Although a biotechnology and biosafety policy that supports the safe development and use of GMOs was passed in 2008, it has taken nearly a decade to debate and pass an explicit biosafety law. While the government recognises the potential benefits of GMOs and the need to regulate GMOs, anti-GMO activism has often slowed down the passage of the biosafety bill. It is also worth noting that most of the GMOs in development in Uganda are food-security crops (bananas, cassava, potatoes), and there is a visibly weak local private sector to advance commercial interests. Where strong trade or commercial prospects are imminent, biosafety regulation would develop much faster. Nevertheless, the Ugandan example illustrates the point that positioning biosafety regulation within the gene technology innovation system can enable progress with GMO development.

Taking a Risk-Based Approach

A risk-based approach to biosafety regulation presupposes that there is a risk source, which follows a causal pathway to potentially bring about harm (Office of the Gene

Technology Regulator, 2005). Therefore, it is important to identify the possible harms GMOs could bring and the likelihood of those harms becoming a reality, and then to develop strategies to mitigate exposure to such harms. The extent of harm or damage should also be assessed against attributes of conventional comparators, and in accordance with international laws such as the CPB. For example, the effect of a Bt gene in cotton on non-target insects can be compared with the conventional practice of spraying synthetic insecticide that also kills non-target insects. A risk-based approach in this situation would therefore compare how much additional risk to non-target organisms Bt cotton would pose compared with a conventional crop that is sprayed with synthetic chemicals. Identifying the risk context is crucial, as is mapping out what the harm could be, and figuring out ways to manage that harm. However, any risk, whether actual or theoretical, also needs to be balanced against the benefits of such a crop. In the case of Bt cotton the benefits may include higher yield and the reduced environmental impact of lower pesticide use, plus the health benefits when farm workers no longer need to spray hazardous chemicals on their crops. Therefore, even if there is a proven risk, this risk may be more than counterbalanced by the benefits of the technology.

This was probably also the spirit behind the CBD and CPB. In fact, Article 16 of the CBD urges state parties to 'facilitate access for and transfer of technologies (which include biotechnology) that are relevant to the conservation and sustainable use of biological diversity, or make use of genetic resources and do not cause significant damage to the environment'. Similarly, Article 2.2 of the CPB urges states to deploy GMOs in a manner that prevents or reduces risks to biological diversity, taking into account also risks to human health. The CPB further provides Annex 3 as a framework for assessing the risks of GMOs. Therefore, an essential element of a biosafety regulation system is the establishment of a risk assessment framework that adequately identifies and manages any risks associated with GMOs. Certainly it is in the best interest of everyone (whether industry or consumers) to have a proper and transparent risk analysis regime to identify and duly manage any adverse effects of GMOs.

In most sub-Saharan African countries, however, risk analysis frameworks have been weak, and most are still evolving. Legislation of most countries adopted a preventative approach to risk, which is an extreme interpretation of the precautionary principle, stated in Principle 15 of the June 1992 Rio Declaration on Environment and Development that, 'where there are threats of serious or irreversible damage, lack of full scientific certainty shall not be used as a reason for postponing cost-effective measures to prevent environmental degradation'. While the CBD and CPB are clear on what the outcome of risk assessment should be, namely minimising risk to biological diversity and human health, in most countries socio-economic, moral and other ethical issues with respect to GMOs appear to be more prominent. However, while the morality of using GMOs and questions about social and economic considerations are important societal concerns, they should, arguably, not be the basis for decision-making when one is following a risk-based approach. In a risk-based approach, which supports biosafety regulation as an integral part of the gene technology innovation system, governments assume the important role of establishing and ensuring a credible and transparent risk analysis framework. Another part of the gene technology innovation system, therefore, should address socio-economic and moral concerns, which in essence are difficult to regulate in practice. Markets would influence and dictate the adoption of GMOs, while societal norms, traditions and beliefs would determine the acceptability of the technology. Issues concerning risks to human health and the environment are different from socio-economic and moral value judgements, and therefore should be less emphasised in a biosafety regulation that follows a risk-based approach. Socio-economic concerns and ethical issues can be better addressed by other actors in the gene technology innovation system, normally by governmental organs dealing with trade and consumer issues. However, in Africa, the practice has been to insert socio-economic issues within the biosafety regulatory frameworks, which in part could be attributed to the CPB, which makes provision for inclusion of socio-economic assessments without giving clear guidance on how countries could pursue such an approach. Consequently, most biosafety regulatory frameworks were conveniently skewed towards socio-economic and moral considerations and became obstacles to the development and use of GMOs rather than mechanisms for judicious assessment of risks to human health and the environment.

Agreeing on an Institutional Model That Best Fits

Most countries in sub-Saharan Africa are grappling with choosing what institutional model to adopt for biosafety regulation. In most countries, it is felt that existing laws are insufficient to address the biosafety of GMOs. Consequently, some countries have opted to have explicit legislation for biosafety, and to establish separate bodies that will regulate GMOs. Kenya, for example, established the National Biosafety Authority in the Ministry of Higher Education, Science and Technology. Tanzania developed specific regulations that are based on the National Environment Act in the Office of the Vice President. Ethiopia made a proclamation under the Prime Minister's Office. Uganda is yet undecided as to which model to use, but currently maintains a distributed model where sectoral agencies have specific responsibilities with respect to GMOs. Each country has developed a system it feels is appropriate in that country's context. However, most of the systems in place lack clear institutional responsibilities among the departments/ministries, which often causes disagreement, leading to difficulties in decision-making on applications for research or general release of GMOs. Adenle *et al.* (2013) mention such disagreements between the Ministry of Environment and Ministry of Agriculture over the regulation and use of GMOs in Egypt. Part of the problem has been that the CPB arose from the CBD, hence ministries responsible for the environment frequently take the lead in biosafety; and yet biosafety is a cross-cutting issue spanning many institutional mandates and priorities, including health and agriculture.

There is no generally perfect institutional model, but each will be tested by the challenges presented by research on GMOs and product development and commercialisation of GMOs. Perhaps, then, an institutional model that best responds to these challenges will evolve in due course. This notwithstanding, it appears likely that, as more familiarity is gained with GMOs, and as private commercial interests in GMOs begin to grow, most countries will adopt distributed systems that enable each responsible sectoral agency to play a part in biosafety regulation. More centralised systems of biosafety regulation are likely to seek increased participation of sister agencies, while the sister agencies will increasingly demand to play their traditional roles in regulating GMOs or products thereof that fall within their respective jurisdictions. It may be prudent, therefore, to promote learning about biosafety regulation as GMO development and use spreads in the continent.

Conclusion

Capturing and creating value from GMOs depends on whether or not a country positions biosafety regulation as an integral part of its gene technology innovation system. It means that different actors can play certain roles within the system. It also means that interaction among actors and learning by actors underpin the functioning of such a system. Such a viewpoint allows other actors within the system, such as trade, industry and consumer organisations, to address more diffuse issues such as value judgements on socio-economics and moral concerns relating to the adoption of GMOs. In the gene technology innovation system, relevant sectoral agencies, such as the food standards agencies, plant quarantine services, regulatory authorities for medicines and environment protection agencies, should possess internal capabilities for dealing with GMOs.

References

Adenle, A. A. *et al.* (2013). Status of development, regulation and adoption of GM agriculture in Africa: views and positions of stakeholder groups. *Food Policy* **43**, 159–166.

African Union Commission (2014). *Science, Technology and Innovation Strategy for Africa 2024*. Addis Ababa: African Union Commission.

African Union Commission (2016). Agenda 2063: The Africa We Want. [Online]. Available from http://agenda2063.au.int/en/home

Alexandrova, N. *et al.* (2005). Biosafety regulations of GMOs: national and international aspects and regional cooperation. *Biotechnology & Biotechnological Equipment* **19**(suppl. 3), 153–172.

Brinckman, D. (2000). The regulation of rBST: the European case. *AgBioForum* **3**(2–3), 164–172.

Chambers, J. A. (2013). *Biosafety of GM Crops in Kenya, Uganda, and Tanzania: An Evolving Landscape of Regulatory Progress and Retreat*. Washington, D.C.: Center for Strategic and International Studies.

Cohen, S. N. (2013). DNA cloning: a personal view after 40 years. *Proceedings of the National Academy of Sciences of the USA* **110**(39), 15521–15529.

Edquist, C. and Johnson, B. (1997). Institutions and organizations in systems of innovation. In *Systems of Innovation: Technologies, Institutions and Organizations*, ed. C. Edquist. London: Routledge, pp. 41–60.

Etzkowitz, H. (2002). Innovation in innovation: the triple helix of university–industry–government relations. *Social Science Information* **42**(3), 293–337.

Jaffe, G. (2004). Regulating transgenic crops: a comparative analysis of different regulatory processes. *Transgenic Research* **13**, 5–19.

Juma, C. (2011). *The New Harvest: Agricultural Innovation in Africa*. New York, NY: Oxford University Press.

Keese, P. *et al.* (2002). Seeds of promise: developing a sustainable agricultural biotechnology industry in sub–Saharan Africa. *Natural Resources Forum* **26**(3), 234–244.

Knight, W. (2002). Zambia bans GM food aid. [Online]. *New Scientist Daily News*, 30 October 2002. Available from www.newscientist.com/article/dn2990-zambia-bans-gm-food-aid/

Koch, M. (2014). Elements of an enabling biosafety and regulatory environment. In *Biotechnology in Africa: Emergence, Initiatives and Future*, ed. F. Wambugu and D. Kamanga. Basel: Springer International Publishing Switzerland, pp. 197–205.

Ladisch, M. R. and Kohlmann, K. L. (1992). Recombinant human insulin. *Biotechnology Progress* **8**(6), 469–478.

Lundvall, B.-A. (2010). *National Innovation Systems: Toward a Theory of Innovation and Interactive Learning*. Copenhagen: Anthem Press.

New Partnership for Africa's Development (2003). *Comprehensive Africa Agriculture Development Programme (CAADP)*. Pretoria: New Partnership for Africa's Development.

Office of the Gene Technology Regulator (2005). *Risk Analysis Framework*. Canberra: Office of the Gene Technology Regulator.

Pingali, P. (2012). Green Revolution: impacts, limits, and the path ahead. *Proceedings of the National Academy of Sciences of the USA* **109**(31), 12302–12308.

Pingali, P. and Raney, T. (2005). *From the Green Revolution to the Gene Revolution: How Will the Poor Fare*? Rome: Food and Agriculture Organization.

Sengooba, T. *et al.* (2005). *Analysis of the Biosafety System in Uganda*. Kampala: National Council for Science and Technology.

United Nations Department of Economic and Social Affairs Population Division (2015). World population prospects: the 2015 revision, key findings and advance tables. Working Paper No. ESA/P/WP.241.

25 Conclusions and Recommendations

E. Jane Morris, Ademola A. Adenle and Denis J. Murphy

The World Needs Agricultural Biotechnology

Global economic growth and agricultural production over the last two decades have lifted millions out of poverty and malnutrition, yet a ninth of the world's population continues to suffer from chronic undernourishment, and most of these people, about 780 million, live in developing countries (World Hunger Education Service, 2015). Looking to the future, the world is facing a 'perfect storm' through a combination of factors, some of which are listed below.

- The most recent estimates of population growth show world population increasing from 7.3 billion to 9.7 billion by 2050, with more than half this growth coming from Africa, where the population is expected to double to 2.5 billion, 100 million more than was estimated 2 years ago (United Nations, 2015)
- The steady increase in yield for some staple crops (especially cereals) since the early twentieth century has now plateaued as they start to reach their yield potential ceiling (Grassini *et al.*, 2013). Although total world food production may perhaps be sufficient to meet projected demand in the years up to 2050, several developing countries and regions have a long history of near-stagnant yields and policy environments that are not very promising (Alexandros and Bruinsma, 2012). Unless this trend can be reversed, those countries will be in a situation of increasing food deficit.
- Land and water resources are increasingly in demand, especially in developing countries, and are subject to overuse, leading to higher levels of environmental and biological stress. This is resulting in soil degradation, salinisation of irrigated areas and competition from uses other than for crops, which is impacting on the capacity for food production.
- Climate change is exacerbating the challenges faced by the agricultural sector, while at the same time agriculture is contributing significantly to the greenhouse-gas (GHG) emissions responsible for climate change. The changing climate is directly affecting crop yields in some regions due to increasing drought, heat waves or severe weather episodes such as flooding. Climatic shifts are also leading to increasing incidence of new pests and diseases that can badly affect crop performance.

In view of these challenges, much has been written about the need for the 'sustainable intensification' of agriculture. This is defined as 'the goal of producing more food with less impact on the environment, intensifying food production while ensuring the natural resource base on which agriculture depends is sustained, and indeed improved, for future generations' (The Montpellier Panel, 2013). These important aspirations are in line with the need for agriculture to address the Sustainable Development Goals as mentioned in Chapter 1. However, it is clear that none of these challenges can be addressed without continued agricultural innovation, of which opportunities through agricultural biotechnology represent a substantial part. The Montpellier Panel report specifically addresses the need for genetic intensification, using both conventional breeding and modern biotechnology to concentrate desirable genes in plants and animals.

Throughout the preceding chapters in this book the authors have addressed the various factors that are impeding, both in developed countries and in developing countries, the adoption of solutions that are available through modern biotechnology, with particular focus on genetically modified organisms (GMOs), but also looking to the future as certain types of new breeding techniques (NBTs) are looking extremely promising. The primary focus of the book is on developing countries since, as mentioned above, these are the countries which are most challenged by malnutrition, low agricultural productivity and food deficits. Now in this final chapter we summarise some of the key messages and make some recommendations as to the way forward.

The Key Messages

It is apparent to all that the debate regarding GMOs has become extremely polarised. Yet, as described in this book, GMOs have enormous potential to contribute to sustainable intensification. Even with the limited, genetically simple traits that are currently available in commercialised GM crops, the benefits have been substantial, not only in economic but also in environmental terms, through reductions in pesticide use and GHG emissions (see Chapter 15 and, for more detail, Brookes and Barfoot (2016)). As the new biotechnologies, such as gene editing, are increasingly applied to crop and livestock improvement (see Chapter 2), the list of traits will expand dramatically and will include more complex traits that will improve yield and quality as well as the ability to adapt to the biotic and abiotic threats of a changing climate.

Over the past two decades, many of the polemical arguments against GMOs have focused not only on perceived risks of the technology but also on the undue influence of the multinational companies (MNCs) as purveyors of the technology, as mentioned in Chapters 4 and 14. In Chapter 14 we heard the view expressed that the anti-GMO movement was merely a proxy for the anti-MNC movement. Yet the extremely stringent regulations imposed by many countries have resulted in escalating regulatory costs, which have imposed such high barriers to entry as to ensure that only the MNCs can afford to bring the technology to market. These stringent regulations prevent the commercialisation of crops (developed usually with public funding) that are locally important in developing countries. Interestingly enough, a recent publication (Schiek *et al.*, 2016) suggests that the costs of developing a GM crop as a public good for deregulation and release in a developing country may be much lower than earlier estimates would suggest. This paper does not state exactly which tests are included in the estimate (e.g., whether provision has been made for any feeding trials, and what compositional analyses are included), and it does make clear that any additional hurdles imposed by regulators could significantly increase the costs. Despite this rather optimistic assessment, the track record of publicly funded projects suggests a more pessimistic picture. A good example is given for Brazil in Chapter 22, which, despite being the country with the world's second-largest production of GM crops, has not been successful in commercialising GM crops developed by its own scientists. A similar situation exists in South Africa (Thomson, 2016) and for many other publicly funded programmes (Huesing *et al.*, 2016).

The high regulatory costs, which are often effectively imposed on producers by importing nations in the developed world, are also cited in Chapters 9 and 20 as barriers to crop improvement. Indeed, it is questionable whether some of the improved GM crops currently undergoing field trials in West Africa (Chapter 23) and Uganda (Chapter 24), as well as the biofortified rice discussed in Chapter 12, will ever make it to market. The message in Chapter 24 that biosafety systems should be regarded as an integral part of a biotechnology innovation system is a key challenge in this regard; the two should not be working against each other.

Many developing countries that are party to the Cartagena Protocol on Biosafety (CPB) have adopted strict risk-averse regulation founded on the precautionary principle (PP); this is also embedded in European legislation. However as described in Chapter 13, the problem with the articulation of the PP in the CPB is that it does not involve any limits, and its unchecked application has led to rejection of safe products as a result of unfounded fears. The inappropriate application of the PP is also cited as a major problem in a number of other chapters, including Chapters 14, 15 and 23, and we believe it to be largely responsible for the current impasse in the introduction of GMOs. Jansen van Rijssen *et al.* (2015) provide a detailed critique of the PP, and amongst others cite Peterson (2007), who stated that the PP 'replaces the balancing of risks and benefits with what might be described as pure pessimism'. We believe that the problems with the CPB (as well as with European legislation) include (1) the fact that consideration of benefits is ignored and (2) the focus on process-based assessment rather than the product-based approach as used in the Canadian regulatory system described in Chapter 5. We discuss these issues further below.

Internationally, the term risk analysis is widely accepted, but the concept of benefit analysis is rarely mentioned. As described in Chapter 3, the risk analysis methodology, which is based on problem formulation, focuses on the harm which might arise from the introduction of a GMO; benefit is considered only at a secondary stage (if it is considered at all). It is difficult to see how (for example) nutritionally enhanced crops would fare in such an analysis. Socio-economic impact analysis as described in Chapters 8 and 10 certainly takes into account benefits, but this is only one dimension of benefit; in any case, socio-economic considerations are not taken into account in all jurisdictions. Chapter 4 focuses on the need for a balanced risk–benefit analysis; it is apparent from analysis of the legislation of many developing countries, as summarised in that chapter, that, in line with the requirements of the CPB, the focus is almost exclusively on risk.

Considering that no risks have materialised during the 20 years since the introduction of GMOs, we consider this excessive focus on risk, as opposed to risk–benefit analysis, to be inappropriate.

The focus on process-based legislation as embedded in the CPB and in many instances of national legislation sets GMOs apart from other breeding techniques. This may have been appropriate when GMOs were first introduced and there was a high level of uncertainty about the technology. However, in the intervening years it has become clear that there is in fact much less uncertainty in genetic modification than in many other breeding techniques. Imprecise techniques such as mutation breeding and embryo rescue are not subject to any regulatory oversight. Meanwhile, the NBTs, including genome editing as described in Chapter 2, have introduced a continuum from conventional breeding to genetic modification, and therefore it is no longer appropriate to separately impose stringent controls on what have traditionally been labelled 'GMOs'. Instead, the assessment needs to be of the product, not of the process by which it was obtained. There are indications that in the future the new gene-editing technologies, where the genetic change may be indistinguishable from naturally occurring variations (termed polymorphisms), might not be subject to the same level of regulation as the earlier, twentieth-century GM technologies (Carroll *et al.*, 2016; Strauss and Sax, 2016). This was reinforced by the conclusion of a report by an impressive list of academic experts that was published in mid 2016 by the US National Academies of Sciences, Engineering and Medicine (US National Academies, 2016).

Despite the product-based approach of the USA, and the US National Academies report, the US Senate has passed a bill that would introduce a national standard for mandatory labelling of GM foods (termed 'bioengineered foods' in the bill) (US Senate, 2016). This is clearly process labelling rather than product labelling. Chapter 6 covers labelling issues in some detail, and makes specific recommendations for labelling regimes in developing countries, in particular recommending voluntary rather than mandatory labelling. It remains to be seen what impact the new approach in the USA will have on developing countries, but the cost implications of labelling should not be ignored.

Issues surrounding socio-economic analysis come under the spotlight in Chapters 8 and 10, but are also relevant in Chapter 12, where the social acceptability of biofortified rice is examined. The authors point out that socio-economic considerations (SEC) have been included in the CPB, as well as in many instances of national legislation, without any definition of the term. Both Chapters 8 and 10 make a strong case for SEC to be evaluated in a scientifically sound manner and with a clear definition of the elements that are being evaluated; considerations should be based on real data rather than suppositions, with the recognition that there are many limitations to what can be determined in an *ex-ante* evaluation. In contrast, as described in Chapter 13, a broad interpretation of SEC has given decision-makers the ability to respond to perceived consumer 'outrage' by blocking regulatory approvals, sometimes just because of assumptions about consumer attitudes. Chapters 18 and 20 provide examples of regulatory decisions that have been overthrown due to consumer pressure in China and India. The EU 'opt-out' clause in Directive 2015/412 broadens the basis for EU member countries to prohibit cultivation of GM crops not only due to ill-defined socio-economic impacts, but also due to a range of other factors, including 'public policy'. We are extremely concerned about these developments, and wish to stress that it is not appropriate for politicians to ignore sound science and block the technology on an arbitrary basis.

If public opinion is to be turned around, there needs to be much better communication at all levels between scientists, the public, policymakers and the media. Chapter 16 stresses the need for more public participation in decision-making, and provides clear indicators as to how biosafety communication should take place, while acknowledging that public participation and communication are currently being neglected in many developing countries. Chapter 17 specifically examines the role of the media. With the rise of social media as an uncensored means of communication, 'good news' messages need to be shared much more strongly, rather than the (often false) scaremongering and anti-MNC messages that have a tendency to go viral. Currently the initiative is in the hands of the anti-GMO activists; a change in approach will require scientists to play a much stronger leadership role in communication and popularisation of the topic.

Several chapters in the book discuss the opportunities for developing countries to address some of their capacity limitations by sharing knowledge and expertise. Harmonisation of regulatory requirements is one important step towards a shared approach by which risk assessments could be undertaken at regional or continental level rather than by individual countries. Chapters 9, 13 and 23 advocate regional or continental harmonisation of risk assessment, while Chapter 11 provides information on OECD consensus documents that are freely available as a resource for regulators to use. Harmonisation needs to go hand in hand with the development of increased capacity, and, as stated in Chapter 7, regional capacity-development programmes could become a conduit to promote collaboration and sharing of resources between countries. Chapter 7 also proposes an innovative, targeted approach to capacity development designed to address the fact that, after 15 years of technical assistance and donor support, many countries have not yet developed a functional biosafety system. We support a harmonised approach, and would encourage countries that lack expertise to be willing to accept the results of risk assessments carried out regionally or internationally.

Particularly in the case of food/feed safety assessments, if agreement on the requirements could be reached at international level, this could pave the way for a single accredited body to undertake the assessment, negating the need for country-specific assessments. This could also help to alleviate the problem of asynchronous approvals of events for commodities, leading to disruptions in international trade, as described in Chapter 21.

Finally, we recognise that modern agricultural biotechnology does not exist in a vacuum but needs to be regarded as an integral part of a holistic agricultural system. The benefits will not be realised unless attention is paid to sustainable agricultural development in a broader context. The governments of developing countries, as well as donor programmes, need to focus on upgrading all aspects of the agricultural system, including seed systems, fertilisers and other agro inputs, as well as markets, so that new technologies are not introduced in a vacuum but are part of a holistic system that contributes to sustainable intensification of agriculture.

The current international regulatory framework for GMOs is unlikely to facilitate the development and adoption of GM products, especially those targeting the nutrition deficit and low agricultural productivity in developing countries. Responding to this challenge, the recommendations in the next section are proposed to address this problem.

Recommendations

Following on from the discussion in the previous section, we conclude the book with some final recommendations.

1. The Cartagena Protocol on Biosafety, largely dating from 2000, is now outmoded and inappropriate, so it needs to be revamped or abandoned.
2. Product-based safety assessment should be the new international norm; process-based assessment is not flexible enough to accommodate rapidly developing technology trends.
3. Regulatory requirements should be in line with actual identified risks, and should be of such a nature as to facilitate the introduction of the improved crops much needed by farmers in the developing world.
4. Risk analysis and benefit analysis should go hand in hand. Benefits should be not only assessed but also communicated at all levels. As with most other technologies, regulation and management of GMOs should reflect the balance of scientifically based risk–benefit assessments.
5. Capacity-building activities should be designed to maximise the opportunities to equip decision-makers with the confidence to move forward with the technology.
6. The requirements for socio-economic considerations, where they are to be included, should be clearly spelled out upfront. They should address only issues that can be adequately considered in an *ex-ante* evaluation.
7. Regional and international efforts at harmonisation should be strongly promoted and encouraged.
8. Broader public participation should be encouraged, and information should be widely shared, while respecting that decision-making needs to be based on sound science.
9. Government officials and policymakers should be actively discouraged from making arbitrary decisions that ignore the input of experts.
10. Developing countries need to invest in agriculture, and, in the particular case of Africa, need to live up to their commitment in the Malabo declaration to allocate at least 10% of public expenditure to agriculture (African Union, 2014).

We urge all concerned parties, namely scientists, government officials, donors, the media and international bodies, to take note of these recommendations and act on them expeditiously. We believe these recommendations could effectively contribute to the successful development of GMOs and increase the adoption of the technology in developing countries.

References

African Union (2014). Malabo declaration on accelerated agricultural growth and transformation for shared prosperity and improved livelihoods. [Online]. Addis Ababa: African Union. Available from http://pages.au.int/sites/default/files/Malabo%20Declaration%202014_11%2026-.pdf

Alexandros, N. and Bruinsma, J. (2012). World agriculture towards 2030/2050: the 2012 revision. FAO Agriculture Development Economics Division. ESA Working Paper No. 12-03.

Brookes, G. and Barfoot, P. (2016). *GM Crops: Global Socio-economic and Environmental Impacts 1996–2014*. Dorchester: PG Economics Ltd.

Carroll, D *et al.* (2016). Regulate genome-edited products, not genome editing itself. *Nature Biotechnology* **34**, 477–479.

Grassini, P. *et al.* (2013). Distinguishing between yield advances and yield plateaus in historical crop production trends. *Nature Communications* **4**, 2918.

Huesing, J. E. *et al.* (2016). Global adoption of genetically modified (GM) crops: challenges for the public sector. *Journal of Agricultural and Food Chemistry* **64**(2), 394–402.

Jansen van Rijssen, F. W. *et al.* (2015). The precautionary principle: making managerial decisions on GMOs is difficult. *South African Journal of Science* **111**(3–4), 1–9.

Peterson, M. (2007). The precautionary principle should not be taken as a basis for decision-making: talking point on the precautionary principle. *EMBO Reports* **8**(4), 305–308.

Schiek, B. *et al.* (2016). Demystification of GM crop costs: releasing late blight resistant potato varieties as public goods in developing countries. *International Journal of Biotechnology* **14**(2), 112–131.

Strauss, S. H. and Sax, J. K. (2016). Ending event-based regulation of GMO crops. *Nature Biotechnology* **34**, 474–477.

The Montpellier Panel (2013). Sustainable Intensification: A New Paradigm for African Agriculture, London. [Online]. Available from www.ag4impact.org/publications/montpellier-panel-report2013/

Thomson, J. A. T. (2016). An African perspective: new biosciences making African agriculture more productive and resilient. In *Creating Sustainable Bioeconomies: The Bioscience Revolution in Europe and Africa*, ed. I. Virgin and E. J. Morris. London: Routledge, pp. 51–61.

United Nations (2015). World population prospects, the 2015 revision. Department of Economic and Social Affairs, Population Division. [Online]. Available from https://esa.un.org/unpd/wpp/

US National Academies (2016). *Genetically Engineered Crops: Experiences and Prospects*. Washington, D.C.: National Academies Press. [Online]. Available from www.nap.edu/catalog/23395/genetically-engineered-crops-experiences-and-prospects

US Senate (2016). Bill S.764 to amend the Agricultural Marketing Act of 1946 to require the Secretary of Agriculture to establish a national disclosure standard for bioengineered foods, and for other purposes. [Online]. Available from www.agriculture.senate.gov/imo/media/doc/Ag%20biotech%20compromise%20proposal.pdf

World Hunger Education Service (2015). 2015 World hunger and poverty facts and statistics. [Online]. Available from www.worldhunger.org/2015-world-hunger-and-poverty-facts-and-statistics/

Index